PUBLIC HEALTH IN THE 21ST CENTURY

SELENIUM

SOURCES, FUNCTIONS AND HEALTH EFFECTS

PUBLIC HEALTH IN THE 21ST CENTURY

Additional books in this series can be found on Nova's website under the Series tab.

Additional E-books in this series can be found on Nova's website under the E-book tab.

NUTRITION AND DIET RESEARCH PROGRESS

Additional books in this series can be found on Nova's website under the Series tab.

Additional E-books in this series can be found on Nova's website under the E-book tab.

PUBLIC HEALTH IN THE 21ST CENTURY

SELENIUM

SOURCES, FUNCTIONS AND HEALTH EFFECTS

CHINATSU AOMORI
AND
MEGUMI HOKKAIDO
EDITORS

Nova Science Publishers, Inc.
New York

For permission to use material from this book please contact us:
Telephone 631-231-7269; Fax 631-231-8175
Web Site: http://www.novapublishers.com

Library of Congress Cataloging-in-Publication Data

Selenium : sources, functions, and health effects / editors, Chinatsu Aomori and Megumi Hokkaido.
p. ; cm.
Includes bibliographical references and index.
ISBN 978-1-61942-061-8 (hardcover)
I. Aomori, Chinatsu. II. Hokkaido, Megumi.
[DNLM: 1. Selenium--pharmacology. 2. Selenium--physiology. QV 138.S5]

572'.555--dc23

2011042423

Published by Nova Science Publishers, Inc. † New York

CONTENTS

PREFACE

Selenium (Se) is an essential trace element of fundamental importance to health, primarily due to its antioxidant activity. Selenium has chemoprotective, anti-inflammatory, and antiviral properties and is related to enhancing immunity, controlling gastrointestinal, and reducing age-related disorders. In this book, the authors present current research in the study of the sources, functions and health effects of selenium. Topics discussed include selenium biorhythms and hormonal regulation; the physiological role of selenium; selenium supplementation of diet effects on antioxidant activity and protects against antibiotic resistance and avian influenza in poultry; selenium in critical illness and tropical diseases; products of animal origin as a source of Se in the human diet; selenium in food materials and stress signaling and proapoptotic pathways induced by selenite in malignant coloncytes.

Chapter 1 - Selenium (Se) is an essential trace element of fundamental importance to health, primarily due its antioxidant activity. The health-related properties of Se include but not limited to protection against cancer, cardiovascular and muscle disorders. Selenium has chemopreventive, anti-inflammatory, and antiviral properties and is related in enhancing immunity, and reducing age-related disorders.

Unlike other metals that interact with proteins in form of cofactors, Se becomes cotranslationally incorporated into the polypeptide chain as part of the amino acid selenocysteine (Sec). Proteins that contain Sec as an integral part of their polypeptide chain are defined as selenoproteins (Sel). Selenoproteins are present in all lineages of life (*i.e.*, bacteria, archaea, and eukarya). In plants, Se occurs as part of an organic compound predominantly selenomethionine (SeMet). Plant absorption of Se principally depends on the concentration and physicochemical forms existing in the soil. Selenium is added to the diet of animals either as an inorganic salt (sodium selenite, calcium selenite or sodium selenate) or as an organo-Se compound more often in the form of Se-yeast. Other forms of organo-Se compounds include SeMet and Se malt. Another source of Se is also available, the Se proteinate formed by the reaction of inorganic Se on a hydrolyzed soya protein. Selenium from organic sources is more thoroughly absorbed and more efficiently metabolized than the inorganic salts, which are poorly absorbed. Absorbed SeMet can be incorporated into tissue proteins in place of methionine or can be further metabolized in the liver and used for the synthesis of specific Sel.

In humans, the Sel family includes the following members: glutathione peroxidases (GPx1-GPx4, GPx6), thioredoxin reductases (TRXR1-TRXR3), iodothyronine deiodinases (D1-D3), selenophosphate synthetase (SPS2), 15-kDa Sel (Sep15), SelH, SelI, SelK, SelM,

SelN, SelO, SelP, MsrB1, SelS, SelT, SelV, SelW. Furthermore, in eukaryotes, the Sel family includes the Fep15, SelJ, SelU1, Sel1-Sel4 and the PDI. Members of this protein family exert many diverse functions, but their synthesis depends on a common set of cofactors and on dietary Se. Although several functions of many Sel are unknown, many disorders are related to changes in Sel structure, activity or expression. Selenium deficiency and mutations or polymorphisms in Sel genes and synthesis cofactors are implicated in a variety of diseases, including muscle and cardiovascular disorders, immune dysfunctions, cancer, neurological disorders and endocrine functions.

In this chapter, the recent progress in the mechanisms of Sel synthesis as well as their health effects is described.

Chapter 2 - Biorhythms caused by Earth rotation are a fundamental phenomenon of Nature embracing all levels of life organization from the cell to the whole organism, to populations, communities, ecosystems and providing optimal levels of adaptation to constantly changing environmental conditions. Many of these rhythms are endogenous cycles of many hormones that are maintained under photoperiodic clamping. Endocrine rhythms span time frames ranging from milliseconds to years, regulating appropriate fluctuations of different biochemical parameters of an organism. At present it is assumed that the pituitary gland is the main «receiver», constantly geared up for the signals coming from the Sun. It regulates all phyllogenetically stabilized biorhythms of an organism via hormonal regulation, being provided by constant blood current. Taking this into account many examples of selenium fluctuations in bio substrates have a logical explanation, based on hormonal regulation of selenium accumulation in tissues and organs.

Selenium is an essential trace element strictly connected with the endocrine system modifying the expression of about 30 selenoproteins. It participates in normal human development, growth, antioxidant defense, male fertility, thyroid hormone metabolism, and in modulating immunity. Biological rhythms of selenium seem to be of great significance due to essentiality of the element to mammal organisms, its antioxidant protective properties and the ability to decrease the risk of cardiovascular diseases and cancer.

Most of the results presented below describe examples of different selenium dynamics without strict proof of the real endogenous biorhythms existence. Nevertheless both give exciting material for conclusions and future research.

Chapter 3 - Seafood and aquaculture products offer a wide diversity of species, a broad range of foods—from whole fresh fish to canned products—and large amounts of several nutrients, invaluable to human health. Consequently, they are considered to have an important role in the human diet. The consumption of these products has been increasing over recent decades. Moreover, the health benefits of a seafood rich diet have been extensively acknowledged in the past decade. In fact, these foods contain various nutritionally beneficial components, such as, readily digested proteins, ω3 polyunsaturated fatty acids (ω3-PUFA), namely, eicosapentaenoic acid (EPA, 20:5ω3) and docosahexaenoic acid (DHA, 22:6ω3) — which are associated with decreased morbidity and mortality from cardiovascular and other diseases as well as with foetal development—, vitamins, and minerals. Regarding the latter nutrients, seafood is a rich source of selenium (Se). Norwegian studies have shown that the quantity of Se in seafood, in average, varies widely from 10 to 290 μg per 100 grams of edible part. The majority of seafood species in the Norwegian diet contains from 20 to 40 μg per 100 grams of edible part. However, some seafood reach higher values, for instance, crab meat is one of the richest sources of this element (84 ± 28 μg/100 g) according to a survey of over 700 food samples representing 100 different types of food product consumed in the United Kingdom. Some estimates based only on Northern European consumers, from Alexander et al., 2007, have shown that seafood account for an average of 31 % of the total Se intake for adults, 24 % for 13-year-olds, 23 % for 9-year-olds, 25 % for 4-year-olds, and 14 % for 2-year-olds. In Japan, which has a very high level of seafood consumption, it was found that the major Se contribution comes from fish (up to 60 % of daily total intake) rather than the staple foods, such as rice and vegetables.

On the other hand, for farmed fish, the amount of Se depends on the content and the form in which Se is present in the feed. According to the literature, Se can be in inorganic form (elemental Se, metal selenides, selenite, selenate) or in organic forms with Se-C bonds. Elemental Se is water-insoluble and, as such, accumulates in anaerobic aquatic sediments. The metal selenides (CdSe, HgSe, PbSe, etc.) are also water-insoluble and display low bioavailability. Moreover, only some organic selenium compounds, such as selenocystine, selenocysteine, selenomethionine or selenoproteins like glutathione peroxidase, have been identified and quantified in seafood and aquaculture products. Up to 91 % of total Se found in aquatic organisms is in the organic form.

In addition, Se content can be reduced by food processing such as cooking (boiling, baking or grilling) due to the volatilization phenomena. Particularly, Se loss has been reported for roasted fish.

For wild and aquaculture fish products, there is a specific advantage of high Se contents. In fact, it is well established that mercury (Hg) and Se bind to form Hg selenides with extremely low solubility, which are thought to be metabolically inert. Moreover, scientific works have found that Se supplementation thwarts the negative effects of Hg exposure in all investigated marine species. Being Hg content high in some marine species, this fact accrues a special importance to Se in seafood and aquaculture products.

Chapter 4 - Selenium (Se) is a widely studied trace element in human and animal due to its role in antioxidant defense system which is needed for the maintenance of health and hormone balance. During last two decades the physiological role of Se in plants has been explored by researchers. Plant roots take up Se from soil water in either the selenate or the selenite ionic forms. In higher plants metabolism of Se is closely related to that of sulfur due to their chemical similarity. Although, Se is not yet confirmed to be required by higher plants, but several studies demonstrate that at low concentrations it may exert diverse beneficial effects, including growth-promoting activities. Moreover, some plant species grown in Se-enriched media have shown enhanced resistance to certain abiotic stresses, e.g. drought, salinity, extreme temperature, metal toxicity and UV-irradiation. Se exerts its beneficial effects on growth and stress tolerance of plants by enhancing their antioxidative capacity. It enhances plants' resistance against oxidative stress caused by reactive oxygen species (ROS). However, agricultural crops plants are sensitive to high Se concentrations which vary among plant species. Although a number of report indicated the protective role of Se in plants, to date the research works conducted on the physiological role of Se is scarce. In this chapter, the authors attempted to review the recent findings related to the physiological role of Se in plants.

Chapter 5 - The worldwide increase in the use of antibiotics as an integral part of the poultry and livestock production industry to treat and prevent infectious bacterial diseases and as growth promoters at subtherapeutic levels in feed has led to the problem of the development of bacterial antibiotic resistance during recent years. These uses promote the selection of antibiotic resistance in bacterial populations. The resistant bacteria from agricultural environments may be transmitted to humans, in whom they cause disease that cannot be treated by conventional antibiotics. Poultry that are given antibiotics often carry antibiotic-resistant strains of *Salmonella* or antibiotic-resistant transposons, which eventually reach humans through poultry meat, eggs and other food. On the other hand, avian influenza is an infection caused by influenza viruses that occur naturally in birds. The virus that causes this infection in birds can change (mutate) to infect humans. Selenoenzyme glutathione blocks the replication of *Salmonella* and influenza virus. Glutathione plays an important role in regulating viral replication and infectivity. Given the current interest in the use of safe alternatives to antibiotics, natural antioxidants including selenium and vitamin E are important for animal health. Regarding nutritional supplementation, selenium and vitamin E can decrease the effects of stress and infection on feed intake and body weight gain in chicken. Nutrition has some effects on both innate and cellular immunities if the feed is severely restricted.

Chapter 6 - The role of Selenium (Se) in critical illness has been an issue of great interest and anxiety for researchers. Sepsis and the systemic inflammatory response syndrome form the pathogenesis of many cases of critical illness, and these are characterised by increased free radical induced oxidative stress. If this inflammatory process is not controlled it can lead to multiple organ dysfunction and contribute to high morbidity and mortality in patients in the intensive care unit (ICU). Therefore it is important to limit the free radical induced oxidative stress process in critical illness. It has been hypothesised that antioxidants, by preventing or removing free radicals, may reduce the oxidative stress in critical illness. Se is an important part of the antioxidant system, and is known to mediate its antioxidant action through the enzyme glutathione peroxidase. Besides this, it also reduces inflammation by inhibiting NF-κB (nuclear factor-kappa B), which is involved in the transcription of several pro inflammatory cytokines such as IL-6 (interleukin-6). Se also improves the immune response of the body by enhancing humoral and cellular immunity. Thus Se may play an important role in improving the clinical outcome in critical illness. Several studies have assessed the role of high dose parenteral Se (as sodium selenite) in critical illness. Although studies have shown an improvement in antioxidant status, there is no clear beneficial effect on overall clinical outcome, in terms of mortality rate in ICU, infection rate, length of stay in ICU or renal replacement therapy. However some studies have shown that high dose Se may improve the mortality rate in patients with particularly severe sepsis or septic shock. Se levels in blood are influenced by the acute phase response, which reduces plasma Se levels through redistribution of selenoproteins. Therefore low plasma Se levels in critical illness may be due to inflammation or severity of illness, rather than Se deficiency. Hence Se levels in blood in critical illness should be interpreted with caution. Further studies are needed to identify Se requirements in critical illness.

Chapter 7 - According to World Health Organization, neglected tropical diseases encompass all diseases that occur solely, or principally, in the tropics. In practice, the term is often taken to refer to infectious diseases that thrive in hot, humid conditions. Clinical and epidemiological studies suggest that selenium (Se) play an important role in tropical diseases, such as tuberculosis, leishmaniasis, filariasis and chagas, acting as preventive agent or in diagnosis and prognosis. Recent studies have evinced the importance of selenium in oxidative status and antioxidant defense capabilities during the course of infection and progression of the illness in human patients and experimental models. For this reason, one of the most relevant mechanism of action proposed involve selenoproteins, i.e. glutathione peroxidase (GPx), an enzyme that protects against oxidative stress and modulates the redox processes. In addition, it was observed that low Se levels were positively correlated with an increased susceptibility to infections. Besides, Se supplementation is proposed as an adjuvant therapy for treatment of these chronic diseases. However, there is a lack in the literature references related to synthesis and biological evaluation of novel derivatives containing selenium moiety against these diseases. During the last years their research group is interested in the design and synthesis of organoselenium compounds as new class of agents for treatment of neglected tropical diseases.

In the present year the authors have reported two general structures with leishmanicidal activity, corresponding both of them to symmetrical compounds. The firsts are alkyl imidoselenocarbamates (alkyl isoselenourea) which possessed a moderate effect *in vitro* and the second ones are selenocyanates and diselenides. It is remarkable that some of them showed stronger *in vitro* antileishmanial activity than edelfosine and miltefosine, used as reference drugs, and combined high potency and low cytotoxicity against Jurkat and THP-1 cells.

Chapter 8 - Selenium consumption in European countries has declined considerably over the last few decades, mainly as a result of eating low-selenium foods. Geographical location is a factor that has a significant effect on Se concentration in products of plant and animal origin. Because selenium content in the environment is low in most European countries, the majority of food products coming from this area contains small amounts of this trace element. Human Se intake depends on the amount and type of ingested food.

Chapter 9 - Sodium selenite (NaSe) continues to be studied as a prospective chemopreventive agent against several types of malignancies including colon cancer. Recent studies demonstrated that in malignant or premalignant cells of digestive system NaSe induces a wide range of effects which may in the end either stabilize or damage these cells. Due to a multifaceted nature of NaSe-dependent mechanisms and their natural crosstalk, their timing and contribution for final observed outcome in the studied model is of paramount importance. Thus the purpose of this work was to investigate the kinetics of selected mechanisms of NaSe in human colon cancer cells HCT-116 during 24 h of exposure. Their results indicate that NaSe has moderate genotoxic effects with subsequent activation of DNA-damage pathway in HCT-116 cells. Furthermore, in thus exposed colonocytes mitogen stress kinase signaling (in particular p38) occurs as well as an early direct mitochondrial changes with resulting cellular degeneration bearing features of apoptosis. These results suggest that in HCT-116 cells NaSe mediates a series of changes of which DNA damage response and direct mitochondrial effects seem to complement each other and to contribute to final observed cell death.

Chapter 10 - Benign prostatic hyperplasia (BPH) and prostate cancer are major sources of morbidity in older men and have an increasing impact on human health in line with the gradual aging of the population. The aetiology of those diseases is still far from being fully understood and, as a direct consequence, it is difficult to identify a rationale long-term strategy to develop an effective therapy. Nutraceuticals are components isolated or purified from food substances currently used for medical or health benefits. Several naturally derived food substances have been studied in BPH and prostate cancer in an attempt to identify alternative therapies for those diseases. Selenium (Se) plays an important role in maintaining equilibrium of a healthy organism. Recent scientific data devoted to investigating the nutraceutic effects of Se confirm a strong correlation between Se supplementation, BPH and prostate cancer. Se is an essential trace element with antioxidative, antimutagenic, antiviral and anticarcinogenic properties and exists naturally in foods, predominantly in the organic form as selenomethionine and selenocysteine. Se has several mechanisms of action, depending on its form. Recent literature suggested that Se could inhibit cell proliferation and induce cell cycle arrest of human prostate cancer cells. Moreover, Se and its derivatives can activate both the intrinsic and extrinsic pathways of apoptosis and inhibit angiogenesis. Despite the positive results obtained with Se supplementation, it is necessary to deeply verify these findings in experimental animal models and controlled clinical studies. In light of this background, their research group recently experimentally demonstrated that the Selenium, in association with Lycopene and *Serenoa repens* is helpful in reducing benign and malign prostate growth. These data confirm the positive effects of Se in prostate disorders suggesting its possible use in therapeutic management of BPH and prostate cancer.

Chapter 11 - Since selenium was first isolated in 1817, there have been many studies on its accumulation and existing forms in food materials as well as its beneficial and toxic effects on human body. The main context of this article reviewed the positive correlation of selenium content in plant food materials with that in their growing soils, the positive correlation of selenium content in animal food materials with that in feedstuffs, the positive correlation of selenium content in food materials with that in human tissues, the association of selenium with proteins in food materials, the organic and inorganic forms of selenium in food materials, selenium deficiency and human diseases such as white muscle, or Keshan disease or other cardiovascular diseases or leading to immotile, deformed sperm and infertility, selenium supplementation including the variation of its beneficial effects caused by the variation of its forms in the supplements for curing or preventing diseases such as cancers and selenium toxicity such as losing hairs or nails to human body.

Chapter 12 - A new approach is used to generate amorphous selenium structures by an ab initiomolecular dynamics method. Crystalline cubic supercells start with 64, 100, 150, 216 and 512 atoms and with the experimental densities of 4.25 g cm^{-3} and 4.45 g cm^{-3}. The samples are amorphized using DMol3 from the suite in Material Studio 3.2 ® by heating the periodic structures to just below the melting point (the undermelt-quench approach) and then cooling them down to 0 K. The structures are relaxed by annealing and quenching, and finally a geometry optimization is carried out. The structural properties: radial distribution functions g(r), bond angle distributions and dihedral angle distributions; electronics properties: electronics density of states; vibrational properties; vibrational density of states and the optical properties: tauc approximation are reported. It is found that the amorphous structure, for both densities, is mainly formed by chains but not at all linear, there are some ring-like structure although not closed, in this chapter gives the first quantitative ratio of chains and rings.

In: Selenium: Sources, Functions and Health Effects
Editors: Chinatsu Aomori and Megumi Hokkaido
ISBN: 978-1-61942-061-8

Chapter 1

THE HEALTH EFFECTS OF SELENOPROTEINS

E. Zoidis and A. C. Pappas
Department of Nutritional Physiology and Feeding,
Faculty of Animal Science and Aquaculture, Agricultural University of Athens,
Athens, Greece

ABSTRACT

Selenium (Se) is an essential trace element of fundamental importance to health, primarily due its antioxidant activity. The health-related properties of Se include but not limited to protection against cancer, cardiovascular and muscle disorders. Selenium has chemopreventive, anti-inflammatory, and antiviral properties and is related in enhancing immunity, and reducing age-related disorders.

Unlike other metals that interact with proteins in form of cofactors, Se becomes cotranslationally incorporated into the polypeptide chain as part of the amino acid selenocysteine (Sec). Proteins that contain Sec as an integral part of their polypeptide chain are defined as selenoproteins (Sel). Selenoproteins are present in all lineages of life (*i.e.,* bacteria, archaea, and eukarya). In plants, Se occurs as part of an organic compound predominantly selenomethionine (SeMet). Plant absorption of Se principally depends on the concentration and physicochemical forms existing in the soil. Selenium is added to the diet of animals either as an inorganic salt (sodium selenite, calcium selenite or sodium selenate) or as an organo-Se compound more often in the form of Se-yeast. Other forms of organo-Se compounds include SeMet and Se malt. Another source of Se is also available, the Se proteinate formed by the reaction of inorganic Se on a hydrolyzed soya protein. Selenium from organic sources is more thoroughly absorbed and more efficiently metabolized than the inorganic salts, which are poorly absorbed. Absorbed SeMet can be incorporated into tissue proteins in place of methionine or can be further metabolized in the liver and used for the synthesis of specific Sel.

In humans, the Sel family includes the following members: glutathione peroxidases (GPx1-GPx4, GPx6), thioredoxin reductases (TRXR1-TRXR3), iodothyronine deiodinases (D1-D3), selenophosphate synthetase (SPS2), 15-kDa Sel (Sep15), SelH, SelI, SelK, SelM, SelN, SelO, SelP, MsrB1, SelS, SelT, SelV, SelW. Furthermore, in eukaryotes, the Sel family includes the Fep15, SelJ, SelU1, Sel1-Sel4 and the PDI. Members of this protein family exert many diverse functions, but their synthesis depends on a common set of cofactors and on dietary Se. Although several functions of many Sel are unknown, many disorders are related to changes in Sel structure, activity or expression. Selenium deficiency and mutations or polymorphisms in Sel genes and synthesis cofactors are implicated in a variety of diseases, including muscle and cardiovascular disorders, immune dysfunctions, cancer, neurological disorders and endocrine functions.

In this chapter, the recent progress in the mechanisms of Sel synthesis as well as their health effects is described.

Keywords: cancer, cardiovascular disease, health, selenium, selenoproteins

INTRODUCTION

Selenoproteins (Sel) is the term used to describe the proteins that contain selenocysteine (Sec). However, proteins that contain selenomethionine (incorporation of selenomethionine (SeMet) into proteins in place of methionine) are not regarded as Sel because of the nonspecific nature of selenium (Se) utilization in these proteins (Heras et al., 2011). Similarly, many homologues of Sel are not considered as Sel since they contain cysteine and not Sec (Fairweather-Tait, et al., 2010). Selenium is incorporated into proteins and is covalently bonded within the amino acid Sec, the 21st amino acid. Currently, the essentiality of Se is beyond doubt and its effect on the appearance of several diseases has been reviewed extensively in the literature (Gladyshev 2006; Pappas et al., 2008).

Selenoproteins are essential for life and several Sel have been characterized as antioxidant enzymes, serving to protect from damage caused by free radicals. Reactive oxygen species (ROS) include not only free radicals like the superoxide anion, or the lipid radicals but also oxidizing non-radical species such as hydrogen peroxide, peroxynitrite and singlet oxygen. Free radical reactions are part of normal human metabolism and can be induced by environmental sources (Halliwell, 1992; Steinbrenner and Sies, 2009). Low levels of ROS modulate signal transduction pathways, and may aid in the defense against infectious agents (Seifried et al., 2007). In contrast, overproduction of these species seems to have detrimental effects and results in oxidative stress (Ramoutar and Brumaghim, 2010). Reactive oxygen species generation by a decrease of coupling of oxidation and phosphorylation in the mitochondria results in an increased electron leakage and overproduction of superoxide radicals.

Table 1. Reactive oxygen and nitrogen species

Radicals	Non-radicals
Alkoxyl, RO*	Hydrogen peroxide, H_2O_2
Hydroperoxyl, HOO*	Hypochlorous acid, HOCl
Hydroxyl, *OH	Ozone, O_3
Peroxyl, ROO*	Singlet oxygen, $^1O_2$
Superoxide, O_2*	Peroxynitrite, $ONOO^-$
Nitric oxide, NO*	Nitroxyl anion, NO^-
Nitrogen dioxide, NO_2*	Nitrous acid, HNO_2

A list of ROS and reactive nitrogen species is presented in Table 1. Once ROS production exceeds the ability of the antioxidant system to neutralize them, lipid peroxidation develops and causes damage to unsaturated lipids in cell membrane and cell integrity is disrupted. Membrane damage is associated with a decreased efficiency of absorption of different nutrients and leads to an imbalance of vitamins, amino acids, inorganic elements and other nutrients in the organism.

Selenoproteins are present in all lineages of life (*i.e.,* bacteria, archaea and eukarya). Much of Se beneficial influence on health is attributed to its presence within at least 25 Sel in humans (Zhang et al., 2005; Ramoutar and Obrumaghim, 2010; Ledesma et al., 2011). In Table 2, Sel of eukaryotes as well as their non Sec containing homologues are presented.

Table 2. Selenoproteins of eukaryotes and brief description of their functions

Sel and non selenocysteine containing homologues	Abbreviation	Cellular distribution/ Major sites of expression	Function
Cytosolic Glutathione peroxidase	GPx1	Cytosol	Antioxidant protection
Gastrointestinal Glutathione peroxidase	GPx2	Gastrointestinal tract	Antioxidant protection
Plasma Glutathione peroxidase	GPx3	Extracellular space and plasma	Maintenance of cellular redox status
Phosholipid hydroperoxide Glutathione peroxidase	GPx4	Cell membrane, many other tissues	Detoxification of lipid hydroperoxides
Epididymal Glutathione peroxidase (non selenocysteine containing protein)	GPx5	Restricted expression to epididymis	Antioxidant protection during spermiogenesis and sperm maturation
Olfactory Glutathione peroxidase	GPx6	Olfactory epithelium, embryonic tissues	Antioxidant protection
Non-selenocysteine containing phospholipid Glutathione peroxidase	GPx7 (NPGPx)	Umbilical cord, ovary	Unknown, possible role in alleviating oxidative stress in breast cancer cells
Non selenocysteine containing glutatione peroxidase	GPX8	Oviduct	Unknown
Thioredoxin reductase Type II	TRXR2	Mitochondria, liver, kidney	Part of the Thioredoxin system. Antioxidant defense, redox regulation, cell signaling
Thioredoxin reductase Type III	TRXR3 (TGR)	Testes	Part of the Thioredoxin system. Antioxidant defense, redox regulation, cell signaling
Iodothyronine deiodinase Type I	D1	Many tissues like liver, kidney, thyroid	Conversion of T4 to T3 and T4 to reverse T3
Iodothyronine deiodinase Type II	D2	Liver, kidney, thyroid, brown adipose tissue	Conversion of T4 to T3
Iodothyronine deiodinase Type III	D3	Placenta, brain, skin, (not in pituitary, thyroid, adult liver)	Conversion of T4 to reverse T3

Table 2. (Continued)

Sel and non selenocysteine containing homologues	Abbreviation	Cellular distribution/ Major sites of expression	Function
Selenophosphate synthetase 1 (non selenocysteine containing protein)	SPS1	Protein that contains a Cys residue in place of Sec	Unknown role in Sel synthesis or any other biological process
Selenophosphate synthetase 2	SPS2	Testes, many other tissues	Role in biosynthesis of Sel. Synthesis of selenophosphate
15-kDa selenoprotein	Sel15	Endoplasmic reticulum, T cells, many other tissues	Role in cell apoptosis and mediation of chemopreventive effects of Se
Selenoprotein M	SelM	Brain and other tissues	Distantly related to Sel15. May be involved in cancer etiology
Selenoprotein H	SelH	Nucleus	Not fully known, possible upregulation of genes involved in glutathione synthesis
Selenoprotein I	SelI	Unknown	Studies with *E. coli* showed specific ethanolamine-phosphotransferase activity
Selenoprotein K	SelK	Cardiomyocytes	Possible antioxidant protection in cardiomyocytes
Selenoprotein N	SelN	Endoplasmic reticulum	It is linked with rigid spine syndrome
Selenoprotein O	SelO	Widely distributed. Only vertebrate homologues of SelO have selenocysteine	Unknown
Selenoprotein P	SelP	Plasma, other tissues	Involved in Se transport, antioxidant defense
Methionine-S-sulfoxide reductase A (non selenocysteine containing protein)	MsrA	Mitochondria	Reduction of methionine-S-sulfoxide
Methionine-R-sulfoxide reductase B1	MsrB1 (previously known as SelR or SelX)	Cytosol, nucleus	Reduction of oxidized methionine residues in damaged proteins
Methionine-R-sulfoxide reductase B2 (non selenocysteine containing protein)	MsrB2	Mitochondria	Reduction of oxidized methionine residues
Methionine-R-sulfoxide reductase B3 (non selenocysteine containing protein)	MsrB3	Endoplasmic reticulum, mitochondria	Reduction of oxidized methionine residues
Selenoprotein S	SelS	Endoplasmic reticulum	Cellular redox balance. Possible influence of inflammatory response

Sel and non selenocysteine containing homologues	Abbreviation	Cellular distribution/ Major sites of expression	Function
Selenoprotein T	SelT	Ubiquitous	Role in regulation of Ca^{2+} homeostasis and neuroendocrine secretion
Selenoprotein V	SelV	Testes	Unknown, possible role in redox regulation
Selenoprotein W	SelW1-SelW2	Heart, muscle and other tissues	Antioxidant protection
Fish 15-kDa Selenoprotein	Fep 15	Endoplasmic reticulum	Fish homologue of Sep15
Selenoprotein J	SelJ	Restricted to actinopterygian fishes and sea urchin	Structural role
Selenoprotein U	SelU1	SelU is found in the Sel form in fish and chicken but in mammals, all three SelU homologues are Cys containing proteins	Unknown
Plasmodium selenoproteins (1-4)	Sel1-Sel4	*Plasmodium* only	Unknown
Protein disulfide isomerase	PDI	Narrowly distributed in eukaryotes	Formation/breakage of disulfide bonds in proteins

Several families of mammalian and human Sel have been cloned and partially characterized with respect to their function (Gladyshev and Hatfield, 1999; Köhrle et al., 2000; Birringer et al., 2002; Kryukov et al., 2003). Mammals contain eight glutathione peroxidase homologues (Chabory et al., 2010), of which five are Sel, including GPxl (also known as cGPx), GPx2 (also known as GI-GPx), GPx3 (also known as pGPx), GPx4 (also known as PHGPx) and GPx6. In detail, glutathione peroxidase family (GPx) has a strong antioxidant role in cell cytosol (GPx1), gastrointestinal tract (GPx2), extracellular space and plasma (GPx3) and in cell membrane and sperm (GPx4). GPx5 is called epididymal GPx due to its restricted expression in the epididymis (Surai, 2006). GPx6 to GPx8 were first identified through large-scale mammalian sequencing programs. The GPx6 is located in olfactory epithelium and embryonic tissues with unknown function. The non-Sec (no selenocysteine) containing phospholipids hydroperoxide glutathione peroxidase (NPGPx or GPx7) is expressed in breast cancer cells (Utomo et al., 2004). Phylogenetic analysis has revealed the presence of a novel member belonging to the GPx family in mammalia and amphibia, and the name GPx8 has been proposed (Toppo et al., 2008). The major function of these peroxidases is considered to be the removal and detoxification of hydrogen peroxide (H_2O_2) and lipid hydroperoxides. Maintenance of cellular redox state is another important function. In addition, GPx are involved in such physiological events as differentiation, signal transduction and regulation of pro-inflammatory cytokine production. Another role of these enzymes is the antioxidant defense during spermiogenesis, maturation of spermatozoa and embryonic development (Ursini, 2000). The thioredoxin reductase (TRXR) family is comprised of 3 members (TRXR1, TRXR2 and TRXR3). The entire thioredoxin system in mammals is dependent on Se (Gladyshev, 2006). The biological role of the system is to provide antioxidant defense, regulate other antioxidant enzymes, control several transcription factors, regulate apoptosis and modulate protein phosphorylation (Surai, 2006). Iodothyronine deiodinase (D) family is comprised of 3 types, namely the D1, D2 and D3 with several defined roles in thyroid metabolism. The three deiodinases, enzymes that activate thyroxine (T4) and inactivate both T4 and T3, are present in all vertebrates. Their importance resides in the fact that T4 must be activated by deiodination to the short-lived biologically active T3 in order to initiate thyroid hormone action (Bianco and Larsen, 2006). Types D1 and D2 convert thyroxin (T4) to bioactive 3,5,3'-tri-iodothyronine (T3), while types D1 and D3 convert T4 to 3',3',5' reverse T3, which is a less bioactive form than T3. Thioredoxin reductases, deiodinases and glutathione peroxidases are all present in the thyroid gland and contribute to thyroid hormone biosynthesis, antioxidant defense and redox control of thyrocytes as well as to thyroid hormone metabolism (Köhrle and Gärtner, 2009).

Other Sel that may not be part of a family include, but not limited to, the selenophosphate synthetase 2 (SPS2), 15-kDa Sel (Sep15), SelH, SelI, SelK, SelM, SelN, SelO, SelP, MsrB1 (SelR/SelX), SelS, SelT, SelV, and SelW. Furthermore, in eukaryotes, the Sel family includes the Fep15, SelJ, SelU1, the *Plasmodium* Sel Sel1-Sel4 and the protein disulfide isomerase (PDI) which is narrowly distributed in eukaryotes. While the role of some of them is still largely unknown, the role of others is clearly understood. The greater attention has been received by the Sel of the humans. More specifically, SPS2 is involved in the synthesis of selenophosphate for Sel synthesis (Becket and Arthur, 2005). Sep15 seems to play a role in cell apoptosis and mediation of chemopreventive effects of Se (Papp et al., 2007). SelH regulates the expression levels of genes involved in de novo glutathione synthesis and phase II detoxification in response to redox status (Panee et al., 2007). SelI function is still elusive with current research pointing that its expression in *Escherichia coli* shows cytidine diphosphate ethanolamine-specific phosphatidyltransferase activity (Horibata and Hirabayashi, 2007). SelK may possess potent antioxidant properties in cardiomyocytes (Lu et al., 2006). SelM is distantly related to Sep15. It may be involved in the early-onset of Alzheimer's disease and play a role in cancer etiology (Kumaraswamy et al., 2000). SelN is a glycoprotein localized within the endoplasmic reticulum. It is directly linked to the rigid spine muscular dystrophy and the classical form of multiminicore disease (Petit et al., 2003). SelO is a widely distributed protein that has homologues in animals, bacteria, yeast and plants, but its function is unknown (Gladyshev, 2006). SelP is an abundant extracellular glycoprotein that is rich in Sec. It is involved in Se transport and antioxidant actions on endothelium. Evidence supports functions of the protein in Se homeostasis, antioxidant defense and transport/delivery of Se to remote tissues (Burk and Hill, 2005). Selenoprotein P is a major Sel in plasma, containing at least 40% of the total plasma Se (Rayman, 2009). Furthermore, plasma SelP seems to be a better indicator of Se nutritional status than the previously used GPx3 (Papp et al., 2007), because full expression of SelP requires a greater Se intake than does full expression of GPx3 (Xia et al., 2005). The methionine-R-sulfoxide reductase B1 (MsrB1, also known as SelR/Sel X) catalyzes the stereospecific reduction of oxidized methionine residues in damaged proteins with thioredoxin as reductant showing antiaging and neurologic properties (Kryukov et al., 2002; Kim and Gladyshev, 2004). SelS is an endoplasmic reticulum protein that participates in the processing and removal of misfolded proteins from the endoplasmic reticulum of mammalian cells to the cytosol where these proteins are further degraded (Rayman, 2009). Additionally, it may be implicated in the control of inflammation response (Curran et al., 2005). SelT plays a role in the regulation of

Ca^{2+} homeostasis and neuroendocrine secretion in response to a cAMP-stimulating trophic factor (Grumolato et al., 2008). SelV has a homology to SelW. It is expressed exclusively in testes. The protein contains specific amino acid sequence motives that predict a role in redox-regulation. Its function is still unknown (Kryukov et al., 2003). Finally, SelW seems to be implicated in antioxidant protection of cardiac and skeletal muscle (Whanger, 2000).

Synthesis of Selenoproteins from Dietary Selenium

Selenomethionine and Sec, are identical to methionine and cysteine except that the sulphur (S) atom is replaced by Se (Combs and Combs, 1984; Whanger, 2000). Yeast and higher plants lost the Sec insertion capability during evolution and therefore, do not possess Sel (Heras, 2011). Plants absorb Se from the soil in the form of selenite or selenate and synthesize SeMet (Rayman, 2004). That means that Se in natural feed ingredients is mainly in the form of SeMet (Combs, 2001). Furthermore, plants express cysteine-containing homologues (Lu and Holmgren, 2009). Vertebrates receive dietary Se in the forms of SeMet and other Se-amino acids, including Sec and its methylated forms, depending on their contents in feed/food components. In addition, currently, the feeds for farm animals are widely supplemented with inorganic Se sources like sodium selenite and sodium selenate as well as with organic form of Se, e.g. selenized yeast. There are principal differences in absorption and metabolism of these forms of Se (Ramoutar and Obrumaghim, 2010). For example, sodium selenite is passively absorbed in the intestine and more efficiently in the ileum segment of the intestine (Pesti, 1976) while, SeMet is actively absorbed in the intestine and thus requires a transport mechanism to actually move the molecules through the enterocyte cell membrane, by all segments of the intestine (Wolffram, 1999). Uptake of Se compounds into cells is assumed to occur *via* anion transporters (Wolffram et al., 1985; Shennan, 1988; Würmli et al., 1989; Huang et al., 1994; Vendeland et al., 1994; Park and Whanger, 1995). Irrespective of the form that Se is received from the diet, it has to be metabolized into Sec in order to be incorporated into Sel.

The incorporation of Se as Sec into a Sel requires a specific mechanism to decode the UGA codon in mRNA, which normally operates in translation termination (Mariotti and Guigó, 2010). Initially, the oxidized forms of inorganic Se (selenite or selenate) undergo reductive metabolism yielding hydrogen selenide (H_2Se) which is converted to Sel (Lu and Holmgren, 2009). Organic sources of Se can also be used for the production of H_2Se. SeMet can be trans-selenated into Sec, similarly to the trans-sulfuration pathway for methionine to cysteine, before lysis by β-lyase to H_2Se. Hydrogen selenide has first to be transformed to Sec. This means that H_2Se has to be metabolized into selenophosphate after catalysis by the SPS2 (Lu and Holmgren, 2009). Selenophosphate reacts with the tRNA specific for serine Ser-$tRNA^{Sec}$ *via* the enzyme Sec-tRNA synthase to give Sec bound tRNA (Sec-$tRNA^{Sec}$) from which Sec is inserted into Sel by the Sec specific UGA codon (Böck, 2000; Combs, 2001; Lacourciere and Stadtmanm, 2001; Rayman, 2004). This incorporation occurs when the mRNA contains a distinct hairpin mRNA sequence downstream of the UGA codon in its 3'-untranslated region (3'-UTR) called Sec insertion sequence (SECIS) or Sec translation element. SECIS prevents termination of the translation by competing for release factors that would otherwise lead to disassembly of the mRNA-ribosomal complex (Chambers et al.,

1986; Böck et al., 1991; Shen et al., 1995; Low and Berry, 1996). Biosynthesis of Sel requires binding of SBP2 (SECIS-binding protein 2) to the SECIS element and recruitment of the Sec tRNA-specific elongation factor (EFsec), connected to $tRNA^{Sec}$ (Bellinger et al., 2009). Namely, SECIS recruits SBP2 and binds the $tRNA^{Sec}$-loaded EFSec, while, several additional proteins bind to SECIS (Fujiwara et al., 1999; Copeland et al., 2000; Copeland and Driscoll, 2001; Schomburg et al., 2004). It seems that all Sel of eukaryotes require a form of the SECIS element for recoding UGA to the Sec codon (Bellinger et al., 2009). Most of the Sel can be classified into two groups according to the location of the Sec in the Sel polypeptide. In the first group, Sec is located on the N-terminal position of the function domain while in the second group Sec is present on the C-terminal (Papp et al., 2007). The first group includes all glutathione peroxidases and iodothyronine deiodinases, SPS2, Sep15, SelH, SelM, SelN, SelP, SelT, SelV, SelW while the second ones includes thioredoxin reductases, SelS, MsrB1, SelO, SelI and Sel K (Lu and Holmgren, 2009).

SELENIUM LEVELS AND REGULATION OF SELENOPROTEIN EXPRESSION

Selenium's essentiality and toxicity have been extensively investigated and well described (Surai, 2006). The disease "selenosis", related to high Se levels, has been reported in humans and in animals in seleniferous areas when intakes are in the range of 3200-6990 μg/day. No "selenosis" observed in the intake range of 240-1510 μg/day (EFSA, 2008). A safe and adequate range for Se intake of between 50 and 200 μg/day has been determined by the US Food and Nutrition Board (Institute of Medicine, 2000) and a Recommended Daily Allowance (RDA) of 55 μg/day Se for adult men and women has been established by the Board. The maximum authorized total Se contents in feed in European Union (EU), background Se plus supplemented one, for farm animals is 0.5 mg/kg of complete feeding stuffs with a moisture content of 12% (EU Commission, 2004). Selenium is supplemented as organo-Se compounds or in the form of inorganic Se. Previous reports reveal that the form of Se (i.e. sodium selenite or yeast products) plays an important role on the appearance or not of adverse toxicological effects on animal growth (Letavayová et al., 2006; Schrauzer, 2000). Selenium enriched yeast products are produced using sodium selenite as the mean for yeast growth. The form of Se in these products is typically that of the seleno-amino acid selenomethionine, accounting for approximately 60-85% of total Se species in the Se-enriched yeast product. Selenocysteine is the second most abundant identified species, approximating to 2-4% of total Se species. Inorganic Se ion is normally found at less than 1% of total, confirming that virtually all of the Se present in the product is organically bound. The remaining proportion is the sum of minor species (EFSA, 2008). Another source of Se is also available, the Se proteinate formed by the reaction of inorganic Se on a hydrolyzed soya protein. Despite the higher bioavailability of Se from organic sources, the toxicity of these organic forms has been shown to be lower than that of inorganic selenite or selenate. This suggests that the increased bioavailability may be counterbalanced by lower toxicity (Surai, 2006). Under this context, studies with ruminants fed at least 10 times the maximum permitted EU Se dietary inclusion rate in the form of Se-enriched yeast derived from a specific strain of *Saccharomyces cerevisiae* revealed no adverse effects on animal health, performance, and feed intake (Juniper et al., 2008). Since, Sel expression is regulated by Se itself it seems that highly controlled mechanisms must be in place to sustain optimal concentrations of Se within cells (Brigelius-Flohe and Banning, 2006).

A. Transcriptional Regulation

Tissue Se concentration and changes in fingernail morphology are regarded as conventional biomarkers of high Se status. However, they are lacking in specificity and sensitivity (Fairweather-Tait et al., 2010). The most promising biomarker appears to be SelP, which appears to reach a plateau after 2–4 wk of supplementation and is well correlated with plasma Se. SelP typically accounts for approximately half of the Se in plasma. It is generally more sensitive than other Sel, such as GPx, in both deficiency and after supplementation and, in addition, the response of SelP to different forms of Se appears to be similar (Fairweather-Tait et al., 2010).

Molecular biomarkers in contrast to conventional ones are potentially better predictors of physiological effects associated with high Se intake. Studies on the transcriptional effects of super-nutritional Se have not identified well-regulated molecular biomarkers of high Se status. In rodents several microarray studies have found 14 to 242 genes altered by a Se intake of 1.0 μg Se/g as compared to Se deficient diets. From all these genes, the ones that were consistently regulated were the Sel. Furthermore, the Se-specific effects detected were primarily caused by Se deficiency and not high Se (Raines and Sunde, 2011). Under this contexts, recent transcriptional studies in chicken revealed that some Sel, like GPx4, are regulated at transcriptional level by high dietary Se. Supranutritional Se level can down-regulate liver GPx4 mRNA levels which means that reserves built by excess of Se may meet antioxidant requirements and no additional GPx4 transcription is necessary (Zoidis et al., 2010).

Transcriptional regulation seems to be a regulatory point of Sel expression. The Nrf2/Keap1 system is one of the major cellular defense mechanisms against oxidative stress. NF-E2-related factor 2 (Nrf2) is the most effective transcription factor that acts through "antioxidant response element" (ARE), a member of the NF-E2 family of basic leucine zipper transcription factors. Kelch-like ECH-associated protein-1 (Keap1) is a cysteine-rich actin associated protein that keeps Nrf2 complexed in the cytosol. Nrf2/Keap1 regulates the expression of phase II detoxification enzymes and redox active proteins, including TRXR1. GPx2 is a target of this transcriptional system and may be up-regulated (Banning et al., 2005). The Nrf2/Keap1 system is important due to its activation by electrophilic compounds, metals, thiol modifiers, and other potential anticarcinogenic compounds derived from dietary sources (Brigelius-Flohe and Banning, 2006). Transcriptional regulation of additional GPx family members extends beyond the Nrf2/Keap1 system and has been reviewed (Brigelius-Flohe, 2006). A complex transcriptional regulation pattern involving interplay between several transcription factors, such as Oct-1, Sp1, Sp3 and multiple transcription start sites in a cell has been reported for the TRXR1 gene (Rundlof and Arner, 2004). Characterization of the promoter region and transcriptional regulation of the GPx4 gene also has been reported (Maiorino et al., 2003; Imai et al., 2006).

B. Post-Transcriptional Regulation

Selenoprotein synthesis is regulated also at post-transcriptional level (Behne and Kyriakopoulos, 2001; Hatfield and Gladyshev, 2002; Driscoll and Copeland, 2003; Schomburg et al., 2004; Caban and Copeland, 2006). A hierarchy in Sel expression during Se deprivation and repletion has been reported (Behne and Kyriakopoulos, 2001). The same authors showed that some tissues and organs are more efficient in maintaining Se levels and the production of certain Sel during Se deprivation compared to other ones. This is indicative of differences in the requirements and biologic roles of Sel in different tissues (Brigelius-Flohe 1999; Behne and Kyriakopoulos, 2001). Selenium deficiency leads to a dramatic loss of activity of Sel, including TRXRs, GPxs, and Ds. The hierarchy during Se deprivation and repletion reveals the significance of specific Sel and in turn determines the priority of the mRNA level and protein expression (Lu and Holmgren, 2009). In detail, in Se-deficient conditions, the activities of most Sel in the liver, kidney, and lung decrease while in the brain remain at a level similar to that during normal Se intake levels (Lu and Holmgren, 2009).

Stability of mRNA may be regulated by Se level since in cases of Se deficiency an increased susceptibility to the nonsense-mediated decay pathway (NMD) and consequently decay of mRNA have been noted (Moriarty et al., 1998; Maquat, 2001; Weiss Sachdev and Sunde, 2001). Selenium may also efficiently control UGA-Sec codon translation (Fletcher et al., 2000; Martin and Berry, 2001), regulate total Sec $tRNA^{Ser(Sec)}$ levels and control the ratio between the methylated and unmethylated Sec $tRNA^{Ser(Sec)}$ isoforms (Hatfield et al., 1991; Chittum et al., 1997; Jameson et al., 2002; Carlson et al., 2005). The methylated isoform of $tRNA^{[Ser]Sec}$ is translationally active and that Se-induced tRNA methylation is a mechanism of regulation of Sel synthesis. It was discovered that the methylated isoform controls the synthesis of Sel involved in the oxidative stress response such as GPx1 and GPx3, whereas the unmethylated form governs synthesis of housekeeping Sel such as TRXR1 and TRXR3 (Jameson and Diamond, 2004).

C. Redox Regulation

Redox regulation has emerged as an essential regulatory process of many pathways in cell biology (Linke and Jakob, 2003; Ghezzi, 2005; Battin and Brumaghim, 2009). Disruption of the intracellular redox balance leads to a state of oxidative stress, during which proteins, nucleic acids, lipids, and other macromolecules can suffer severe damage (Surai, 2006). Oxidative stress appears to be a major factor in aging. It has been implicated in numerous diseases such as Alzheimer's, diabetes and cancer (Berlett and Stadtman, 1997; Kovacic and Jacintho, 2001; Aliev et al., 2002).

TRXRs and GPxs, through the action of Sec within their catalytic sites, serve housekeeping redox functions. This is mediated by controlling the activity of cellular proteins and scavenging free radicals. On the other hand, these antioxidant enzymes respond to oxidative stress by inducing their gene expression and by changing their activity and subcellular localization (Hirota et al., 1997; Karimpour et al., 2002; Hattori et al., 2005). In response to cellular stress conditions, protein translation is reduced allowing the cells to conserve resources in order to initiate a reconfiguration of gene expression (Holcik and

Sonenberg, 2005). It can be concluded that regulation of Sel synthesis is a complicated process involving redox regulation through the thioredoxin and glutaredoxin systems.

HEALTH EFFECTS

Selenium is fundamental for life and adequate amounts of this element are required for optimal animal and human health. The population of many countries in Europe and other parts of the world still have a dietary Se intake below of that of 55 μg/day (RDA, recommended dietary allowance (USA); PRI, population reference intake (EU)) recommended by health regulatory bodies such as the Institute of Medicine in USA and the European Food Safety Authority (Institute of Medicine, 2000; Combs, 2001; Rayman, 2005; EFSA, 2009; Fairweather-Tait, 2011). As mentioned previously, these recommendations were based on the GPx3 optimal enzyme activity (Thomson et al., 1993; Duffield et al., 1999), although recent studies showed that SelP may be a better indicator of Se nutritional status (Xia et al., 2005). It seems that the recommended dietary intake may need to be revised.

The precise molecular mechanisms behind the effects of Se in physiological and in pathological conditions remain elusive. Most of its physiological roles are directly attributed to its presence within Sel. Moderate Se deficiency has been linked to many disorders, such as increased cancer and infection risk, male infertility, decrease in immune and thyroid function, and several neurologic conditions, such as epilepsie, Alzheimer's- and Parkinson's disease (Rayman, 2000). For some of these disorders, the evidence is rather insufficient, lacks consensus and should be further established. At times of low Se intake, the reserves built up in the tissues begin to diminish. Among different tissues the rate of depletion is different. Brain, reproductive and endocrine organs maintain the Se levels longer compared to liver, muscle and skin that rapidly lose their Se content (Behne et al., 1988). The hierarchy of Se retention during Se depletion exists not only among tissues and organs but also among Sel. During depletion, Se is rapidly mobilized from GPx1 stores, whereas expression of other Sel such as GPx4, GPx2, D2, D3 and TRXR is hardly affected or may even be increased, like in case of D1. This hierarchy is reflected and in the level of mRNA with some mRNAs to be preferentially translated into Sel (Grunder-Culeman et al., 1999; Low et al., 2000; Zoidis et al., 2010). Those proteins residing high in the hierarchy of Se retention during Se depletion also seem to lead in the priority for repletion (Hill et al., 1992; Bermano et al., 1995; Gross et al., 1995; Lei et al., 1995; Michell et al., 1998; Driscoll and Copeland, 2003).

A. Cardiovascular Diseases

Selenoprotein function in cardiovascular disease has been investigated primarily by analysis of oxidative stress under conditions of Se supplementation and/or deficiency (Steinbrenner and Sies, 2009). Oxidative stress impairs vascular endothelial cells and provokes cardiovascular diseases such as atherosclerosis, hypertension, and congestive heart failure (Lum and Roebuck, 2001). Selenoproteins are crucially involved in the antioxidant defense system of the organism and the use of Se to prevent or even treat cardiovascular diseases has been under investigation for many years (Bellinger et al., 2009). Although there are several epidemiological and clinical studies, the present chapter will focus on research studies. Selenium supplementation elevates mRNA expression and activity of GPx1, GPx4 and TRXR1 in vascular endothelial or smooth muscle cells and thus inhibits oxidative stress, cell damage and apoptosis from oxidized LDL (low-density lipoprotein) or triol, a cytotoxic hydroxylated cholesterol derivative found in blood, cells, tissues and atherosclerotic plaques in humans (Thomas et al., 1993; Miller et al., 2001; Tang et al., 2005; Steinbrenner et al., 2006). Similarly, in rodents, long-term Se deficiency strongly decreases GPx activity and mRNA expression and increases both physiological and cholesterol oxide-induced damage to the heart and vasculature. Dietary Se supplementation can reverse these effects (Huang et al., 2002; Wu and Huang, 2004; Stranges et al., 2010). Furthermore, Se supplemented animals and their offspring exhibit reduced ischaemia-induced oxidative damage to the heart and improved recovery of cardiac function (Ostadalova et al., 2007; Venardos et al., 2004).

The precise role of specific Sel in cardiovascular disease has been partially elucidated, particularly with the GPx enzymes. GPx1 has been shown to inhibit ischemia/reperfusion induced apoptosis of cardiac myocytes in mice (Maulik et al., 1999). Genetic deletion of GPx1 in mice produces heart and vascular dysfunction and tissue irregularities (Forgione et al., 2002). Moreover, GPx1-overexpressing mice are more resistant than wild-type to doxorubicin-induced cardiac dysfunction as measured by heart contractility, blood flow rate and heartbeat rate (Xiong et al., 2006). GPx3 is expressed mainly in plasma and probably regulates redox dependent functions of the vasculature. Excess ROS due to decreased GPx3 activity results in inadequate nitric oxide (NO) levels, which, in turn, disrupts platelet inhibitory mechanisms and increases arterial thrombosis (Kenet et al., 1999). Additionally, hypoxia regulates GPx3 expression (Bierl et al., 2004), and there is evidence of an association between polymorphisms in the GPx3 promoter and increased risk of ischemic stroke (Voetsch et al., 2007; Voetsch et al., 2008). Overexpression of GPx4 reduces the atherogenic effects of lysophosphatidylcholine and 7-oxocholesterol, including necrosis and apoptosis of endothelial cells (Guo et al., 2001). Furthermore, overexpression of mitochondrial GPx4 protects against simulated ischemia/reperfusion in neonatal cardiomyocytes in vitro (Hollander et al., 2003). Mice heterozygous for GPx4 exhibit massive lipid peroxidation that produces cell death, which is dependent on 12/15-lipoxygenase and is mediated by apoptosis-inducing factor (Seiler et al., 2008). Taken together, these studies suggest that GPx4 inhibits atherosclerosis by reducing lipid and lipoprotein oxidation and downstream damaging mechanisms. Several reviews implicate the TRXR/TRX system in regulating processes of the cardiovascular system (Ago and Sadoshima, 2006; Berndt et al., 2007; World et al., 2006). Changes in the intracellular redox environment alter inter- and intra-cellular signaling (Arner and Holmgren, 2000; Maulik and Das, 2008), including activation of hypertrophic and apoptotic pathways in cardiac myocytes (Nakamura et al., 1998; Amin et al., 2001; Tanaka et al., 2001; Remondino et al., 2003; Pimentel et al., 2006). Furthermore, the TRXR/TRX system contributes in regulating myocardial remodeling through the reversible oxidation of signaling molecules (Ago and Sadoshima, 2006; Berndt et al., 2007). For example, adrenergic receptor activation induced hypertrophy of adult rat cardiomyocytes is affected by the oxidation of cysteine thiols of Ras oncogene that can be reduced by TRXR1 (Kuster et al., 2005). It is noteworthy that TRXRs directly reduce substrates other than TRX (Andersson et al., 1996), which may have relevant effects on heart and vascular function. SelK is a transmembrane protein localized to endoplasmic reticulum that has an antioxidant function in cardiomyocytes and high mRNA expression in the heart (Lu et al., 2006). Plasma SelP supplies Se to cells (Burk and Hill, 2009), probably supporting optimal expression of GPxs, TRXRs and other Se dependent enzymes. Additionally, SelP reduces peroxynitrite-induced protein oxidation and nitration, as well as lipid and LDL peroxidation (Arteel et al., 1998), at the expense of oxidizing TRX (Takebe et al., 2002). It can be concluded that there is need for a mechanistic understanding of specific Sel function in the cardiovascular system in order to clearly determine the therapeutic benefits of Se (Bellinger et al., 2009).

B. Cancer

A brief overview of the role of Sel in cancer will be provided in the present piece of work. There is a lot of available data about the role of Sel in cancer etiology (Squires and Berry, 2006; Jackson and Combs, 2008; Brigelius-Flohe, 2008; Hatfield et al., 2009; Heras et al., 2011) and the cancer-preventing properties of Se (Diwadkar-Navsariwala and Diamond, 2004; Squires and Berry, 2006; Brigelius-Flohe, 2008). Several mechanisms through which Sel would exert their protective effect against cancer have been proposed, namely modulation of cellular division rate, metabolic alteration of some carcinogenic agents with a decrease in the formation of carcinogenic metabolites, cell antioxidant protection against oxidative damage, stimulation of the immune system, inhibition of angiogenesis and induction of apoptosis in cancer cells, inhibition of the activity of hepatic enzymes and activation of detoxifying enzymes (Navarro-Alarcon and Lopez-Martinez, 2000; Brozmanová et al., 2010; Ramoutar and Brumaghim, 2010; Suzuki et al., 2010; Valdiglesias et al., 2010; Fairweather-Tait et al., 2011). There are many studies that indicate that Se could reduce the risk of different forms of cancer (Squires and Berry, 2006; Selenius et al., 2010). Supplemental Se has been shown to have cancer-protective effects in a variety of experimental settings and clinical studies (reviewed by Whanger, 2004) and to reduce the incidence and mortality of total cancer, prostate cancer, liver cancer, and stomach cancer in human interventional trials (Valdiglesias et al., 2010). The anticarcinogenic effect of Se against leukemia and cancers of the colon, rectum, pancreas, breast, ovaries, prostate, bladder, lung, and skin seems clear at least under some conditions (Rayman, 2000; Whanger, 2004; Valdiglesias et al., 2010).

On the contrary, there are some studies that indicate that Se may be implicated, under certain conditions (i.e. form of supplemental Se), in Type 2 diabetes and other diseases including cancer itself. In detail, secondary analysis of the NPC (Nutritional Prevention of Cancer) study (Stranges et al., 2007) and the recently reported SELECT (Selenium and Vitamin E Cancer Prevention Trial) study (Ledesma et al., 2011; Lippman et al., 2009) also raised the possibility that Se increased the risk of Type 2 diabetes. Many Sel gene polymorphisms have been linked to risk of cancer. Polymorphisms of GPx1 have been linked to various forms of cancer, including lung, breast, prostate, head and neck cancer (Hu et al., 2003; Hu et al., 2005; Foster et al., 2006). Polymorphisms in GPx2, GPx4 and SelP have been implicated in colorectal cancer (Al-Taie et al., 2004; Bermano et al., 2007), Sep15 polymorphisms may increase lung cancer risk (Jablonska et al., 2008), whereas SelS promoter polymorphisms have been linked to gastric cancer (Shibata et al., 2009). Recently, epistasis between polymorphisms of SelP and mitochondrial superoxide dismutase (SOD) were shown to confer risk of prostate cancer (Cooper et al., 2008). Additionally, changes in expression of GPx1, GPx2, Sep15, SelP and TRXR1 have been observed in different forms of cancer (Diwadkar-Navsariwala and Diamond, 2004; Squires and Berry, 2006).

The NPC trials originally sought to determine whether Se supplementation could reduce the risk of skin carcinomas. Although skin cancer incidence did not differ between groups, the original study found decreases in total incidence of cancer and of prostate, lung and colorectal cancers (Clark et al., 1996). Follow-up studies confirmed the protective effect of Se in preventing prostate cancer (Duffield-Lillico et al., 2003). Thus it is surprising that the SELECT study found no significant reduction in prostate cancer with Se supplementation (Lippman et al., 2009). However, the supplementation of trial participants was terminated early because of concerns about diabetes and increased prostate cancer from vitamin E, although the subjects in the study are still being monitored for possible health benefits. It is possible that the trial was terminated too early to observe changes similar to those seen with the NPC trial. The SELECT study had several design differences from the earlier NPC trial, including the use of purified selenomethionine in supplements as opposed to selenized yeast used in the earlier trial (Lippman et al., 2009). It seems that the source of the Se supplement in SELECT and the relatively high initial levels of Se in the enrolled men have contributed to the negative results obtained in this trial (Hatfield and Gladyshev, 2009). Interestingly, the combination of Se and vitamin E did not seem to increase either diabetes or prostate cancer. The incidences of these disorders with the dual-supplement group were notably lower than the increases found with either supplement alone. The direct roles of Sel should be examined in order to assess whether supplementation is advisable for treatment or prevention of a specific disease (Sunde et al., 2008; Barnes et al., 2009; Bellinger et al., 2009; Rayman, 2009, Brozmanová et al., 2010; Novonty et al., 2010).

C. Endocrine Disorders

Selenoproteins are of central importance in the production of hormones and growth factors, especially for thyroid hormone production (Köhrle et al., 2005; Köhrle and Gärtner, 2009; Fairweather-Tait et al., 2011). The energy demands of endocrine tissue as well as redox reactions involved in the production and release of factors require Sel such as GPxs and TRXRs to prevent accumulation of ROS. Activation of thyroid hormone is dependent upon the iodothyronine deiodinases class of Sel. These enzymes, as discussed earlier catalyze the deiodination of the prohormone T4 to the active hormone T3, and to the inactive metabolites rT3 (reverse tri-iodothyronine) and T2 (di-iodothyronine) (Larsen and Berry, 1995; Visser, 1996). Mutations in SBP2 were found to be responsible for deficiencies in thyroid function in the case of two families (Dumitrescu et al., 2005). One of these mutations impaired exon splicing, leading to an intron retention that changed the reading frame, resulting in a truncated SBP2. The altered SBP2 led to decreased levels of D2 (Beckett and Arthur, 2005). Thus mutations in the machinery essential for Sel production can negatively affect health status (Papp et al., 2007). Nevertheless, the mutations in SBP2 do not have greater consequences. Targeted disruption of the genes encoding several Sel, including TRXR1 and TRXR2, and GPx4, or of the gene encoding $tRNA^{Sec}$, leads to embryonic or early postnatal lethality (Schweizer et al., 2004). Thus impairing the Sel synthesis machinery could be expected to have more severe consequences than those seen with SBP2 mutations. Part of the answer may be in the complexity of SBP2 and its varying affinities for different Sel mRNAs. The RNA binding domain is intact in the SBP2 mutations, but selectivity for differing SECIS elements is impaired (Squires et al., 2007). Two major forms of SECIS elements, forms 1 and 2 (numbered in order of discovery), have been described, although unique elements exist in some Sel messages (Small-Howard and Berry, 2005). The wild-type form of SBP2 has greater affinity for form 2; one mutation discussed above renders SBP2 less selective as well as reducing overall affinity (Squires et al., 2007). Of course, it is likely that mutations in SBP2 that completely prevented its function would be lethal and never detected in adults and families. The mutations in SBP2 make up an interesting example of how synthesis of Sel can be selectively impaired to cause a specific disorder (Bellinger et al., 2009).

Diabetes mellitus is a disorder resulting in impaired control of blood glucose levels by either impaired insulin release (Type 1) or impaired insulin function or insulin resistance (Type 2). The resulting hyperglycaemia increases ROS production, which may contribute to the progression of this disorder (Roberts and Sindhu, 2009). Diabetic men and women have a 2-5-fold higher risk of coronary heart disease, stroke and peripheral vascular disease than matched non-diabetic individuals (Navarro-Alarcon et al., 1999). Therefore, reducing risk of vascular disease requires, among other, improved glycemic control. Additionally, the erythrocyte and serum Se concentrations, as well as Sel and SOD activities are frequently reduced. The epidemiological studies performed in erythrocytes of diabetic patients have indicated an enhanced lipidic peroxidation and GPx1 activity as compared with healthy controls. Nevertheless, the SOD activity was impaired as much as that the enhancement of the GPx1 activity was not able to compensate it. The final result was an increase in peroxidation (Navarro-Alarcon and Lopez-Martinez, 2000). Some studies have suggested that Se may be beneficial in treating diabetes (Battell et al., 1998; Faure, 2003; Ozdemir et al., 2005; Aydemir-Koksoy and Turan, 2008). However, recent clinical trials such as the SELECT study have suggested a possible risk of developing Type 2 diabetes resulting from Se supplementation (Bleys et al., 2007; Stranges et al., 2007; Lippman et al., 2009). Selenium possesses insulin-mimetic properties in vitro and in vivo that appear to be independent of insulin release (McNeill et al., 1991; Ghosh et al., 1994; Beckett and Arthur, 2005), which could potentially accelerate development of insulin resistance. SelS is glucose-regulated, and was originally discovered in a rodent model for diabetes (Walder et al., 2002; Karlsson et al., 2004). Mice overexpressing GPx1 develop insulin-resistance, a hallmark of Type 2 diabetes (McClung et al., 2004). GPx1 is increased by Se supplementation (Sunde et al., 2005), and thus may have a role in the apparent increased risk of diabetes reported in recent Se-supplementation studies (Stranges et al., 2007; Lippman et al., 2009). Further research on the function of Sel and diabetes is required in order to clarify their role in the onset, progression and treatment of the disorder.

D. Neurological Diseases

Selenium is retained within the brain even under conditions of dietary Se deficiency, implying its potential role in neurological disorders (Behne et al., 1988; Nakayama et al., 2007). Damage from ROS takes place in neurodegenerative disorders such as Alzheimer's disease (AD), Parkinson's disease (PD), ischemic damage, exposure to environmental toxins and drugs of abuse, and brain tumors (Chen and Berry, 2003).

Oxidative damage to macromolecules is an early indication of AD that can appear before clinical symptoms (Moreira et al., 2006). Alzheimer's disease patients suffer memory loss, impaired cognitive function and changes in behavior and personality (Reddy and Beal, 2005).

The brains of AD patients can be identified by their characteristic extracellular plaques consisting of the protein, amyloid β, as well as by intracellular neurofibrillary tangles. Most cases of AD are 'late-onset', progressing with age, and the causes are unclear. However, several autosomal dominant mutations have been identified that can result in 'early-onset' AD. One of these is a mutation in presenilin-2, an enzyme involved in processing amyloid precursor protein (Kowalska et al., 2004). A mouse model overexpressing the human mutation has reduced levels of brain SelM, an ER-specific Sel with antioxidant and ER-protein folding functions (Hwang et al., 2005). Thus, SelM may have a protective role in AD. Although originally identified as a plasma protein, SelP is abundant in neurons and ependymal cells in the human brain (Scharpf et al., 2007). Expression of SelP in brain increases with aging, suggesting it may play a role in ameliorating oxidative stress (Lu et al., 2004). Genetic deletion of SelP impairs synaptic function in the hippocampus; a region involved in memory, and reduces spatial learning as well as long-term potentiating, a cellular model for learning and memory (Peters et al., 2006). A recent analysis of expression data indicated that SelP was also increased in AD beyond that found in aging (Miller et al., 2008). Although a specific role for SelP in AD is uncertain, the co-localization of SelP with plaques and neurofibrillary tangles suggests that it could play a role in mitigating the oxidation accompanying plaques (Bellinger et al., 2008). Serum SelP is greatly influenced by dietary Se, and thus Se supplementation may have a direct neuroprotective role by increasing SelP expression. Recent studies have suggested that Se supplementation can decrease amyloid toxicity in cell culture and animal models (Lovell et al., 2009; Strozyk et al., 2009). An ancillary study of the SELECT study, PREADVISE (Prevention of Alzheimer's Disease by Vitamin E and Selenium), is currently in progress (expected completion in August 2013) to examine the possible benefits of increased dietary Se on preventing AD (Kryscio et al., 2004). Although participants of the SELECT study have been advised to discontinue their supplements because of the possibility that Se may increase the risk of Type 2 diabetes or that vitamin E may increase the risk of prostate cancer, the cohort is still being monitored (Lippman et al., 2009). A decrease in the risk of AD could possibly justify any increase in the risk of diabetes with Se supplementation for individuals with family history or early signs of the disease (Bellinger et al., 2009).

Dopamine is a neurotransmitter that controls many important brain functions despite being released from only 2% of neurons in the brain (Chinta and Andersen, 2005). Severe loss of dopamine-releasing neurons in the substantia nigra is central to the neurodegenerative disorder, PD (Iversen and Iversen, 2007). Symptoms of PD include rigidity, tremor and loss of movement control, with mood changes and cognitive impairments found in later stages of the disease (Fahn, 2003). Parkinson's disease is characterized by loss of dopamine terminals in putamen and caudate within the striatum from neurons projecting from the substantia nigra (the nigrostriatal pathway). The dopaminergic neurons in substantia nigra exhibit lesions termed 'Lewy bodies', made up of aggregates of ubiquitinated α-synuclein (Galvin, 2006). Several findings suggest an involvement of Sel in preserving the nigrostriatal pathway. The substantia nigra and putamen have higher concentrations of Se than other brain regions (Chen and Berry, 2003). Selenium deficiency increases pathology in mouse models of the disease (Imam et al., 1999; Kim et al., 1999; Imam and Ali, 2000; Virmami et al., 2003). Parkinson's disease patients have an approx. 50% decrease in glutathione, suggesting impaired GPx function (Zeevalk et al., 2008). Chemical lesions of dopaminergic terminals and neurons are greatly exacerbated in Se-deficient animals (Kim et al., 1999; Kim et al., 2000), whereas Se supplementation was protective to dopamine neurons and up-regulated GPx activity (Islam et al., 2002; Zafar et al., 2003). A recent report demonstrated that GPx1 is associated with microglia in PD pathology (Power and Blumberg, 2009). Knockout of GPx1 in mice greatly potentiates dopamine loss and pathology in a rodent PD model (Klivenyi et al., 2004), whereas overexpression of GPx1 has a protective role (Bensadoun et al., 1998; Ridet et al., 2006). Thus, GPxs and other Sel may play important roles in protecting dopaminergic transmission and preventing PD. However, to date, no changes in Sel expression or function have been reported to correlate directly with this disease.

Epilepsy is a chronic neurological disorder characterized by seizures which cause interruptions in normal brain function (Fisher et al., 2005). There are many classifications of epilepsy syndromes, with each seizure type presenting unique problems, and thus treatment options. Owing to the variations within this disorder, additional treatments for epilepsy are being explored. A clinical study performed in infants showed that low blood levels of Se lead to infant seizures and neurological conditions (Ashrafi et al., 2007). Epilepsy, ischemia and brain trauma cause a signal cascade of free radicals and activation of pro-apoptotic transcription factors, resulting in neuronal loss (Savaskan et al., 2003). Rats on Se-deficient diets had increased susceptibility to kainate-induced seizures and cell loss (Savaskan et al., 2003). Another study combining Se and topiramate (TPM) (an anti-epileptic drug which inhibits voltage gated sodium and calcium channels), showed protective effects following PTZ (pentylentetrazol)-induced seizures (Naziroglu et al., 2008). GPx and plasma membrane calcium ATPase activity were increased following PTZ challenge in rats treated with Se and TPM, thus inhibiting free radical production and regulating calcium-dependent processes (Kutluhan et al., 2009). SelP-knockout mice develop neurological seizures and movement disorders when raised on restricted Se diets (Hill et al., 2003; Schomburg et al., 2003), providing further evidence for a possible role for Sel in preventing epilepsy.

CONCLUSION

Selenoproteins require a common set of cofactors for their synthesis, and are dependent upon dietary Se intake. The energy cost of the organism to produce and maintain these cofactors and synthesize Sel suggests the collective importance of this protein family to cell function. However, the functions of these proteins are quite heterogeneous. Selenoproteins play important roles in numerous diseases and conditions, including cardiovascular disorders, cancer, neurodegeneration and endocrine disorders. In view of the diverse roles of Sel, strategies to target expression and/or function of specific Sel could be considered for therapeutic treatment and prevention of disorders. Different dietary forms of Se may selectively increase specific Sel. Pharmaceuticals could also target specific Sel or factors involved in their synthesis. The functions of many Sel remain unknown, that is why understanding the function of each member of the Sel family will be an important tool in determining the health benefits of Se.

REFERENCES

Ago, T. and Sadoshima, J. (2006). Thioredoxin and ventricular remodeling. *Journal of Molecular and Cellular Cardiology.* 41, 762–773.

Aliev, G., Smith, M.A., Seyidov, D., Neal, M.L., Lamb, B.T., Nunomura, A., Gasimov, E.K., Vinters, H.V., Perry, G., LaManna, J.C., Friedland, R.P. (2002). The role of oxidative stress in the pathophysiology of cerebrovascular lesions in Alzheimer's disease. *Brain Pathology.* 12, 21–35.

Al-Taie, O.H., Uceyler, N., Eubner, U., Jakob, F., Mork, H., Scheurlen, M., Brigelius-Flohe, R., Schottker, K., Abel, J., Thalheimer, A., Katzenberger, T., Illert, B., Melcher, R., Köhrle, J. (2004). Expression profiling and genetic alterations of the selenoproteins GI-GPx and SePP in colorectal carcinogenesis. *Nutrition and Cancer.* 48, 6–14.

Amin, J.K., Xiao, L., Pimental, D.R., Pagano, P.J., Singh, K., Sawyer, D.B., Colucci, W.S. (2001). Reactive oxygen species mediate α-adrenergic receptor-stimulated hypertrophy in adult rat ventricular myocytes. *Journal of Molecular and Cellular Cardiology.* 33, 131–139.

Andersson, M., Holmgren, A., Spyrou, G. (1996). NK-lysin, a disulfide-containing effector peptide of T-lymphocytes, is reduced and inactivated by human thioredoxin reductase: implication for a protective mechanism against NK-lysin cytotoxicity. *Journal of Biological Chemistry.* 271, 10116–10120.

Arner, E.S. and Holmgren, A. (2000). Physiological functions of thioredoxin and thioredoxin reductase. *European Journal of Biochemistry.* 267, 6102–6109.

Arteel, G.E., Mostert, V., Oubrahim, H., Briviba, K., Abel, J., Sies, H. (1998). Protection by selenoprotein P in human plasma against peroxynitrite-mediated oxidation and nitration. *Biological Chemistry.* 379, 1201–1205.

Ashrafi, M.R., Shabanian, R., Abbaskhanian, A., Nasirian, A., Ghofrani, M., Mohammadi, M., Zamani, G.R., Kayhanidoost, Z., Ebrahimi, S., Pourpak, Z. (2007). Selenium and intractable epilepsy: is there any correlation? *Pediatric Neurology.* 36, 25–29.

Aydemir-Koksoy, A. and Turan, B. (2008). Selenium inhibits proliferation signaling and restores sodium/potassium pump function of diabetic rat aorta. *Biological Trace Element Research.* 126, 237–245.

Banning, A., Deubel, S., Kluth, D., Zhou, Z., Brigelius-Flohe, R. (2005). The GI-GPx gene is a target for Nrf2. *Molecular and Cellular Biology.* 25, 4914–4923.

Barnes, K.M., Evenson, J.K., Raines, A.M., Sunde, R.A. (2009). Transcript analysis of the selenoproteome indicates that dietary selenium requirements of rats based on selenium-regulated selenoprotein mRNA levels are uniformly less than those based on glutathione peroxidase activity. *Journal of Nutrition.* 139, 199–206.

Battell, M.L., Delgatty, H.L., McNeill, J.H. (1998). Sodium selenate corrects glucose tolerance and heart function in STZ diabetic rats. *Molecular and Cellular Biochemistry.* 179, 27–34.

Battin, E.E. and Brumaghim, J.L. (2009). Antioxidant activity of sulfur and selenium: A review of reactive oxygen species scavenging, glutathione peroxidase, and metal-binding antioxidant mechanisms. *Cell Biochemistry and Biophysics.* 55, 1–23.

Beckett, G.J. and Arthur, J.R. (2005). Selenium and endocrine systems *Journal of Endocrinol*ogy. 184, 455-465.

Behne, D., Hilmert, H., Scheid, S., Gessner, H., Elger, W. (1988). Evidence for specific selenium target tissues and new biologically important selenoproteins. *Biochimica et Biophysica Acta*. 966, 12–21.

Behne, D. and Kyriakopoulos, A. (2001). Mammalian selenium-containing proteins. *Annual Review of Nutrition*. 21, 453–473.

Bellinger, F.P., He, Q.P., Bellinger, M.T., Lin, Y., Raman, A.V., White, L.R., Berry, M.J. (2008). Association of selenoprotein P with Alzheimer's pathology in human cortex. *Journal of Alzheimers Disease*. 15, 465–472.

Bellinger, F.P., Raman, A.V., Reeves, M.A., Berry, M.J. (2009). Regulation and function of selenoproteins in human disease. *Biochemical Journal*. 422, 11–22.

Bensadoun, J.C., Mirochnitchenko, O., Inouye, M., Aebischer, P., Zurn, A.D. (1998). Attenuation of 6-OHDA-induced neurotoxicity in glutathione peroxidase transgenic mice. *European Journal of Neuroscience*. 10, 3231–3236.

Berlett, B.S. and Stadtman, E.R. (1997). Protein oxidation in aging, disease and oxidative stress. *Journal of Biological Chemistry*. 272, 20313–20316.

Bermano, G., Nicol, F., Dyer, J.A., Sunde, R.A., Beckett, G.J., Arthur, J.R., Hesketh, J.E. (1995). Tissue-specific regulation of selenoenzyme gene expression during selenium deficiency in rats. *Biochemical Journal*. 311, 425–430.

Bermano, G., Pagmantidis, V., Holloway, N., Kadri, S., Mowat, N.A., Shiel, R.S., Arthur, J.R., Mathers, J.C., Daly, A.K., Broom, J., Hesketh, J.E. (2007). Evidence that a polymorphism within the 3'-UTR of glutathione peroxidase 4 is functional and is associated with susceptibility to colorectal cancer. *Genes and Nutrition*. 2, 225–232.

Berndt, C., Lillig, C.H., Holmgren, A. (2007). Thiol-based mechanisms of the thioredoxin and glutaredoxin systems: implications for diseases in the cardiovascular system. *American Journal of Physiology*. 292, H1227–H1236.

Bianco, A.C. and Larsen, P.R. (2006). Selenium, deiodinases and endocrine function. In: Hatfield, D.L., Berry, M.J., Gladyshev, V.N. (Eds.), Selenium: Its molecular biology and role in human health. Second Edition, Springer science + business media LLC, New York, USA, pp. 207-219.

Bierl, C., Voetsch, B., Jin, R. C., Handy, D.E., Loscalzo, J. (2004). Determinants of human plasma glutathione peroxidase (GPx3) expression. *Journal of Biological Chemistry*. 279, 26839–26845.

Birringer, M., Pilawa, S., Flohe, L. (2002). Trends in selenium biochemistry. *Natural Product Reports*. 19, 693–718.

Bleys, J., Navas-Acien, A., Guallar, E. (2007). Serum selenium and diabetes in U.S. adults. *Diabetes Care*. 30, 829–834.

Böck, A. (2000). Biosynthesis of selenoproteins–an overview. *Biofactors*. 11, 77–78.

Böck, A., Forchhammer, K., Heider, J., Leinfelder, W., Sawers, G., Veprek, B., Zinoni, F. (1991). Selenocysteine: the 21st amino acid. *Molecular Microbiology*. 5, 515–520.

Brigelius-Flohe, R. (2008). Selenium compounds and selenoproteins in cancer. *Chemistry and Biodiversity*. 5, 389–395.

Brigelius-Flohe, R. (1999). Tissue-specific functions of individual glutathione peroxidases. *Free Radical Biology and Medicine*. 27, 951–965.

Brigelius-Flohe, R. (2006). Glutathione peroxidases and redox-regulated transcription factors. *Biological Chemistry*. 387, 1329–1335.

Brigelius-Flohe, R. and Banning, A. (2006). Part of the series: from dietary antioxidants to regulators in cellular signaling and gene regulation: sulforaphane and selenium, partners in adaptive response and prevention of cancer. *Free Radical Biology and Medicine*. 40, 775–787.

Brozmanová, J., Máníková, D., Vlčková, V., Chovanec, M. (2010). Selenium: a double-edged sword for defense and offence in cancer. *Archives of Toxicology*. 84, 919-938.

Burk, R.F. and Hill, K.E. (2009). Selenoprotein P: expression, functions, and roles in mammals. *Biochimica et Biophysica Acta*. 1790, 1441-1447.

Burk, R.F. and Hill, K.E. (2005). Selenoprotein P: An extracellular protein with unique physical characteristics and a role in selenium homeostasis. *Annual Review of Nutrition*. 2, 215-235.

Caban, K. and Copeland, P.R. (2006). Size matters: a view of selenocysteine incorporation from the ribosome. *Cellular and Molecular Life Sciences*. 63, 73–81.

Carlson, B.A., Xu, X.M., Gladyshev, V.N., Hatfield, D.L. (2005). Selective rescue of selenoprotein expression in mice lacking a highly specialized methyl group in selenocysteine tRNA. *Journal of Biological Chemistry*. 280, 5542–5548.

Chabory, E., Damon, C., Lenoir, A., Henry-Berger, J., Vernet, P., Cadet, R., Saez, F., Drevet, J. R. (2010). Mammalian glutathione peroxidases control acquisition and maintenance of spermatozoa integrity. *Journal of Animal Science*. 88, 1321-1331.

Chambers, I., Frampton, J., Goldfarb, P., Affara, N., McBain, W., Harrison, P.R. (1986). The structure of the mouse glutathione peroxidase gene: the selenocysteine in the active site is encoded by the 'termination' codon, TGA. *EMBO Journal*. 5, 1221–1227.

Chen, J. and Berry, M. J. (2003). Selenium and selenoproteins in the brain and brain diseases. *Journal of Neurochemistry*. 86, 1–12.

Chinta, S. J. and Andersen, J. K. (2005). Dopaminergic neurons. *International Journal of Biochemistry and Cell Biology*. 37, 942–946.

Chittum, H.S., Baek, H.J., Diamond, A.M., Fernandez-Salguero, P., Gonzalez, F., Ohama, T., Hatfield, D.L., Kuehn, M., Lee, B.J., Hatfield, D., Hampton, L. (1997). Selenocysteine tRNA[Ser]Sec levels and selenium-dependent glutathione peroxidase activity in mouse embryonic stem cells heterozygous for a targeted mutation in the $tRNA^{[Ser]Sec}$ gene: selenium induces changes in the selenocysteine $tRNA^{[Ser]Sec}$ population in mammalian cells. *Biochemistry*. 36, 8634–8639.

Clark, L.C., Combs, Jr, G.F., Turnbull, B.W., Slate, E.H., Chalker, D.K., Chow, J., Davis, L.S., Glover, R.A., Graham, G.F., Gross, E.G., Krongrad, A., Lesher, Jr, J.L., Park, H.K., Sanders, Jr, B.B., Smith, C.L., Taylor, J.R. (1996). Effects of selenium supplementation for cancer prevention in patients with carcinoma of the skin: a randomized controlled trial. Nutritional Prevention of Cancer Study Group. *Journal of the American Medical Association*. 276, 1957–1963.

Combs, G.F. Jr. (2001). Selenium in global food systems. *British Journal of Nutrition*. 85, 517–547.

Combs, G.F. and Combs, S.B. (1984). The Nutritional Biochemistry of Selenium. *Annual Review of Nutrition*. 4, 257-280.

Cooper, M.L., Adami, H.O., Gronberg, H., Wiklund, F., Green, F.R. and Rayman, M.P. (2008). Interaction between single nucleotide polymorphisms in selenoprotein P and mitochondrial superoxide dismutase determines prostate cancer risk. *Cancer Research*. 68, 10171–10177.

Copeland, P.R. and Driscoll, D.M. (2001). RNA binding proteins and selenocysteine. *Biofactors*. 14, 11–16.

Copeland, P.R., Fletcher, J.E., Carlson, B.A., Hatfield, D.L., Driscoll, D.M. (2000). A novel RNA binding protein, SBP2, is required for translation of mammalian selenoprotein mRNAs. *EMBO Journal*. 19, 306–314.

Curran, J.E., Jowett, J.B.M., Elliott, K.S., Gao, Y., Gluschenko, K., Wang, J., Azim, D.M.A., Cai, G., Mahaney, M.C., Comuzzie, A.G., Dyer, T.D., Walder, K.R., Zimmet, P., MacCluer, J.W., Collier, G.R., Kissebah, A.H., Blangero, J. (2005). Genetic variation in selenoprotein S influences inflammatory response. *Nature Genetics*. 37, 1234-1241.

Diwadkar-Navsariwala, V. and Diamond, A.M. (2004). The link between selenium and chemoprevention: a case for selenoproteins. *Journal of Nutrition*. 134, 2899–2902.

Driscoll, D.M. and Copeland, P.R. (2003). Mechanism and regulation of selenoprotein synthesis. *Annual Review of Nutrition*. 23, 17–40.

Duffield, A.J., Thomson, C.D., Hill, K.E., Williams, S. (1999). An estimation of selenium requirements for New Zealanders. *American Journal of Clinical Nutrition*. 70, 896–903.

Duffield-Lillico, A.J., Dalkin, B.L., Reid, M.E., Turnbull, B.W., Slate, E.H., Jacobs, E.T., Marshall, J.R., Clark, L.C. (2003). Selenium supplementation, baseline plasma selenium status and incidence of prostate cancer: an analysis of the complete treatment period of the Nutritional Prevention of Cancer Trial. *British Journal of Urology International*. 91, 608–612.

Dumitrescu, A.M., Liao, X-H., Abdullah, M.H.S., Lado-Abeal, J., Majed, F.A., Moeller, L.C., Boran, G., Schomburg, L., Weiss, R.E., Refetoff, S. (2005). Mutations in the *SBP2* gene produce abnormal thyroid hormone metabolism in man. *Nature Genetics*. 37, 1247-1252.

EFSA, 2008. Selenium-enriched yeast as source for selenium added for nutritional purposes in foods for particular nutritional uses and foods (including food supplements) for the general population. Scientific Opinion of the Panel on Food Additives. *The EFSA Journal*. 766, 1-42.

EFSA, 2009. EFSA Panel on Dietetic Products, Nutrition and Allergies (NDA). *The EFSA Journal*. 7, 1220-1244.

European Union Commission (2004). List of the authorised additives in feedingstuffs published in application of Article 9t (b) of Council Directive 70/524/EEC concerning additives in feedingstuffs (2004/C 50/01). *Official Journal of the European Union*. 47 (C50), 1–144.

Fahn, S. (2003) Description of Parkinson's disease as a clinical syndrome. *Annals of the New York Academy of Sciences*. 991, 1–14.

Fairweather-Tait S.J., Bao Y., Broadley M.R., Collings R., Ford D., Hesketh J.E., Hurst R. (2011). Selenium in Human Health and Disease. *Antioxidants and Redox Signaling*. 14, 1337-1383.

Fairweather-Tait, S.J., Collings, R., Hurst. R. (2010). Selenium bioavailability: current knowledge and future research requirements. *American Journal of Clinical Nutrition*. 91, 1484–1491.

Faure, P. (2003). Protective effects of antioxidant micronutrients (vitamin E, zinc and selenium) in type 2 diabetes mellitus. *Clinical Chemistry and Laboratory Medicine*. 41, 995–998.

Fisher, R. S., van Emde Boas, W., Blume, W., Elger, C., Genton, P., Lee, P. and Engel, Jr, J. (2005). Epileptic seizures and epilepsy: definitions proposed by the International League Against Epilepsy (ILAE) and the International Bureau for Epilepsy (IBE). *Epilepsia.* 46, 470–472.

Fletcher, J.E., Copeland, P.R., Driscoll, D.M. (2000). Polysome distribution of phospholipid hydroperoxide glutathione peroxidase mRNA: evidence for a block in elongation at the UGA/selenocysteine codon. *RNA.* 6, 1573–1584.

Forgione, M.A., Cap, A., Liao, R., Moldovan, N.I., Eberhardt, R.T., Lim, C.C., Jones, J., Goldschmidt-Clermont, P.J., Loscalzo, J. (2002). Heterozygous cellular glutathione peroxidase deficiency in the mouse: abnormalities in vascular and cardiac function and structure. *Circulation.* 106, 1154–1158.

Foster, C.B., Aswath, K., Chanock, S.J., McKay, H.F., Peters, U. (2006). Polymorphism analysis of six selenoprotein genes: support for a selective sweep at the glutathione peroxidase 1 locus (3p21) in Asian populations. *BMC Genetics.* **7**, 56-76.

Fujiwara, T., Busch, K., Gross, H.J., Mizutani, T. (1999). A SECIS binding protein (SBP) is distinct from selenocysteyl-tRNA protecting factor. *Biochimie.* 81, 213–218.

Galvin, J.E. (2006) Interaction of α-synuclein and dopamine metabolites in the pathogenesis of Parkinson's disease: a case for the selective vulnerability of the substantia nigra. *Acta Neuropathologica.* 112, 115–126.

Ghezzi, P. (2005). Regulation of protein function by glutathionylation. *Free Radical Research.* 39, 573–580.

Ghosh, R., Mukherjee, B., Chatterjee, M. (1994). A novel effect of selenium on streptozotocin-induced diabetic mice. *Diabetes Research.* 25, 165–171.

Gladyshev, V.N. (2006). Selenoproteins and selenoproteomes. In: Hatfield, D.L., Berry, M.J., Gladyshev, V.N. (Eds.), Selenium: Its molecular biology and role in human health. Second Edition, Springer science+business media LLC, New York, USA, pp. 99-110.

Gladyshev, V.N. and Hatfield, D.L. (1999). Selenocysteine-containing proteins in mammals. *Journal of Biomedical Science.* 6, 151–160.

Gross, M., Oertel, M., Köhrle, J. (1995). Differential selenium-dependent expression of type I 5'-deiodinase and glutathione peroxidase in the porcine epithelial kidney cell line LLC-PK1. *Biochemical Journal.* 306, 851-856.

Grumolato, L., Ghzili, H., Montero-Hadjadje, M., Gasman, S., Lesage, J., Tanguy, Y., Galas, L., Ait-Ali, D., Leprince, J., Guérineau, N.C., Elkahloun, A.G., Fournier, A., Vieau, D., Vaudry, H., Anouar, Y. (2008). Selenoprotein T is a PACAP-regulated gene involved in intracellular Ca^{2+} mobilization and neuroendocrine secretion. *FASEB Journal.* 22, 1756–1768.

Grundner-Culeman, E., Martin, 3rd G.W., Harney, J.W., Berry, M.J. (1999). Two distinct SECIS structures capable of directing selenocysteine incorporation in eukaryotes. *RNA.* 5, 625–635.

Guo, Z., Van Remmen, H., Yang, H., Chen, X., Mele, J., Vijg, J., Epstein, C. J., Ho, Y. S. and Richardson, A. (2001). Changes in expression of antioxidant enzymes affect cell-mediated LDL oxidation and oxidized LDL-induced apoptosis in mouse aortic cells. *Arteriosclerosis, Thrombosis, and Vascular Biology.* 21, 1131–1138.

Halliwell, B. (1992). Reactive Oxygen Species and the Central Nervous System. *Journal of Neurochemistry.* 59, 1609-1623.

Hatfield, D.L. and Gladyshev, V,N, (2009). The outcome of Selenium and Vitamin E Cancer Prevention Trial (SELECT) reveals the need for better understanding of selenium biology. *Molecular Interventions.* 9, 18–21.

Hatfield, D.L., Yoo, M.H., Carlson, B.A., Gladyshev, V.N. (2009). Selenoproteins that function in cancer prevention and promotion. *Biochimica et Biophysica Acta.* 1790, 1541-1545.

Hatfield, D.L., Lee, B.J., Hampton, L., Diamond, A.M. (1991). Selenium induces changes in the selenocysteine tRNA[Ser]Sec population in mammalian cells. *Nucleic Acids Research.* 19, 939–943.

Hatfield, D.L. and Gladyshev, V.N. (2002). How selenium has altered our understanding of the genetic code. *Molecular and Cellular Biology.* 22, 3565–3576.

Hattori, H., Imai, H., Furuhama, K., Sato, O., Nakagawa, Y. (2005). Induction of phospholipid hydroperoxide glutathione peroxidase in human polymorphonuclear neutrophils and HL60 cells stimulated with TNF-alpha. *Biochemical and Biophysical Research Communications.* 337, 464–473.

Heras, I.L., Palomo, L., Madrid, Y. (2011). Selenoproteins: the key factor in selenium essentiality. State of the art analytical techniques for selenoproteins studies. *Analytical and Bioanalytical Chemistry.* 400, 1715-1727.

Hill, K.E., Zhou, J., McMahan, W.J., Motley, A.K., Atkins, J.F., Gesteland, R.F., Burk, R.F. (2003). Deletion of selenoprotein P alters distribution of selenium in the mouse. *Journal of Biological Chemistry.* 278, 13640–13646.

Hill, K.E., Lyons, P.R., Burk, R.F. (1992). Differential regulation of rat liver selenoprotein mRNAs in selenium deficiency. *Biochemical and Biophysical Research Communications.* 185, 260–263.

Hirota, K., Matsui, M., Iwata, S., Nishiyama, A., Mori, K., Yodoi, J. (1997). AP-1 transcriptional activity is regulated by a direct association between thioredoxin and Ref-1. Proceedings of the National Academy of Sciences of the United States of America. 94, 3633–3638.

Holcik, M. and Sonenberg, N. (2005). Translational control in stress and apoptosis. *Nature Reviews Molecular Cell Biology.* 6, 318-327.

Hollander, J.M., Lin, K.M., Scott, B.T., Dillmann, W.H. (2003). Overexpression of PHGPx and HSP60/10 protects against ischemia/reoxygenation injury. *Free Radical Biology and Medicine.* 35, 742–751.

Horibata, Y. and Hirabayashi, Y. (2007). Identification and characterization of human ethanolaminephosphotransferase1. *Journal of Lipid Research.* 48, 503-508.

Hu, Y.J. and Diamond, A.M. (2003). Role of glutathione peroxidase 1 in breast cancer: loss of heterozygosity and allelic differences in the response to selenium. *Cancer Research.* 63, 3347–3351.

Hu, Y., Benya, R.V., Carroll, R.E., Diamond, A.M. (2005) Allelic loss of the gene for the GPx1 selenium-containing protein is a common event in cancer. *Journal of Nutrition.* 135, 3021S–3024S.

Huang, K., Lauridsen, E., Clausen, J. (1994). The uptake of Na-selenite in rat brain. Localization of new glutathione peroxidases in the rat brain. *Biological Trace Element Research.* 46, 91-102.

Huang, K., Liu, H., Chen, Z., Xu, H. (2002). Role of selenium in cytoprotection against cholesterol oxide-induced vascular damage in rats. *Atherosclerosis.* 162, 137–144.

Hwang, D.Y., Cho, J.S., Oh, J.H., Shim, S.B., Jee, S.W., Lee, S.H., Seo, S.J., Lee, S.K., Lee, S.H., Kim, Y.K. (2005). Differentially expressed genes in transgenic mice carrying human mutant presenilin-2 (N141I): correlation of selenoprotein M with Alzheimer's disease. *Neurochemical Research*. 30, 1009–1019.

Imai, H., Saito, M., Kirai, N., Hasegawa, J., Konishi, K., Hattori, H., Nishimura, M., Naito, S., Nakagawa, Y. (2006). Identification of the positive regulatory and distinct core regions of promoters, and transcriptional regulation in three types of mouse phospholipids hydroperoxide glutathione peroxidase. *Journal of Biochemistry*. 140, 573–590.

Imam, S.Z. and Ali, S.F. (2000). Selenium, an antioxidant, attenuates methamphetamine-induced dopaminergic toxicity and peroxynitrite generation. *Brain Research*. 855, 186–191.

Imam, S.Z., Newport, G.D., Islam, F., Slikker, Jr, W., Ali, S.F. (1999). Selenium, an antioxidant, protects against methamphetamine-induced dopaminergic neurotoxicity. *Brain Research*. 818, 575–578.

Institute of Medicine. 2000. Selenium. In: Dietary reference intakes for vitamin C, vitamin E, selenium, and carotenoids. *National Academy Press, Washington*. pp 284-324.

Islam, F., Zia, S., Sayeed, I., Zafar, K.S., Ahmad, A.S. (2002). Selenium-induced alteration of lipids, lipid peroxidation, and thiol group in circadian rhythm centers of rat. *Biological Trace Element Research*. 90, 203–214.

Iversen, S.D. and Iversen, L.L. (2007). Dopamine: 50 years in perspective. *Trends in Neuroscience*. 30, 188–193.

Jablonska, E., Gromadzinska, J., Sobala, W., Reszka, E., Wasowicz, W. (2008). Lung cancer risk associated with selenium status is modified in smoking individuals by Sep15 polymorphism. *European Journal of Nutrition*. 47, 47–54.

Jackson, M.I. and Combs, Jr, G.F. (2008). Selenium and anticarcinogenesis: underlying mechanisms. *Current Opinion in Clinical Nutrition and Metabolic Care*. 11, 718–726.

Jameson, R.R., Carlson, B.A., Butz, M., Esser, K., Hatfield, D.L., Diamond, A.M. (2002). Selenium influences the turnover of selenocysteine $tRNA^{([Ser]Sec)}$ in Chinese hamster ovary cells. *Journal of Nutrition*. 132, 1830–1835.

Jameson, R.R. and Diamond, A.M. (2004). A regulatory role for Sec $tRNA^{[Ser]Sec}$ in selenoprotein synthesis. *RNA*. 10, 1142–1152.

Juniper, D.T., Phipps, R.H., Givens, D.I., Jones, A.K., Green, C., Bertin, G. (2008). Tolerance of ruminant animals to high dose in-feed administration of a selenium-enriched yeast. *Journal of Animal Science*. 86, 197-204.

Karimpour, S., Lou, J., Lin, L.L., Rene, L.M., Lagunas, L., Ma, X., Karra, S., Bradbury, C.M., Markovina, S., Goswami, P.C., Spitz, D.R., Hirota, K., Kalvakolanu, D.V., Yodoi, J., Gius, D. (2002). Thioredoxin reductase regulates AP-1 activity as well as thioredoxin nuclear localization *via* active cysteines in response to ionizing radiation. *Oncogene*. 21, 6317-6327.

Karlsson, H.K., Tsuchida, H., Lake, S., Koistinen, H.A., Krook, A. (2004). Relationship between serum amyloid A level and Tanis/SelS mRNA expression in skeletal muscle and adipose tissue from healthy and type 2 diabetic subjects. *Diabetes*. 53, 1424–1428.

Kenet, G., Freedman, J., Shenkman, B., Regina, E., Brok-Simoni, F., Holzman, F., Vavva, F., Brand, N., Michelson, A., Trolliet, M., Loscalzo, J., Inbal, A. (1999). Plasma glutathione peroxidase deficiency and platelet insensitivity to nitric oxide in children with familial stroke. *Arteriosclerosis, Thrombosis, and Vascular Biology.* 19, 2017–2023.

Kim, H.C., Jhoo, W.K., Choi, D.Y., Im, D.H., Shin, E.J., Suh, J.H., Floyd, R.A., Bing, G. (1999). Protection of methamphetamine nigrostriatal toxicity by dietary selenium. *Brain Research.* 851, 76–86.

Kim, H., Jhoo, W., Shin, E., Bing, G. (2000). Selenium deficiency potentiates methamphetamine-induced nigral neuronal loss: comparison with MPTP model. *Brain Research.* 862, 247–252.

Kim, H.Y. and Gladyshev, N.V. (2004). Methionine sulfoxide reduction in mammals: characterization of methionine-R-sulfoxide reductases. *Molecular Biology of the Cell.* **15**, 1055-1064.

Klivenyi, P., Starkov, A.A., Calingasan, N.Y., Gardian, G., Browne, S.E., Yang, L., Bubber, P., Gibson, G.E., Patel, M.S., Beal, M.F. (2004). Mice deficient in dihydrolipoamide dehydrogenase show increased vulnerability to MPTP, malonate and 3-nitropropionic acid neurotoxicity. *Journal of Neurochemistry.* 88, 1352–1360.

Köhrle, J. and Gärtner, R. (2009). Selenium and thyroid. Best Practice and Research Clinical Endocrinology and Metabolism. 23, 815–827.

Köhrle, J., Brigelius-Flohe, R., Böck, A., Gartner, R., Meyer, O., Flohe, L. (2000). Selenium in biology: facts and medical perspectives. *Biological Chemistry.* 381, 849-864.

Köhrle, J., Jakob, F., Contempré, B., Dumont, J.E. (2005). Selenium, the thyroid, and the endocrine system. *Endocrine Reviews.* 26, 944-984.

Kovacic, P. and Jacintho, J.D. (2001). Mechanisms of carcinogenesis: focus on oxidative stress and electron transfer. *Current Medicinal Chemistry.* 8, 773-796.

Kowalska, A., Pruchnik-Wolinska, D., Florczak, J., Modestowicz, R., Szczech, J., Kozubski, W., Rossa, G., Wender, M. (2004). Genetic study of familial cases of Alzheimer's disease. *Acta Biochimica Polonica.* 51, 245–252.

Kryscio, R.J., Mendiondo, M.S., Schmitt, F.A., Markesbery, W.R. (2004). Designing a large prevention trial: statistical issues. *Statistics in Medicine.* 23, 285–296.

Kryukov, G.V., Castellano, S., Novoselov, S.V., Lobanov, A.V., Zehtab, O., Guigo, R., Gladyshev, V.N. (2003). Characterization of mammalian selenoproteomes. *Science.* 300, 1439-1443.

Kryukov, G.V., Kumar, R.A., Koc, A., Sun, Z., Gladyshev, N.V. (2002). Selenoprotein R is a zinc-containing stereo-specific methionine sulfoxide reductase. Proceedings of the National Academy of Sciences of the United States of America. 99, 4245-4250.

Kumaraswamy, E., Malykh, A., Korotkov, K.V., Kozyavkin, S., Hu, Y., Kwon, S.Y., Moustafa, M.E., Carlson, B.A., Berry, M.J., Lee, B.J., Hatfield, D.L., Diamond, A.M., Gladyshev, V.N. (2000). Structure-expression relationships of the 15-kDa selenoprotein gene: possible role of the protein in cancer etiology. *Journal of Biological Chemistry.* 275, 35540-35547.

Kuster, G.M., Pimentel, D.R., Adachi, T., Ido, Y., Brenner, D.A., Cohen, R.A., Liao, R., Siwik, D.A., Colucci, W.S. (2005). α-Adrenergic receptor-stimulated hypertrophy in adult rat ventricular myocytes is mediated via thioredoxin-1-sensitive oxidative modification of thiols on Ras. *Circulation.* 111, 1192–1198.

Kutluhan, S., Naziroglu, M., Celik, O., Yilmaz, M. (2009). Effects of selenium and topiramate on lipid peroxidation and antioxidant vitamin levels in blood of pentylentetrazol-induced epileptic rats. *Biological Trace Element Research*. 129, 181–189.

Lacourciere, G.M. and Stadtmanm, T.C. (2001). Utilization of selenocysteine as a source of selenium for selenophosphate *biosynthesis*. *Biofactors*. 14, 69-74.

Larsen, P.R. and Berry, M.J. (1995). Nutritional and hormonal regulation of thyroid hormone deiodinases. *Annual Review of Nutrition*. 15, 323–352.

Ledesma, M.C., Jung-Hynes, B. Schmit, T.L., Kumar, R., Mukhtar, H. Ahma, N. (2011). Selenium and vitamin E for prostate cancer: Post-SELECT (Selenium and Vitamin E Cancer Prevention Trial) status. *Molecular Medicine*. 17, 134-143.

Lei, X.G., Evenson, J.K., Thompson, K.M., Sunde, R.A. (1995). Glutathione peroxidase and phospholipid hydroperoxide glutathione peroxidase are differentially regulated in rats by dietary selenium. *Journal of Nutrition*. 125, 1438-1446.

Letavayová, L., Vlčková, V., Brozmanová, J. (2006). Selenium: From cancer prevention to DNA damage. *Toxicology*. 227, 1–14.

Linke, K. and Jakob, U. (2003). Not every disulfide lasts forever: disulfide bond formation as a redox switch. *Antioxidants and Redox Signaling*. 5, 425–434.

Lippman, S.M., Klein, E.A., Goodman, P.J., Lucia, M.S., Thompson, I.M., Ford, L.G., Parnes, H.L., Minasian, L.M., Gaziano, J.M., Hartline, J.A. et al. (2009). Effect of selenium and vitamin E on risk of prostate cancer and other cancers: the Selenium and Vitamin E Cancer Prevention Trial (SELECT). *Journal of the American Medical Association*. 301, 39–51.

Lovell, M.A., Xiong, S., Lyubartseva, G., Markesbery, W.R. (2009). Organoselenium (Sel-Plex diet) decreases amyloid burden and RNA and DNA oxidative damage in APP/PS1 mice. *Free Radical Biology and Medicine*. 46, 1527–1533.

Low, S.C. and Berry, M.J. (1996). Knowing when not to stop: selenocysteine incorporation in eucaryotes. *Trends in Biochemical Sciences*. 21, 203-208.

Low, S.C., Grundner-Culemann, E., Harney, J.W., Berry, M.J. (2000). SECIS-SBP2 interactions dictate selenocysteine incorporation efficiency and selenoprotein hierarchy. *EMBO Journal*. 19, 6882-6890.

Lu, C., Qiu, F., Zhou, H., Peng, Y., Hao, W., Xu, J., Yuan, J., Wang, S., Qiang, B., Xu, C. (2006). Identification and characterization of selenoprotein K: An antioxidant in cardiomyocytes. *FEBS Letters*. 580, 5189-5197.

Lu, J. and Holmgren, A. (2009). Selenoproteins. *Journal of Biological Chemistry*. 284, 723-727.

Lu, T., Pan, Y., Kao, S.Y., Li, C., Kohane, I., Chan, J., Yankner, B.A. (2004). Gene regulation and DNA damage in the ageing human brain. *Nature*. 429, 883–891.

Lum, H. and Roebuck, K.A. (2001). Oxidant stress and endothelial cell dysfunction. *American Journal of Physiology*. 280, C719–C741.

Maiorino, M., Scapin, M., Ursini, F., Biasolo, M., Bosello, V., Flohe, L. (2003). Distinct promoters determine alternative transcription of GPx4 into phospholipid-hydroperoxide glutathione peroxidase variants. *Journal of Biological Chemistry*. 278, 34286-34290.

Maquat, L.E. (2001). Evidence that selenium deficiency results in the cytoplasmic decay of GPx1 mRNA dependent on pre-mRNA splicing proteins bound to the mRNA exon-exon junction. *Biofactors*. 14, 37-42.

Mariotti, M. and Guigó, R. (2010). Selenoprofiles: profile-based scanning of eukaryotic genome sequences for selenoprotein genes. *Bioinformatics*. 26, 2656–2663.

Martin, G.W. 3rd and Berry, M.J. (2001). Selenocysteine codons decrease polysome association on endogenous selenoprotein mRNAs. *Genes to Cells.* 6, 121-129.

Maulik, N. and Das, D.K. (2008). Emerging potential of thioredoxin and thioredoxin interacting proteins in various disease conditions. *Biochimica et Biophysica Acta.* 1780, 1368–1382.

Maulik, N., Yoshida, T., Das, D.K. (1999). Regulation of cardiomyocyte apoptosis in ischemic reperfused mouse heart by glutathione peroxidase. *Molecular and Cellular Biochemistry.* 196, 13–21.

McClung, J.P., Roneker, C.A., Mu, W., Lisk, D.J., Langlais, P., Liu, F., Lei, X.G. (2004). Development of insulin resistance and obesity in mice overexpressing cellular glutathione peroxidase. Proceedings of the National Academy of Sciences of the United States of America, 101, 8852–8857.

McNeill, J.H., Delgatty, H.L., Battell, M. L. (1991). Insulinlike effects of sodium selenate in streptozocin-induced diabetic rats. *Diabetes.* 40, 1675–1678.

Miller, J.A., Oldham, M.C., Geschwind, D.H. (2008). A systems level analysis of transcriptional changes in Alzheimer's disease and normal aging. *Journal of Neuroscience.* 28, 1410–1420.

Miller, S., Walker, S.W., Arthur, J.R., Nicol, F., Pickard, K., Lewin, M.H., Howie, A.F., Beckett, G.J. (2001). Selenite protects human endothelial cells from oxidativedamage and induces thioredoxin reductase. *Clinical Science.* 100, 543–550.

Mitchell, J.H., Nicol, F., Beckett, G.J., Arthur, J.R. (1998). Selenoprotein expression and brain development in preweanling selenium- and iodine-deficient rats. *Journal of Molecular Endocrinology.* 20, 203-210.

Moreira, P.I., Honda, K., Zhu, X., Nunomura, A., Casadesus, G., Smith, M.A., Perry, G. (2006). Brain and brawn: parallels in oxidative strength. *Neurology.* 66, S97–S101.

Moriarty, P.M., Reddy, C.C., Maquat, L.E. (1998). Selenium deficiency reduces the abundance of mRNA for Se-dependent glutathione peroxidase 1 by a UGA-dependent mechanism likely to be nonsense codon-mediated decay of cytoplasmic mRNA. *Molecular and Cellular Biology.* 18, 2932-2939.

Nakamura, K., Fushimi, K., Kouchi, H., Mihara, K., Miyazaki, M., Ohe, T., Namba, M. (1998). Inhibitory effects of antioxidants on neonatal rat cardiac myocyte hypertrophy induced by tumor necrosis factor-α and angiotensin II. *Circulation.* 98, 794–799.

Nakayama, A., Hill, K.E., Austin, L.M., Motley, A.K., Burk, R.F. (2007). All regions of mouse brain are dependent on selenoprotein P for maintenance of selenium. *Journal of Nutrition.* 137, 690–693.

Navarro-Alarcon, M., Lopez-G de la Serrana, H., Perez-Valero, V., Lopez-Martinez, M.C. (1999). Serum and urine selenium concentrations as indicators of body status in patients with diabetes mellitus. *Science of the Total Environment.* 228, 79-85.

Navarro-Alarcon, M. and Lopez-Martinez, M.C. (2000). Essentiality of selenium in the human body: relationship with different diseases. *Science of the Total Environment.* 249, 347-371.

Naziroglu, M., Kutluhan, S., Yilmaz, M. (2008). Selenium and topiramate modulates brain microsomal oxidative stress values, Ca^{2+}-ATPase activity, and EEG records in pentylentetrazol-induced seizures in rats. *Journal of Membrane Biology.* 225, 39–49.

Novotny, L., Rauko, P., Kombian, S.B., Edafiogho, I.O. (2010). Selenium as chemoprotective anti-cancer agent: reality or wishful thinking? *Neoplasma.* 57, 383–391.

Ostadalova, I., Vobecky, M., Chvojkova, Z., Mikova, D., Hampl, V., Wilhelm, J., Ostadal, B. (2007). Selenium protects the immature rat heart against ischemia/ reperfusion injury. *Molecular and Cellular Biochemistry.* 300, 259–267.

Ozdemir, S., Ayaz, M., Can, B., Turan, B. (2005). Effect of selenite treatment on ultrastructural changes in experimental diabetic rat bones. *Biological Trace Element Research.* 107, 167–179.

Panee, J., Stoytcheva, Z.R., Liu, W., Berry, M.J. (2007). Selenoprotein H is a redox-sensing high mobility group family DNA-binding protein that up-regulates genes involved in glutathione synthesis and phase II detoxification. *Journal of Biological Chemistry.* 282, 23759-23765.

Papp, L.V., Lu, J., Holmgren, A., Khanna, K.K. (2007). From selenium to selenoproteins: Synthesis, identity, and their role in human health. *Antioxidants and Redox Signaling.* 9, 775-806.

Pappas, A.C., Zoidis, E., Surai, P.F., Zervas, G. (2008). Selenoproteins and maternal nutrition. *Comparative Biochemistry and Physiology.* – B, 151, 361-372.

Park, Y.C. and Whanger, P.D. (1995). Toxicity, metabolism and absorption of selenite by isolated rat hepatocytes. *Toxicology.* 100, 151-162.

Pesti, G.M. and Combs, G.F. (1976). Studies on the enteric absorption of selenium in the chick using localized coccidial infections. *Poultry Science.* 55, 2265-2274.

Peters, M.M., Hill, K.E., Burk, R.F., Weeber, E.J. (2006). Altered hippocampus synaptic function in selenoprotein P deficient mice. *Molecular Neurodegeneration.* 1, 12.

Petit, N., Lescure, A., Rederstorff, M., Krol, A., Moghadaszadeh, B., Wewer, U.M., Guicheney, P. (2003). Selenoprotein N: an endoplasmic reticulum glycoprotein with an early developmental expression pattern. *Human Molecular Genetics.* 12, 1045-1053.

Pimentel, D.R., Adachi, T., Ido, Y., Heibeck, T., Jiang, B., Lee, Y., Melendez, J.A., Cohen, R.A., Colucci, W.S. (2006). Strain-stimulated hypertrophy in cardiac myocytes is mediated by reactive oxygen species-dependent Ras S-glutathiolation. *Journal of Molecular and Cellular Cardiology.* 41, 613–622.

Power, J.H. and Blumbergs, P.C. (2009). Cellular glutathione peroxidase in human brain: cellular distribution, and its potential role in the degradation of Lewy bodies in Parkinson's disease and dementia with Lewy bodies. *Acta Neuropathologica.* 117, 63–73.

Raines, A.M. and Sunde, R.A. (2011). Selenium toxicity but not deficient or supernutritional selenium status vastly alters the transcriptome in rodents. *BMC Genomics.* 12, 26.

Ramoutar, R.R. and Brumaghim, J.L. (2010). Antioxidant and anticancer properties and mechanisms of inorganic selenium, oxo-sulfur and oxo-selenium compounds. *Cell Biochemistry and Biophysics.* 58, 1–23.

Rayman, M.P. (2000). The importance of selenium to human health. *Lancet.* 356, 233-241.

Rayman, M.P. (2004). The use of high-selenium yeast to raise selenium status: how does it measure up? *British Journal of Nutrition.* 92, 557-573.

Rayman, M.P. (2005). Selenium in cancer prevention: a review of the evidence and mechanism of action. *Proceedings of the Nutrition Society.* 64, 527-542.

Rayman, M.P. (2009). Selenoproteins and human health: Insights from epidemiological data. *Biochimica et Biophysica Acta.* 1790, 1533–1540.

Reddy, P.H. and Beal, M.F. (2005). Are mitochondria critical in the pathogenesis of Alzheimer's disease? *Brain Research Reviews.* 49, 618–632.

Remondino, A., Kwon, S.H., Communal, C., Pimentel, D.R., Sawyer, D.B., Singh, K., Colucci, W.S. (2003). *β*-Adrenergic receptor-stimulated apoptosis in cardiac myocytes is mediated by reactive oxygen species/c-Jun NH2-terminal kinase-dependent activation of the mitochondrial pathway. *Circulation Research.* 92, 136–138.

Ridet, J.L., Bensadoun, J.C., Deglon, N., Aebischer, P., Zurn, A.D. (2006). Lentivirus-mediated expression of glutathione peroxidase: neuroprotection in murine models of Parkinson's disease. *Neurobiology of Disease.* 21, 29–34.

Roberts, C.K. and Sindhu, K.K. (2009). Oxidative stress and metabolic syndrome. *Life Sciences.* 84, 705–712.

Rundlof, A.K. and Arner, E.S. (2004). Regulation of the mammalian selenoprotein thioredoxin reductase 1 in relation to cellular phenotype, growth, and signaling events.

Savaskan, N.E., Brauer, A.U., Kuhbacher, M., Eyupoglu, I.Y., Kyriakopoulos, A., Ninnemann, O., Behne, D., Nitsch, R. (2003). Selenium deficiency increases susceptibility to glutamate-induced excitotoxicity. *FASEB Journal.* 17, 112–114.

Scharpf, M., Schweizer, U., Arzberger, T., Roggendorf, W., Schomburg, L., Köhrle, J. (2007). Neuronal and ependymal expression of selenoprotein P in the human brain. *Journal of Neural Transmission.* 114, 877–884.

Schomburg, L., Schweizer, U., Holtmann, B., Flohe, L., Sendtner, M., Köhrle, J. (2003). Gene disruption discloses role of selenoprotein P in selenium delivery to target tissues. *Biochemical Journal.* 370, 397–402.

Schomburg, L., Schweizer, U., Köhrle, J. (2004). Selenium and selenoproteins in mammals: extraordinary, essential, enigmatic. *Cellular and Molecular Life Sciences.* 61, 1988-1995.

Schrauzer, G.N. (2000). Selenomethionine: a review of its nutritional significance, metabolism and toxicity. *Journal of Nutrition.* 130, 1653-1656.

Schweizer, U., Schomburg, L., Savaskan, N.E. (2004). The neurobiology of selenium: lessons from transgenic mice. *Journal of Nutrition.* 134, 707–710.

Seifried, H.E., Anderson, D.E., Fisher, E.I., Milner, J.A. (2007). A review of the interaction among dietary antioxidants and reactive oxygen species. *Journal of Nutritional Biochemistry.* 18, 567–579.

Seiler, A., Schneider, M., Forster, H., Roth, S., Wirth, E. K., Culmsee, C., Plesnila, N., Kremmer, E., Radmark, O., Wurst, W., Bornkamm, G.W., Schweizer, U., Conrad, M. (2008). Glutathione peroxidase 4 senses and translates oxidative stress into 12/15-lipoxygenase dependent- and AIF-mediated cell death. *Cell Metabolism.* 8, 237–248.

Selenius, M., Rundlof, A.K., Olm, E., Fernandes, A.P., Bjornstedt, M. (2010). Selenium and selenoproteins in the treatment and diagnostics of cancer. *Antioxidants and Redox Signaling.* 12, 867–880.

Shen, Q., Leonard, J.L., Newburger, P.E. (1995). Structure and function of the selenium translation element in the 3'-untranslated region of human cellular glutathione peroxidase mRNA. *RNA.* 1, 519-525.

Shennan, D. (1988). Selenium (selenate) transport by human brush border membrane vesicles. *British Journal Nutrition.* 59, 13-19.

Shibata, T., Arisawa, T., Tahara, T., Ohkubo, M., Yoshioka, D., Maruyama, N., Fujita, H., Kamiya, Y., Nakamura, M., Nagasaka, M., Iwata, M., Takahama, K., Watanabe, M., Hirata, I. (2009). Selenoprotein S (SEPS1) gene −105G>A promoter polymorphism influences the susceptibility to gastric cancer in the Japanese population. *BMC Gastroenterology.* 9, 2.

Small-Howard, A.L. and Berry, M.J. (2005). Unique features of selenocysteine incorporation function within the context of general eukaryotic translational processes. *Biochemical Society Transactions.* 33, 1493–1497.

Squires, J. and Berry, M.J. (2006) Selenium, selenoproteins, and cancer. *Hawaii Medical Journal.* 65, 239–240.

Squires, J.E., Stoytchev, I., Forry, E.P., Berry, M.J. (2007). SBP2 binding affinity is a major determinant in differential selenoprotein mRNA translation and sensitivity to, nonsense-mediated decay. *Molecular and Cellular Biology.* 27, 7848–7855.

Steinbrenner H. and Sies, H. (2009). Protection against reactive oxygen species by selenoproteins. *Biochimica et Biophysica Acta.* 1790, 1478–1485.

Steinbrenner, H., Alili, L., Bilgic, E., Sies, H., Brenneisen, P. (2006). Involvement of selenoprotein P in protection of human astrocytes from oxidative damage. *Free Radical Biology and Medicine.* 40, 1513–1523.

Stranges, S., Marshall, J.R., Natarajan, R., Donahue, R.P., Trevisan, M., Combs, G.F., Cappuccio, F.P., Ceriello, A., Reid, M.E. (2007). Effects of long-term selenium supplementation on the incidence of type 2 diabetes: a randomized trial. *Annals of Internal Medicine.* 147, 217–223.

Stranges, S., Navas-Acien, A., Rayman, M.P., Guallar, E. (2010). Selenium status and cardiometabolic health: state of the evidence. *Nutrition, Metabolism and Cardiovascular Diseases.* 20, 754-60.

Strozyk, D., Launer, L.J., Adlard, P.A., Cherny, R.A., Tsatsanis, A., Volitakis, I., Blennow, K., Petrovitch, H., White, L.R., Bush, A.I. (2009). Zinc and copper modulate Alzheimer Aβ levels in human cerebrospinal fluid. Neurobiol. *Aging.* 30, 1069–1077.

Sunde, R.A., Evenson, J.K., Thompson, K.M., Sachdev, S.W. (2005). Dietary selenium requirements based on glutathione peroxidase-1 activity and mRNA levels and other Se-dependent parameters are not increased by pregnancy and lactation in rats. *Journal of Nutrition.* 135, 2144-2150.

Sunde, R.A., Raines, A.M., Barnes, K.M., Evenson, J. K. (2008). Selenium status highly-regulates selenoprotein mRNA levels for only a subset of the selenoproteins in the selenoproteome. Bioscience Reports, 29, 329–338.

Surai, P.F. (2006). Selenium in Nutrition and Health. Nottingham University Press, Nottingham, UK.

Suzuki, M., Endo, M., Shinohara, F., Echigo, S., Rikiishi, H. (2010). Differential apoptotic response of human cancer cells to organoselenium compounds. *Cancer Chemotherapy and Pharmacology.* 66, 475–484.

Takebe, G., Yarimizu, J., Saito, Y., Hayashi, T., Nakamura, H., Yodoi, J., Nagasawa, S., Takahashi, K. (2002). A comparative study on the hydroperoxide and thiol specificity of the glutathione peroxidase family and selenoprotein P. *Journal of Biolgical Chemistry.* **277**, 41254–41258.

Tanaka, K., Honda, M., Takabatake, T. (2001). Redox regulation of MAPK pathways and cardiac hypertrophy in adult rat cardiac myocyte. *Journal of the American College of Cardiology.* 37, 676–685.

Tang, R., Liu, H., Wang, T., Huang, K. (2005). Mechanisms of selenium inhibition of cell apoptosis induced by oxysterols in rat vascular smooth muscle cells. *Archives of Biochemistry and Biophysics.* 441, 16–24.

Thomas, J.P., Geiger, P.G., Girotti, A.W. (1993). Lethal damage to endothelial cells by oxidized low density lipoprotein: role of selenoperoxidases in cytoprotection against lipid hydroperoxide- and iron-mediated reactions. *Journal of Lipid Research.* 34, 479–490.

Thomson, C.D., Robinson, M.F., Butler, J.A., Whanger, P.D. (1993). Longterm supplementation with selenate and selenomethionine: selenium and glutathione peroxidase (EC 1.11.1.9) in blood components of New Zealand women. *British Journal of Nutrition.* 69, 577-588.

Toppo, S., Vanin, S., Bosello, V., Tosatto, S.C. (2008). Evolutionary and structural insights into the multifaceted glutathione peroxidase (Gpx) superfamily. *Antioxidants and Redox Signaling.* 10, 1501-1514.

Ursini, F. (2000). The world of glutathione peroxidases. *Journal of Trace Elements in Medicine and Biology.* 14, 116.

Utomo, A., Jiang, X., Furuta, S., Yun, J., Levin, D.S., Wang, Y.C.J., Desai, K.V., Green, J.E., Chen, P.L., Lee, W.H. (2004). Identification of a novel putative non-selenocysteine containing phospholipid hydroperoxide glutathione peroxidase (NPGPx) essential for alleviating oxidative stress generated from polyunsaturated fatty acids in breast cancer cells. *Journal of Biologial Chemistry.* 279, 43522-43529.

Valdiglesias, V., Pasaro, E., Mendez, J., Laffon, B. (2010). In vitro evaluation of selenium genotoxic, cytotoxic, and protective effects: a review. *Archives of Toxicology.* 84, 337–351.

Venardos, K., Harrison, G., Headrick, J. and Perkins, A. (2004). Effects of dietary selenium on glutathione peroxidase and thioredoxin reductase activity and recovery from cardiac ischemia-reperfusion. *Journal of Trace Elements in Medicine and Biology.* 18, 81–88.

Vendeland, S.C., Deagen, J.T., Butler, J.A., Whanger, P.D. (1994). Uptake of selenite, selenomethionine and selenate by brush border membrane vesicles isolated from rat small intestine. *Biometals.* 7, 305-312.

Virmani, A., Gaetani, F., Imam, S., Binienda, Z., Ali, S. (2003). Possible mechanism for the neuroprotective effects of L-carnitine on methamphetamine-evoked neurotoxicity. Annals of the New York Academy of Sciences. 993, 197–207.

Visser, T.J. (1996). Pathways of thyroid hormone metabolism. *Acta Medica Austriaca.* 23, 10–16.

Voetsch, B., Jin, R.C., Bierl, C., Benke, K.S., Kenet, G., Simioni, P., Ottaviano, F., Damasceno, B.P., Annichino-Bizacchi, J.M., Handy, D.E., Loscalzo, J. (2007). Promoter polymorphisms in the plasma glutathione peroxidase (GPx-3) gene: a novel risk factor for arterial ischemic stroke among young adults and children. *Stroke.* 38, 41–49.

Voetsch, B., Jin, R.C., Bierl, C., Deus-Silva, L., Camargo, E.C., Annichino-Bizacchi, J. M., Handy, D.E., Loscalzo, J. (2008). Role of promoter polymorphisms in the plasma glutathione peroxidase (GPx-3) gene as a risk factor for cerebral venous thrombosis. *Stroke.* 39, 303–307.

Walder, K., Kantham, L., McMillan, J. S., Trevaskis, J., Kerr, L., De Silva, A., Sunderland, T., Godde, N., Gao, Y., Bishara, N., Windmill, K., Tenne-Brown, J., Augert, G., Zimmet, P.Z., Collier, G.R. (2002). Tanis: a link between type 2 diabetes and inflammation? *Diabetes.* 51, 1859–1866.

Weiss Sachdev, S. and Sunde, R.A. (2001). Selenium regulation of transcript abundance and translational efficiency of glutathione peroxidase-1 and -4 in rat liver. *Biochemistry Journal.* 357, 851-858.

Whanger, P.D. (2000). Selenoprotein W: a review. *Cellular and Molecular Life Sciences*. 57, 1846-1852.

Whanger P.D. (2004). Selenium and its relationship to cancer: an update. *British Journal of Nutrition*. 91, 11–28.

Wolffram, S. (1999). Absorption and metabolism of selenium: differences between inorganic and organic sources. In: Lyons, T.P. and Jacques, K.A.(Eds.), Proceedings of Alltech's 15th Annual Symposium, Nottingham University Press, Nottingham, UK, pp. 547-566.

Wolffram, S., Arduser, F., Scharrer, E. (1985). In vivo intestinal absorption of selenate and selenite by rats. *Journal of Nutrition*. 115, 454-459.

World, C.J., Yamawaki, H., Berk, B.C. (2006). Thioredoxin in the cardiovascular system. *Journal of Molecular Medicine*. 84, 997–1003.

Wu, Q. and Huang, K. (2004). Effect of long-term Se deficiency on the antioxidant capacities of rat vascular tissue. *Biological Trace Element Research*. 98, 73–84.

Würmli, R., Wolffram, S., Stingelin, Y., Scharrer, E. (1989). Stimulation of mucosal uptake of selenium from selenite by L-cystein in sheep small intestine. *Biological Trace Element Research*. 20, 75-85.

Xia, Y., Hill, K.E., Byrne, D.W., Xu, J., Burk, R.F. (2005). Effectiveness of selenium supplements in a low-selenium area of China. *American Journal of Clinical Nutrition*. 81, 829-834.

Xiong, Y., Liu, X., Lee, C.P., Chua, B.H., Ho, Y.S. (2006). Attenuation of doxorubicin-induced contractile and mitochondrial dysfunction in mouse heart by cellular glutathione peroxidase. *Free Radical Biology and Medicine*. 41, 46–55.

Zafar, K.S., Siddiqui, A., Sayeed, I., Ahmad, M., Salim, S., Islam, F. (2003). Dose-dependent protective effect of selenium in rat model of Parkinson's disease: neurobehavioral and neurochemical evidences. *Journal of Neurochemistry*. 84, 438–446.

Zeevalk, G.D., Razmpour, R., Bernard, L.P. (2008). Glutathione and Parkinson's disease: is this the elephant in the room? *Biomedicine and Pharmacotherapy*. 62, 236–249.

Zhang, Y., Fomenko, D.E., Gladyshev, V.N. (2005). The microbial selenoproteome of the Sargasso Sea. *Genome Biology*. 6, 37-53.

Zoidis E., Pappas A.C., Georgiou C.A., Komaitis E., Fegeros K. (2010). Selenium affects the expression of GPx4 and catalase in the liver of chicken. *Comparative Biochemistry and Physiology – B*. 155, 294-300.

In: Selenium: Sources, Functions and Health Effects
Editors: Chinatsu Aomori and Megumi Hokkaido
ISBN: 978-1-61942-061-8

Chapter 2

SELENIUM BIORHYTHMS AND HORMONAL REGULATION

*N. A. Golubkina**
Institute of Nutrition RAMS, Moscow, Russia

INTRODUCTION

Biorhythms caused by Earth rotation are a fundamental phenomenon of Nature embracing all levels of life organization from the cell to the whole organism, to populations, communities, ecosystems and providing optimal levels of adaptation to constantly changing environmental conditions. Many of these rhythms are endogenous cycles of many hormones that are maintained under photoperiodic clamping (Walker et al, 2010). Endocrine rhythms span time frames ranging from milliseconds to years, regulating appropriate fluctuations of different biochemical parameters of an organism. At present it is assumed that the pituitary gland is the main «receiver», constantly geared up for the signals coming from the Sun. It regulates all phyllogenetically stabilized biorhythms of an organism via hormonal regulation, being provided by constant blood current. Taking this into account many examples of selenium fluctuations in bio substrates have a logical explanation, based on hormonal regulation of selenium accumulation in tissues and organs.

Selenium is an essential trace element strictly connected with the endocrine system modifying the expression of about 30 selenoproteins. It participates in normal human development, growth, antioxidant defense, male fertility, thyroid hormone metabolism, and in modulating immunity (BeckettandArthur, 2005). Biological rhythms of selenium seem to be of great significance due to essentiality of the element to mammal organisms, its antioxidant protective properties and the ability to decrease the risk of cardiovascular diseases and cancer.

Most of the results presented below describe examples of different selenium dynamics without strict proof of the real endogenous biorhythms existence. Nevertheless both give exciting material for conclusions and future research.

---

* E-mail: segolubkina@rambler.ru

CIRCADIAN BIORHYTHMS

Biological cycles that repeat approximately every 24 hours are called circadian rhythms. At present circadian biorhythms are considered to be the most studied among natural rhythmicity phenomena. They compose diurnal oscillations of multiple biochemical processes, regulated by endogen clocks. The circadian timing system comprises peripheral oscillators, located in most of the tissues of the body and central pacemaker, located in the suprachiazmatic nucleus of the hypothalamus (Schibler et al, 2003). Circadian genes and the proteins, produced by these genes, constitute the molecular components of the circadian oscillator which form positive/negative feedback loops and generate circadian rhythms. Besides these genes circadian regulation is achieved via genes, controlling clocks, including different genes of cell cycle. A wide range of biological processes are regulated by the circadian clock including sleep-wake cycles, body temperature, energy metabolism, cell cycle and hormone secretion (Lowrey and Takahashi, 2004; Kondratov et al, 2007, Refinetti, 2006), reflecting genetic adaptation of metabolism in an organism to life conditions on the Earth (Hahlberg, 1974).

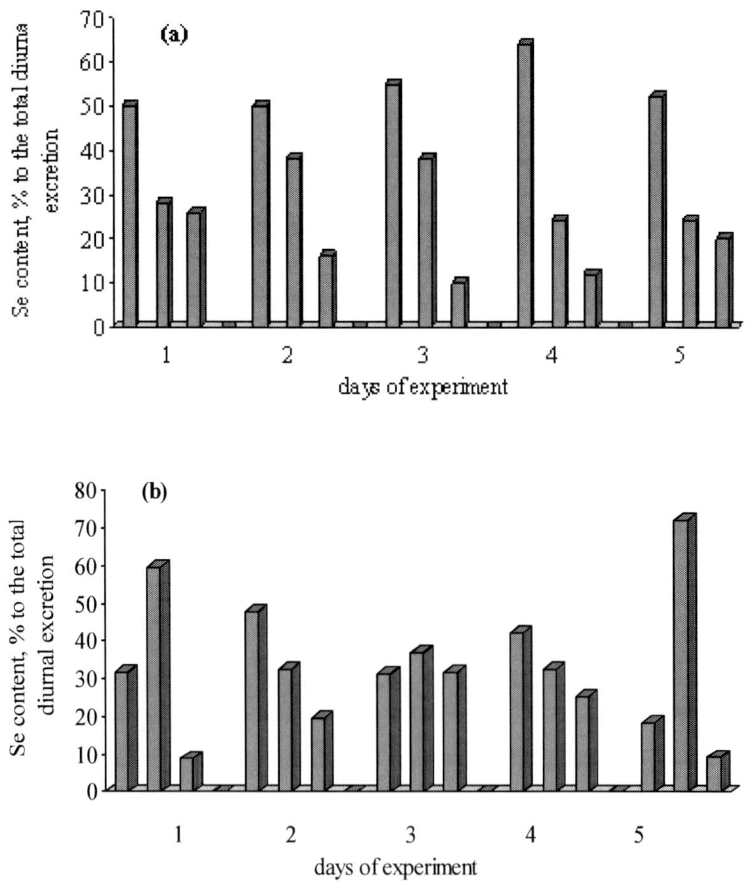

Figure 1. Biorhythms of selenium excretion with urine: (a) sodium selenate and b) selenium enriched yeast (Selena, Finland).

In evaluation of the human selenium status an amazing variability of selenium concentration in urine was revealed that made this parameter uninformative. But in most cases this fact was considered only as an annoying phenomenon, preventing utilization of a noninvasive evaluation method widely used in medicine for estimation of other biochemical parameters. The answer to the problem was found in calculating diurnal selenium excretion with urine, or, with certain limitations, the element's concentration per creatinine content. Such an approach leaves aside the decision of the main question: lack or existence of temporal correlations and the reason and mechanism of significant oscillations.

Taking into account the different metabolisms of organic and inorganic forms of selenium (Schrauser, 2006), it seems probable that the value of urine selenium excretion should be initially determined by the chemical form of the element. Thereby pharmacological loading was used for the selenium biorhythms investigation using inorganic selenium (sodium selenate) and selenomethyonine (incorporated into proteins in selenium enriches yeast "Selena" preparation, Finland) (Golubkina and Papazyan, 2006). Dynamics of selenium excretion was evaluated by three temporal periods: morning (from 6 to 12 hours), day (from 12 to 19 hours) and evening (from 19 hours to 6 hours in the morning) during 5 days with constant increase of selenium loading: the first day – 0 μg of Se, the second– 50 μg, the third – 100 μg, the fourth – 200 μg, the fifth – 400 μg, the sixth – 800 μg. In all cases preparation administration was achieved at 22 o'clock *per os* during all five days of the experiment. The consumption level of selenium with food was 68 μg/day.

The results reveal accurate dynamics, typical to each chemical form of selenium (Figure 1). Thus, administration of sodium selenate results in the value of fluctuation period for selenium excretion equal to 24 hours with the maximal excretion in the morning regardless of the dose of inorganic selenium. As the moment of selenium administration starts up the biological clocks of selenium accumulation and excretion, the period of Se (+6) metabolic cycle will also be equal to 24 hours. Other time of selenate administration may cause a shift in urine excretion rhythm.

Selenomethionine supplementation in a form of selenium enriched yeast changes significantly the excretion biorhythm: the period of these oscillations happens to be twice as large as for selenate - 48 hours with two excretion maximums, corresponding to 18 and 30 hours. In other words maximal value of selenium concentration in urine is registered alternately in the morning of the first day and at day time of the second one.

Revealed peculiarities of excretion dynamics for organic and inorganic forms of selenium are undoubtedly connected with principal differences in metabolism of these derivatives. As can be seen from Figure 2, an increased dose of sodium selenate leads to a multifold elevation of selenium in urine, while selenium enriched yeast supplementation causes saturation of the selenium level in urine at a dose of 400 μg/day and higher. The main chemical form of selenium in selenium enriched yeast is selenomethionine of proteins. This compound is known to be actively absorbed in the intestine via a sodium-dependant transport system of methionine and substitutes methionine in proteins with the formation of the so called "selenium depot". Selenate (analogous to selenite) passes through intestine walls via passive diffusion and becomes a substrate for selenocystein synthesis, providing a very limited and non specific incorporation of selenocystein in proteins.

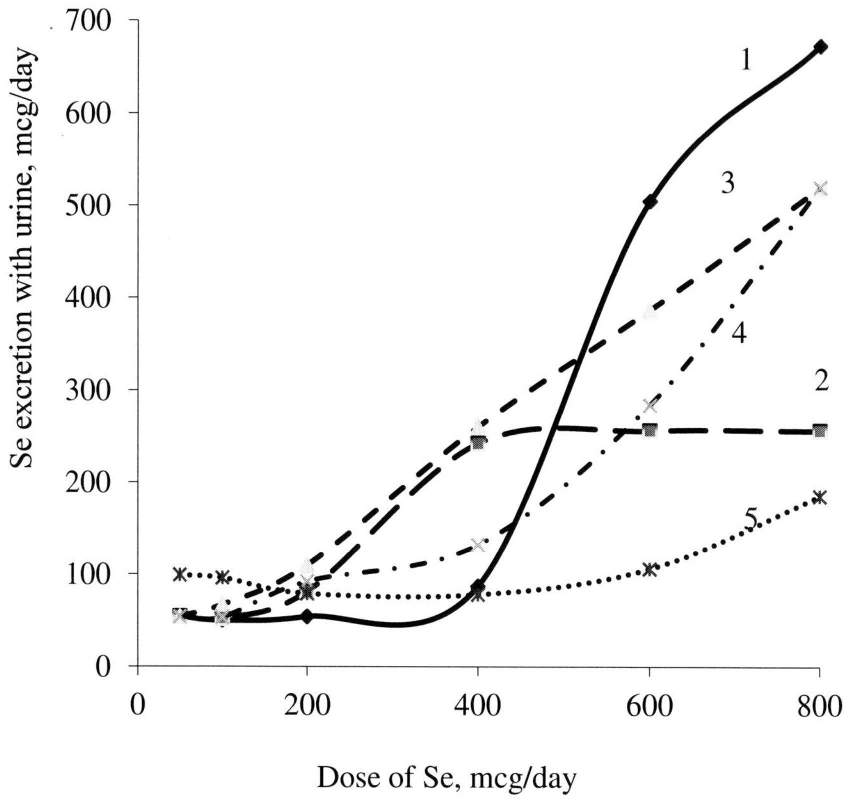

Figure 2. Selenium excretion with urine after different chemical forms administration: 1) sodium selenate, 2) selenium enriched yeast, 3) selenopyran, 4) dimethylpyrasolyl selenide, 5) selenium enriched spirulina.

Inorganic selenium may lead to a formation of selenothrisulphides, which are fluently oxidized and eliminated from proteins (Schrauzer, 2003). Thus quicker metabolism of Se(+6) is in positive agreement with the lower value of the selenium excretion period compared to that of selenomethionine in proteins. Nevertheless this does not explain the existence of two maximums of selenium excretion in the case of organic form supplementation and this phenomenon needs further investigation.

Though the above mentioned forms of selenium are wide spread in nature the question of metabolism and excretion dynamics for other selenium forms is still open. That is especially urgent for minor derivatives such as selenium analog of allicin (Tsuneyoshi et al, 2006) or for polysaccharides containing selenium (Ferri et al, 2007), and also for synthetic derivatives with low toxicity, such as selenopyran and 3-dimethylpirasolyl selenide (Golubkina and Papazyan, 2006).

Figure 2 reveals that the latter two synthetic analogs possess dynamics of selenium excretion that differ from the appropriate curves for sodium selenate and selenium enriched yeast: a linear correlation is registered between selenium excretion value and dose of selenopyran and a hyperbolic curve – for 3-dimethylpirasolyl selenide excretion. This fact supposes different metabolic ways for selenium accumulation and excretion of the appropriate synthetic derivatives and as a consequence the possibility of other types of oscillations for these chemical forms.

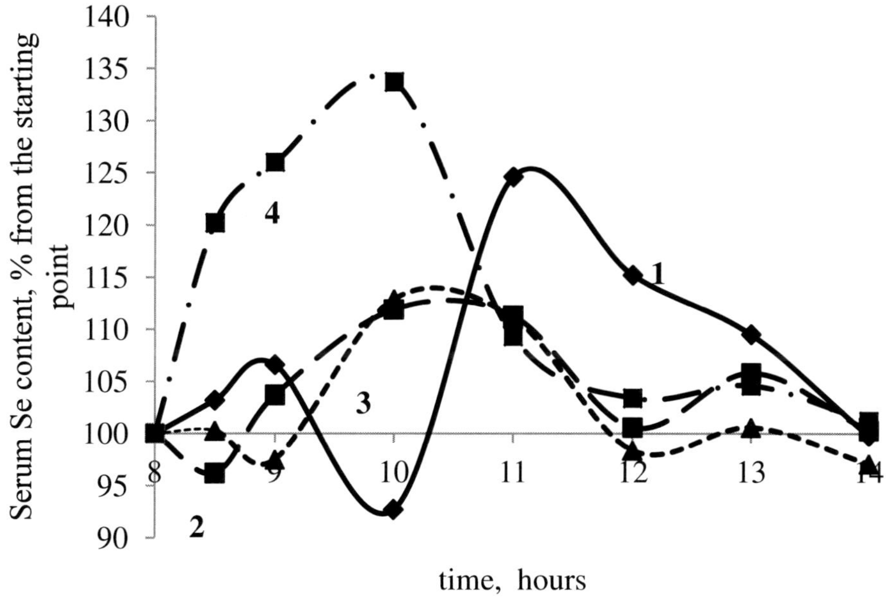

Figure 3. Serum selenium fluctuations: 1) control, 2) Se-vitasil, 3)Se-spirulina, 4) Selmevit (sodium selenite).

Such data should be considered an important characteristic of selenium metabolism that is especially significant for new supplements with specific chemical forms of selenium or plants fortified with the element (for instance, selenium enriched potatoes or *Brassica, Allium, Capsicum* species fortified with the element).

Investigation of selenium concentration dynamics in serum seems to be more complex due to limited time intervals available. The present data do not give the full characteristic of circadian fluctuations but prove the existence of dynamic oscillations of the element (Kukes et al, 2003).

The pharmacokinetic study was conducted on healthy volunteers using the following preparations: Selmevit (Bivitech Inc, Russia), a polyvitamin supplement, containing sodium selenite, vitamins and methionine, Se-Vitasil (Eney-Vitl Inc., Russia) – auto-lysate of selenium enriched yeast, and Spirulina-Se (Agro-Victoria Inc., Russia) – cianobacteria *Spirulina*, fortified with selenium. During one month the same volunteers took the above supplements with weekly intervals: the first week (control)-no selenium containing supplements, the second week- Selmevit, the third week- Se-Vitasil, the fourth week- Spirulina-Se, using one and the same dose- 75 μg Se. Blood samples were taken from the cubital vein at 8 a.m. on an empty stomach and immediately after that the corresponding supplement was administered. Blood was then sampled from all investigated persons at the terminal intervals of 0.5-1 hour over 6 hours. Within the following 6 hours volunteers did not eat (Kukes et al, 2003).

As can be seen from Figure 3, during 6 hours the endogenous selenium level demonstrates a statistically significant maximum, corresponding to 11 o'clock (P<0.001

compared to the data for 8 o'clock) and a minimum at 10 o'clock ($P<0.002$ compared to the data for 8 o'clock).

The most significant rise of serum selenium concentration was found after sodium selenite supplementation: already 30 minutes after the Selmevit administration the selenium level significantly exceeded the selenium content after Se-Vitasil supplementation ($P<0.05$). An hour later statistically significant differences were demonstrated both with Se-Vitasil ($P<0.01$), and Spirulina-Se ($P<0.05$).

At the same time the maximum serum selenium concentration was detected two hours after the intake regardless of the type of supplement. That becomes obvious from the curve resulted from the extraction of endogenous selenium levels from that of pharmacological loading (Figure 4).

One can indicate accurate fluctuations of serum selenium accumulation with statistically significant minimums at 11 o'clock for sodium selenite and 11.30 - for organic forms of selenium ($P<0.002$ compared to the data at 8 o'clock). Besides this sodium selenite (Selmevit) demonstrated the second concentration peak in serum corresponding to 12 o'clock ($P<0.002$ compared to the data at 8 o'clock).

Temporal differences in serum selenium concentration maximums with and without administration of selenium derivatives (organic selenium and sodium selenate): 11 and 10 o'clock respectively - seem to be the result of the synchronizing effect of preparations as an exogenous factor, affecting the rhythmicity of physiological processes.

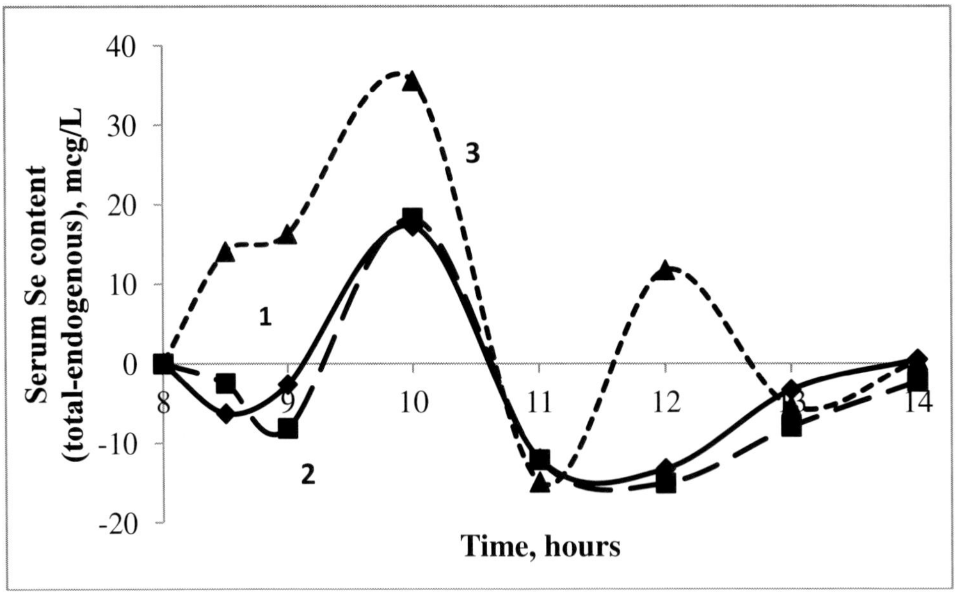

Figure 4. Dynamics of plasma selenium accumulation: 1)Se-vitasil, 2)Se-spirulina, 3)Selmevit (sodium selenite).

The results of selenium concentration oscillations in serum during fasting suppose that the process may be regulated by growth hormone (GH) fluctuations.

Indeed as has been demonstrated in 1988 (Ho et al, 1988), fasting induces enhancement of GH release through combined frequency and amplitude modulation contrary to the mixed nutrients effect forming an extremely low basal level of GH at day time and unpredictable effect on GH release. Thus the first maximum of GH concentration in serum during fasting lies at approximately 11 o'clock that corresponds to the maximum of endogenous selenium excretion described above. The correlation seems to be possible due to a significant increase of serum selenium concentration as a result of GH administration (Golubkina et al, 1997, see below).

On the other hand it should be taken into account that melatonin plays one of the most prominent roles in circadian biorhythms being known as a circadian neuroendocrine transducer determining activity/sleep phases, blood biochemical characteristics etc (Richter et al, 2004). The above experiment does not allow for a correlation of detected fluctuations with the activity of this hormone due to a very narrow temporal interval. Barlow-Walden et al (1995) investigations have shown that exogenously administered melatonin causes a 2-fold rise of glutathione peroxidase activity within 30 minutes in the brain of rats. Furthermore, brain glutathione peroxidase activity is higher at night than during the day and is correlated with high night-time tissue melatonin levels. Glutathione peroxidase is thought to be the principal enzyme eliminating peroxides in the brain. This antioxidant enzyme reduces the formation of hydroxyl radicals formed via iron-catalyzed Fenton-type reactions from hydrogen peroxide by reducing this oxidant to water. Since the hydroxyl radical is the most toxic oxygen radical known, induction of brain glutathione peroxidase might be an important mechanism by which melatonin exerts its potent neuroprotective effects.

Only three years later in 1998 circadian biorhythms of glutathione peroxidase and glutathione reductase activity were described in chicken brains (Pablos et al, 1998). The authors revealed that glutathione peroxidase exhibited a marked 24 h rhythm with peak activity which had acrophases about 8 hours after lights off and about 4 hours after the serum melatonin peak was detected. Glutathione reductase activity exhibited similar robust rhythms with the peaks occurring roughly 2 hours after those of glutathione peroxidase. The authors suggested that neural glutathione peroxidase activity increased due to the rise of nocturnal melatonin levels while glutathione reductase activity rose slightly later possibly due to an increase of its substrate, oxidized glutathione. The exposure of chicks to constant light for 6 days eliminated the melatonin rhythm as well as the peaks in both glutathione peroxidase and glutathione reductase activities. These findings suggested that the melatonin rhythm may be related to the night time increases in enzyme activities, although high concentration of GH at night may strengthen the effect. In healthy individuals biochemical variables (melatonin, thyroid stimulating hormone, cortisol, serotonin etc) exhibit circadian rhythm. But in patients with affective disorders many of these circadian rhythms are disturbed in phase and amplitude (Schulzand and Steimer, 2009; McClung, 2007; Dallaspezia and Benedetti. 2009).

Several investigations indicate that mistakes in circadian clocks genes expression may have serious consequences on trans-activation of targets controlling cell cycle and on cells ability to undergo apoptosis with the onset of genomic instability, increase of cell proliferation and promotion of cancer development (Sobia Rana, Saqib Mahmood, 2010).

This raises a question whether disturbances in circadian selenium biorhythms may be connected with cancer. Genetic or functional disturbances in molecular circadian clocks are registered for different forms of cancer, such as breast, colorectal, endometrial and prostate cancer (Mulero-Navatto and Esreller, 2008; Saharand and Sassone-Corsi, 2009; Schernhammer et al, 2003). Persons working rotating night shifts, that is in conditions preventing the increase of GSHPx activity at night, are known to exhibit an elevated risk of intestinal (Tokunaga et al, 2008), prostate (Klevecz et al, 1987), and breast cancer development (Zhu et al, 2005). Although diurnal fluctuations of selenium in cancer patients have not been investigated earlier, several forms of cancer are characterized by specific distribution of the element between serum and erythrocytes (Table 1), the differences with healthy persons being not only in absolute values but in serum/erythrocytes concentration ratio.

Table 1. Selenium distribution in blood of patients with different forms of cancer

Diagnosis	Se content, % to healthy Persons		Se eryth/ serum ratio	References
	Serum	Erythrocytes		
Head and neck cancer	71.3	63.2	1.93	Goodwin et al, 1983
All forms of cancer	79.2	95.9	1.87	Robinson et al, 1979
Liver cirrhosis	77.8	87.4	1.81	Miller et al, 1983
Breast cancer	100	111.7	1.47	Vernie et al, 1983
Stomach cancer	86.4	53.0	1.28	Tutelyan et al, 2002

Ultradian Rhythms

Circadian rhythms are a basic and well recognized feature of human physiology and behavior. However biological rhythms that repeat more or less frequently than every 24 hours are also fundamental to the body's function. In general, ultradian rhythms (those with length of less than 24 hours) and intradian rhythms (those with a length greater than 24 hours) do not coincide with conspicuous environmental cues, and how they are generated is not well understood.

The theory of a basic rest-activity cycle has led to many studies of ultradian cycles in alertness-sleepness, hunger, heart function, sexual excitement, urine formation and other functions.

Hormones are also released in ultradian cycles. Many are secreted in a more or less regular pattern every few hours. More frequent cycles of release, every few minutes, have also been documented. Although the mechanisms of hormone secretion have not been uncovered, patterned release has been shown to be extremely important for proper functioning. Although these cycles do not appear to be tightly coordinated, it is clear, that ultradian rhythms with a cycle of 90 minutes as well as with cycles of a few minutes to several hours are a basic component of many human functions.

The first description of ultradian rhythm seems to be the discovery of the dynamics of selenium distribution between serum and erythrocytes after radioactive selenium administration, the so called "erythrocytes pump" (McMurray and Davidson, 1979). Already 1-2 minutes after selenium administration erythrocytes concentrate about 50-70% of the whole blood selenium. 15-20 minutes are necessary for all selenium to be removed from erythrocytes.

At first the element binds to plasma albumin and later to globulins. Here selenium is supposed to form a complex with glutathione that gives it an opportunity to bind with plasma proteins.

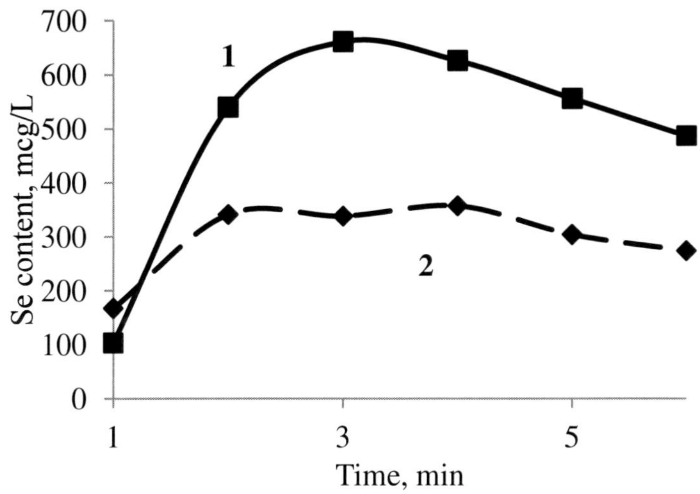

Figure 5. Dynamics of selenium accumulation in serum (1) and erythrocytes (2).

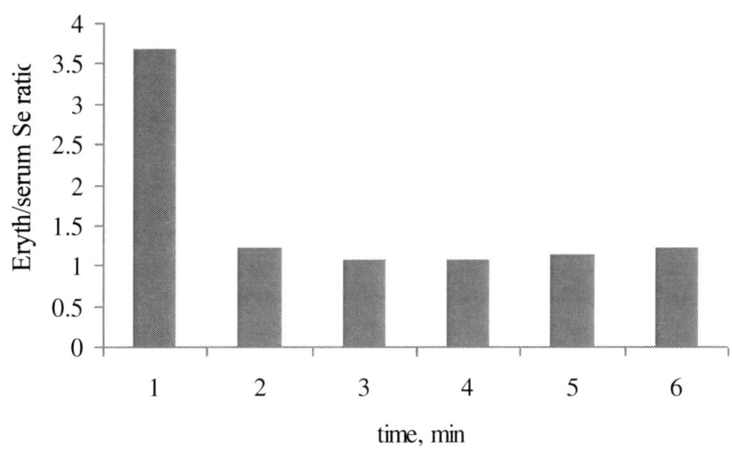

Figure 6. Dynamics of serum/erythrocytes selenium ratio.

Analogous quick incorporation of selenium into erythrocytes is demonstrated for sodium selenite, the time appearance of selenium in plasma varying for different mammal species (the lowest for rats and the highest for human beings; Sandholm, 1975). It is important that redistribution of selenium between plasma and erythrocytes is also accompanied by changes in humoral immunological parameters.

Investigation of selenium distribution dynamics between serum and erythrocytes was achieved on volunteers using sodium selenate. Blood was sampled in a single-step from vein into the test tube with heparin. Immediately after that 0.1 mL 0.01 N sodium selenate solution was added to each tube and samples were incubated at 37°C, taking probes each minute

(Golubkina et al, 2003). One drop of sodium selenate solution results in a serum selenium increase reaching maximum at the third minute after the beginning of incubation (Figure 5).

By this moment selenium redistribution between serum and erythrocytes is over as the selenium concentration ratio in serum/erythrocyte is stabilized (Figure 6).

The decrease of total selenium concentration in erythrocytes and serum after 3 minutes of incubation may be connected with the formation of volatile hydrogen selenide. The latter is known to be the key regulated form of selenium in plasma (Levander et al, 1998) and in conditions *in vitro* it seems possible to lose some of this volatile compound.

The above described dynamics of selenium redistribution between serum and erythrocytes is accompanied by synchronous redistribution of humoral immunity parameters: immunoglobulins A, G and M (Figure 7). Exactly at the third minute of incubation maximal accumulation of IgA and IgM is registered in erythrocytes, while Ig G concentration reaches its maximum at the second minute of incubation (Figure 7).

Changes of immunoglobulins concentration in serum demonstrate similar dynamics: the lowest IgA and IgM levels are found at the third minute that corresponds to the highest concentration of selenium.

Thus, selenium concentration ratio in serum and erythrocytes seems to be an extremely informative parameter, characterizing both the immune and selenium status of an organism.

Indeed, up to the present time the most prevailing criterion of the human selenium status evaluation has been serum selenium content.

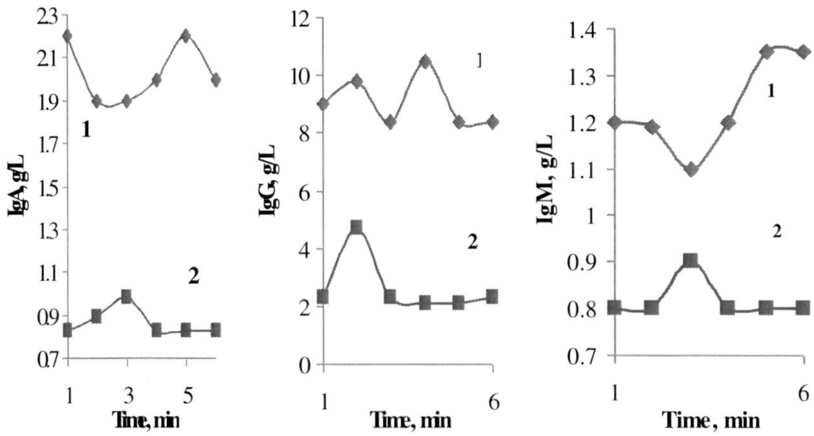

Figure 7. Dynamics of a) Ig A, b)IgG, c)IgM in serum (1) and erythrocytes (2).

The *in vitro* results described above and the data of selenium distribution between blood components for patients with different pathologies show that the endogenous regulation of selenium accumulation is connected first of all with redistribution of an element between serum and erythrocytes.

It is known that many diseases are characterized by changes of selenium homeostasis in the organism, leading to either the development of selenium deficiency or, in general, to redistribution of the element between plasma and erythrocytes (Table 2).

Such redistribution is not always accompanied by a decrease of both parameters. For instance, in patients with chronic pancreatitis selenium concentrations in serum and erythrocytes are decreased (Bowrey, 1999). On the contrary, children with Down syndrome demonstrate the decrease of GSHPx activity in erythrocytes, but increase of serum selenium concentration parameter (Anneren et al, 1989). AIDS patients possess very low levels of plasma selenium and significantly higher selenium content in erythrocytes compared to healthy persons (see Table 2).

The results allow assigning clinically the most significant group of pathologies, connected with selenium redistribution from plasma to erythrocytes. They are immunodeficiency cases of chemical and radiation nature, infections. It seems reasonable that improvement of patient health with the help of selenium containing supplements so popular at present, may become questionable and real improvement may need better redistribution of selenium in blood components than selenium saturation.

That may be demonstrated by the data of factors affecting the selenium status changes during infection (Golubkina et al, 1997). Serum selenium concentration is known to be decreased in patients with these pathologies (Sammalcorpi et al, 1988).

Thus hepatitis A and B patients demonstrate a decrease of serum selenium level during the development of the disease. Growth hormone and prolactin (PL) concentrations also decrease during this period. The highest decrease of serum selenium is detected in patients with severe clinical conditions, and GH and PL concentration decrease is more significant for patients with light stage (Table.3).

Analogous correlation between GH and PL levels with the selenium status is found also in laboratory animals. Thus, infection of mice with *Salmonella thyphimurium* results in an almost doubling decrease of serum selenium concentration compared to intact control ($P<0.01$, Figure 8).

GH administration causes a sharp increase of serum selenium concentration while parlodel - semi synthetic dopamine agonist, being the most effective drug for suppression of endogenous secretion of GH and PL by the pituitary gland (Jasmin et al, 1959), decreases the selenium level almost three fold ($P<0.05$).

Table 2. Selenium distribution between serum and erythrocytes in patients with different diseases (GolubkinaandPapazyan, 2006)

Disease	Serum Se	Erythrocyte Se	Se eryth/ /Se serum	Number of patients
Control (Moscow residents)	103±12	199±24	1.93±0.4	120
Control (the Urals residents)	100±13	187±17	1.90±0.3	55
AIDS	60±7	269±25	4.48±0.4	10
Syphilis	71±16	150±40	2.11±0.3	20
Neurocytoma	63±6	161±14	2.56±0.2	18
Surgical patients	80±9	150±12	1.88±0.2	20
CHD*	118±13	221±25	1.87±0.3	15
Atrophic gastritis	87±10	165±17	1.87±0.4	16
Infertility	91±8	169±17	1.86±0.3	20
Impotence and sexual weakness	87±10	161±18	1.85±0.4	23
Malabsorption	83±7	151±14	1.82±0.3	14
Atrophic gastroduodenitis	87±8	151±16	1.74±0.2	12
Partial gastrectomy	95±8	122±14	1.28±0.3	16

* coronary heart disease.

Table 3. Concentration of growth hormone, prolactin and selenium in serum of patients with viral hepatitis A and B .(M±m)

Heap titis	Severity degree	Growth hormone ng/ml (n)	Growth hormone % to the control	Prolactin µg/g (n)	Prolactin % to the control	Serum Se µg/L (n)	Serum Se % to the Control
A	Light	0.25± 0.03 (32)	17	81± 15 (34)	38	80.8±2.5 (21)	74
	Mode rately severe	2.60± 0.23 (15)	173	331.0± 26.2 (15)	153	69.7±4.1 (7)	64
B	Light	0.25± 0.04 (26)	17	66.0± 9.4 (26)	31	83.8±5.2 (22)	77
	Mode rately severe	1.98± 0.98 (22)	53	119.0± 20.7 (22)	55	81.2±4.9 (14)	75
	Severe	0.79± 0.41 (11)	53	102.0± 16.9 (10)	47	59.3±9.5 (5)	54

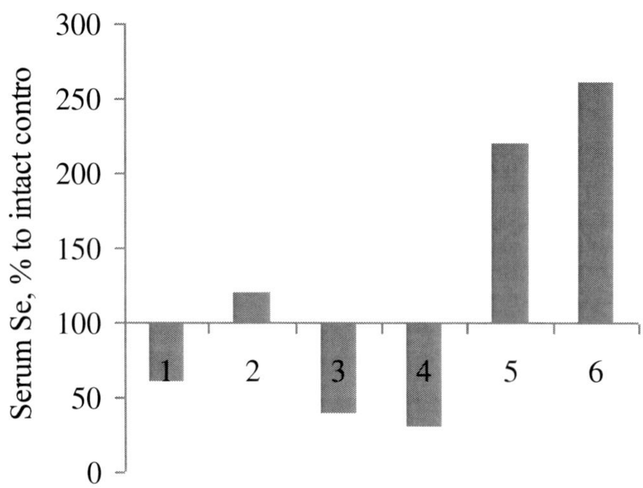

Figure 8. Modulation of selenium concentration in mice serum with selenium, GH and parlodel 1) infected control, 2) GH 0.2 U/mouse intravenously 3 times during 8 days; 3) parlodel 3 mg/mouse per os daily during 8 days; 4) parlodel 5 mg/mouse intramascular 3 times during 8 days; 5) sodium selenate solution (0.66 mg/L) daily during 8 days; 6) selenium enriched yeast in dry bread during 8 days (12.4 mg of selenium in 1 kg of dry bread).

Results on mice show that lifespan of animals during sharp infection decreases after GH and selenium injection (sodium selenate or selenium enriched yeast), while parlodel administration increases the time of survival (Figure 9).

Thus, one can see the dual effect of GH and its antagonist parlodel both on the serum selenium level and on time of survival after infection.

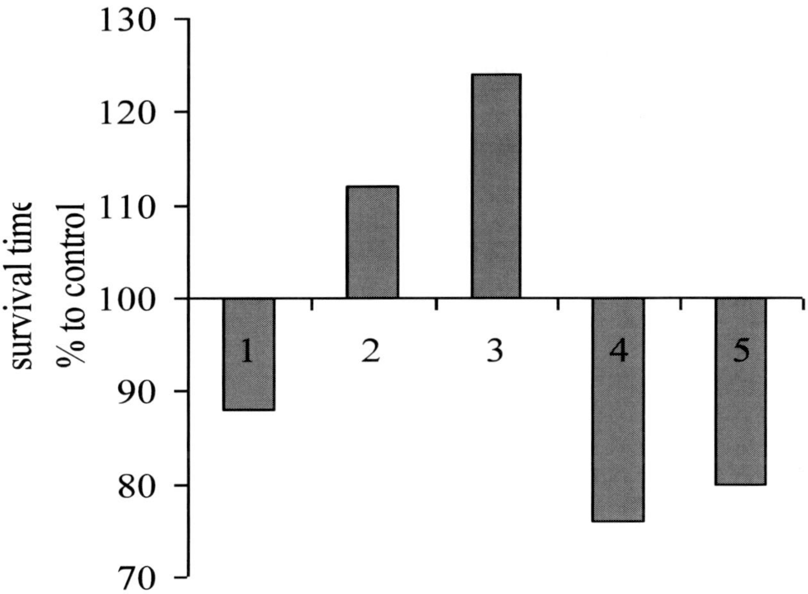

Figure 9. Lifespan of mice after *S. Typhimurium* infection. 1-GH 0.2 U/mouse twice a week 3 times before infection and later up to the death; 2-0.05 mg of parlodel/mouse per os daily during 8 days; 3-0.05 mg of parlodel/mouse intramuscular twice a week 3 times before infection and later beginning from the 8th day before infection; 4-sodium selenate, beginning from the 8th day before infection; 5-Se-containing yeast in bread.

GH is known to suppress cell immunity, to stimulate mortality from cancer and to induce nonspecific polyarthritis (Jasmin et al, 1959). A well known immunoproliferative effect of GH is expressed in a significant increase of the relative size of thymus, liver and lymphatic nodules (Kimball et al, 1968). Bacterial polysaccharides are shown to cause GH secretion to blood (Kimball et al, 1968). The phenomenon is in positive agreement with the known data about the protective effect of antisomatotropic serum in experimental tetanus in mice (Lazarev et al, 1976). Parlodel, being a GH antagonist, naturally increases the lifespan of animals, but it is not the fact of protection that attracts attention but its value (Figure 9).

In the above experiment selenium does not increase but decreases the survival of mice during infection. Taking into account the protective and selenium decreasing effect of parlodel, one should suppose the existence of an active mechanism of natural decrease in selenium, GH and PL concentrations in blood as a physiological principle of survival during infection.

It is of significance, that utilization of organic (selenium enriched yeast) and inorganic forms of selenium (50 μg/day) by patients with viral hepatitis A and B does not affect the clinical current of the disease, though the level of lipids peroxides decreases by 14-16% compared to the control group. These data indicate the existence of intensive hormonal regulation of the selenium status and reveals the possibility of the existence of two regulation mechanisms - antioxidant and hormonal, composing a dual pair.

Infradian Rhythms of Selenium in Living Organism

Among works devoted to the biological functions of selenium, a significant part discusses age-related changes of the human selenium status. Study of the dynamics of selenium accumulation in serum reveals relatively high stability of the parameter during the major part of human life from 18 years old and up to 60.

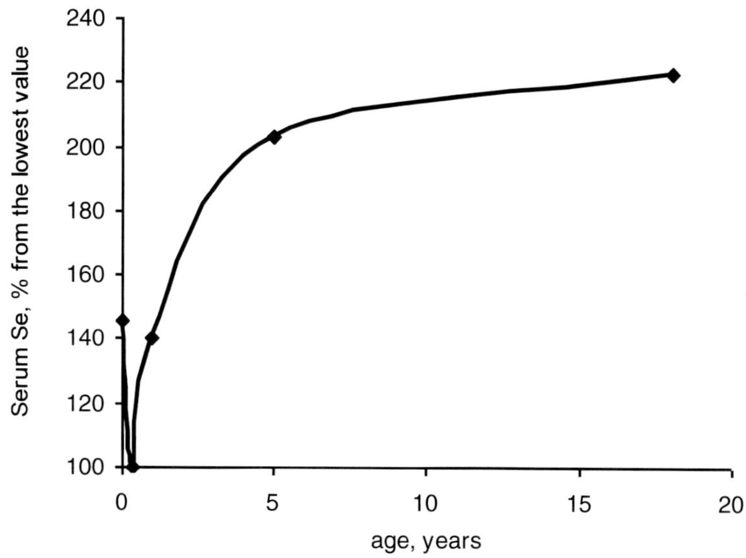

Figure 10 Dynamics of serum selenium accumulation during life.

The highest fluctuation is found for the first year of life, especially the first three month after birth and to a lesser extent for elderly persons (Figure 10). The sharp decrease of serum selenium concentration during 1-3 months of human life (Muntau et al, 2002) corresponds to the hormonal crisis of neonates, when estrogen concentration decreases substantially.

Xiaodong Zhou investigations (2007) revealed an intensive positive effect of estrogens on the selenium status. Female Sprague Dawley rats were bilaterally ovariectomised and implanted with either a placebo pellet (OVX) or pellet with estradiol (OVX+E2) at 7 weeks of age. At 12 weeks of age, 60 μCi (43 ng total) of ^{75}Se as selenite was orally administered to each rat.

As can be seen from Figure 11, exposure to estrogens significantly increases the selenium status of animals as measured by selenium concentration and GSHPx activity in the plasma, liver and brain. Selenium content in erythrocytes is also increased by estrogen treatment.

Additionally estradiol upregulates the hepatic levels of GSHPx mRNA and Sel P mRNA (Figure 12).

According to the suggested hypothetical model of selenium metabolism regulation by estrogens, estrogens may activate selenoprotein P (SelP) gene transcription in the liver by binding to half-site estrogen response elements in the promoter region of SelP gene, or by activation of transcription factors. After synthesis in the liver, SelP may deliver selenium to extrahepatic tissues via a receptor-mediated mechanism.

The testosterone suppressing effect on the selenium status takes place in puberty when an intensive hormonal rearrangement of an organism is achieved. Epidemiological data reveal that male serum selenium concentration and liver GSHPx activity is significantly decreased during this period whereas female serum selenium concentrations remain relatively constant throughout puberty (Marano et al, 1991; Capel and Smallwood, 1983). These findings support the role of testosterone in causing a redistribution and accumulation of selenium to the testis during male puberty. In general erythrocyte GSHPx activity is greater in females than in males (Guemouri et al, 1991) while renal GSHPx activity is greater in male than in female rats (Finley and Kincaid, 1991).

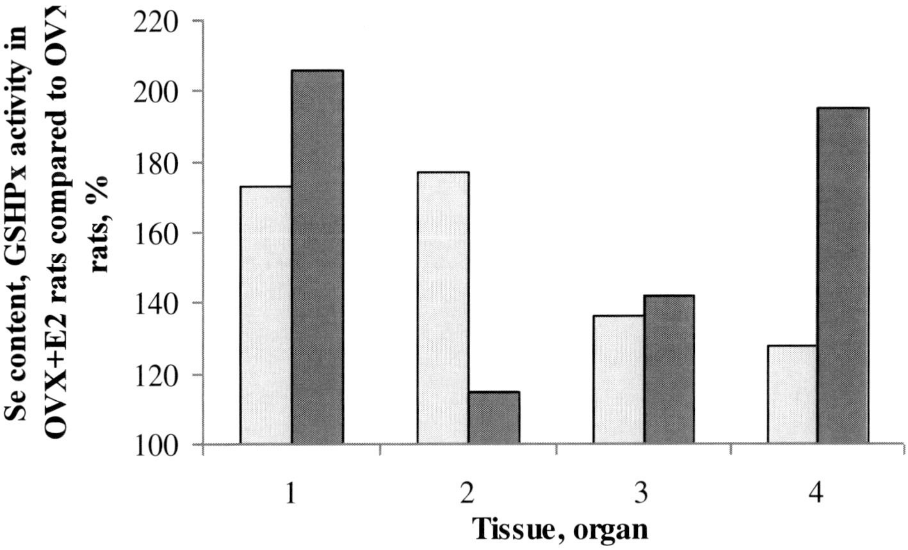

Figure 11. Effects of estrogen status on tissue Se level (light columns) and GSHPx activity (dark columns) in female rats: 1) plasma, 2) erythrocytes, 3) liver, 4) brain.

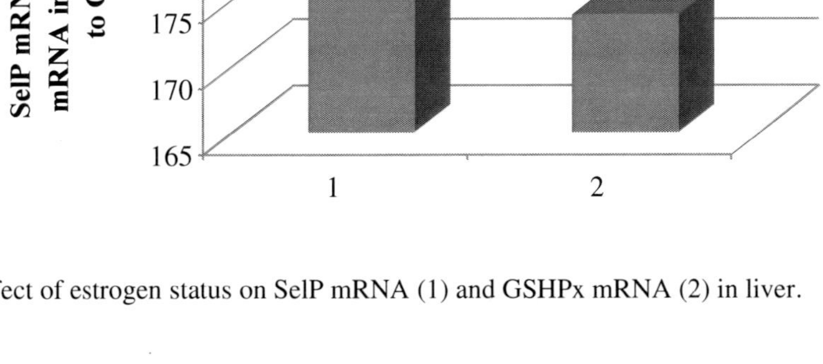

Figure 12. Effect of estrogen status on SelP mRNA (1) and GSHPx mRNA (2) in liver.

The decline in estrogen secretion during menopause appears to affect selenium metabolism. Unexpectedly erythrocyte and plasma GSHPx activity is shown to be significantly greater in post-menopausal women compared to that of premenopausal women (Smith et al, 2000; Ha and Smith, 2009) possibly in response to oxidative processes increasing with age. On the contrary serum selenium concentrations do not differ between these groups (Ha and Smith, 2009), while the selenium level in erythrocytes of post-menopausal women may be either equal (Ha and Smith, 2009), or lower (Karita et al, 2008) than the appropriate values for pre-menopausal women.

At the same time a positive correlation between serum estrogen levels and erythrocytes GSHPx activity in pre- and post-menopausal women receiving estrogen replacement therapy is demonstrated (Massafra et al, 2002). The use of estrogen-containing contraceptives also increases erythrocyte GSHPx activity in premenopausal women (Massafra et al, 1993).

Serum selenium concentration of elderly persons is known to be statistically lower than that of healthy adults (Ducros et al, 2000), that may be connected both with partial changes in nutrition character and a decrease of antioxidant defense. Indeed investigation of plasma fractional composition of middle age and elderly women reveals a significant decrease in GSHPx activity in serum of elderly persons (Bortoli et al, 1991; Ito et al, 1998; Martin et al, 1991).

Table 4. Plasma albumin, Se status and Se content of the plasma selenoproteins of the 3 groups of women (Ducros et al, 2000)

Parameter	Young adults	Free-living elderly	Institutionalized Elderly
Total Se, μg/L	84.3 ± 14.1	80.6±12.5	65.8±6.6 *,**
Albumin, g/L	44±6	42±2	37±3 *,**
GSHPx, U/L	391±64	303±45 [2]	282±87 [*]
Se in albumin, μg/L	36.9±7.1	32.4±5.9	28.5±5.2*,***
Se in Sel P, μg/L	28.8±11.8	33.1±9.6	23.2±5.0 [**]
Dietary Se intake, μg/day	35.1±6.6	33.5±8.6	31.6±5.4

[***] $p<0.05$ vs mean for free-living elderly.

This proves the fact that aging is accompanied not only by a decrease of total serum selenium content but also by lowering the antioxidant defense. Human plasma selenium is associated with three proteins: SelP, GSHPx and albumin.

The investigation has shown that SelP concentration in plasma is not related to age (Hill et al, 1996; Marchaluk et al, 1995, Ducros et al, 2000). On the contrary aging decreases selenium bound to albumen (Table.4). Epidemiological investigation of the human selenium status in Dagestan proves the decreased values of serum selenium concentration in elderly persons. Unexpectedly centenarians (89-104 years old) demonstrate significantly higher levels of serum selenium (Golubkina and Dibirova, 2011) (Table 5).

This amazing fact is in positive agreement with the Chinese data (Yonghua Li et al, 2011) of centenarians' hair selenium concentration increase compared to persons of middle age (Figure 13). At present the explanation of this phenomenon is absent but one can suppose that 89 years old opens a new period of infradian selenium biorhythms.

Table 5. Serum selenium concentration of Dagestan inhabitants (Gerontological Centre, Makhachkala, Dagestan)

Parameter	Young adults	Free-living elderly	Institutionalized elderly
Total Se, µg/L	84.3 ± 14.1	80.6±12.5	65.8±6.6 *,**
Albumin, g/L	44±6	42±2	37±3 *,**
GSHPx, U/L	391±64	303±45 ^{2}	282±87 *
Se in albumin, µg/L	36.9±7.1	32.4±5.9	28.5±5.2*,***
Se in Sel P, µg/L	28.8±11.8	33.1±9.6	23.2±5.0 **
Dietary Se intake, µg/day	35.1±6.6	33.5±8.6	31.6±5.4

* p<0.05 ** p<0.01 compared to persons of 20-59 years old, ***-blood donors.

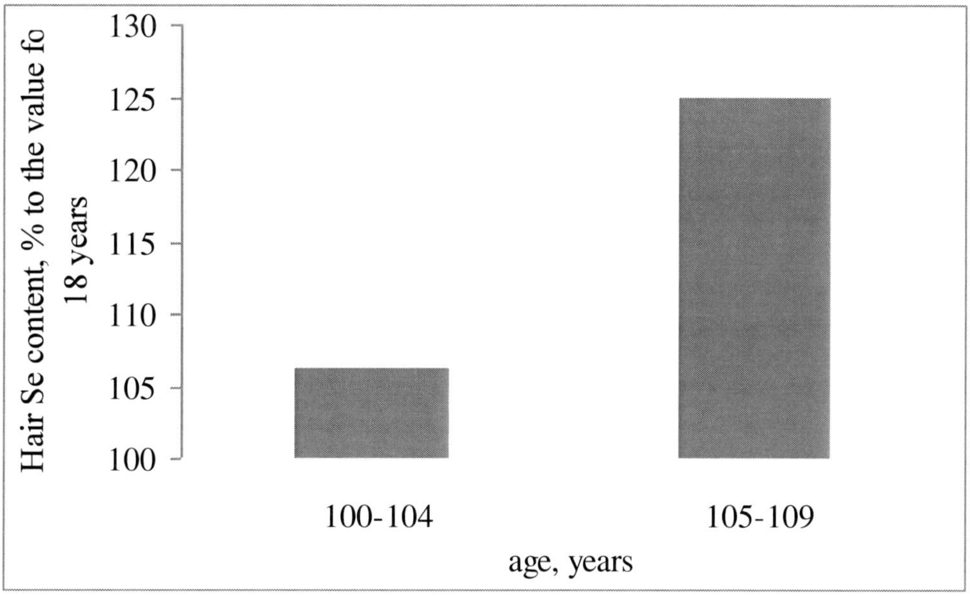

Figure 13. Selenium content in hair of Chinese centenarians (Yonghua Li et al, 2011).

Indeed most of the above data coincide with the well-known periods of human life, described by Fibonacci numbers (Livio, 2003): the first year of life (infancy) is characterized by a sharp serum selenium decrease, 1-2, 2-3, 3- 5, 5-8 and 8-13 years periods are accompanied by an exponential increase of this parameter, puberty from 13 to 21 years results in plasma selenium saturation, 54-89 years interval reveals the decline of the human selenium status, while centenarians (older than 89 years) demonstrate the selenium status increase.

The same associations with Fibonacci numbers are characteristic to age-related changes of selenium concentration in toe- and finger-nails (Figure 14) (Alfthan et al, 1992). The dynamics of selenium accumulation in toe- and fingernails associated with age demonstrates a curve twist at the age of about 13, the lowest level at the end of puberty (about 21 years), reaching the highest values of selenium concentrations by 34 and a significant decrease in selenium concentration by 55.

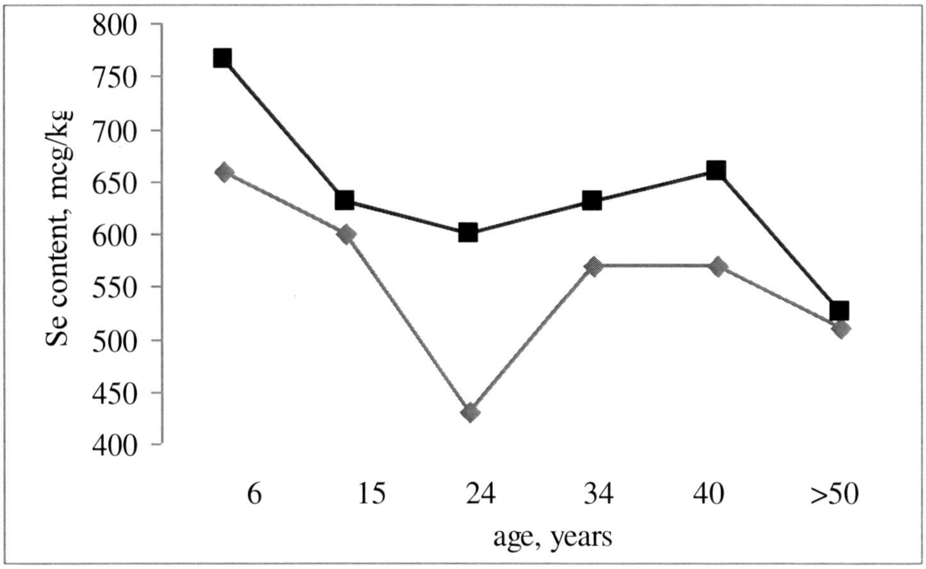

Figure 14. Changes in selenium concentration in fingernails (1) and toenails (2) during human life (Alfthan et al, 1992).

Biorhythms are known to reflect the phenomenon of symmetry in living organisms that makes the correlation between age-related changes of biochemical parameters of the selenium status and the Golden ratio, reflected in Fibonacci numbers, to be highly valuable (Livio, 2003). Toenails selenium concentration is known to exceed that for fingernails, which may be connected with the higher growth rate of fingernails. These differences in growth rate may determine a certain discrepancy between selenium concentration fluctuations for toenails and fingernails, being more expressed in the latter case. Comparing age-related dynamics of selenium accumulation by serum and toe/finger-nails one can indicate that childhood and puberty periods reflect contra phase changes of selenium concentration in these tissues: a significant increase of selenium accumulation in serum and a decrease in the latter case. The age of 20 is the time of plasma selenium stabilization while in toe/finger-nails this is the beginning of intensive selenium accumulation. Nevertheless at present the toe/fingernail dynamics of accumulation of the element has no explanation and the question needs further investigation.

Seasonal Fluctuations

Very little data exist about seasonal fluctuations of selenium levels in blood though intensive hormonal changes in spring-summer-autumn-winter suppose appropriate changes in the human selenium status.

The work of Reyes et al (2000) supports the hormonal regulation theory by demonstrating higher levels of the element in plasma of women at delivery in summer compared to other seasons of the year. The authors have paid attention to the fact that the morbidity of intrahepatic cholestasis of pregnancy is the lowest only in summer.

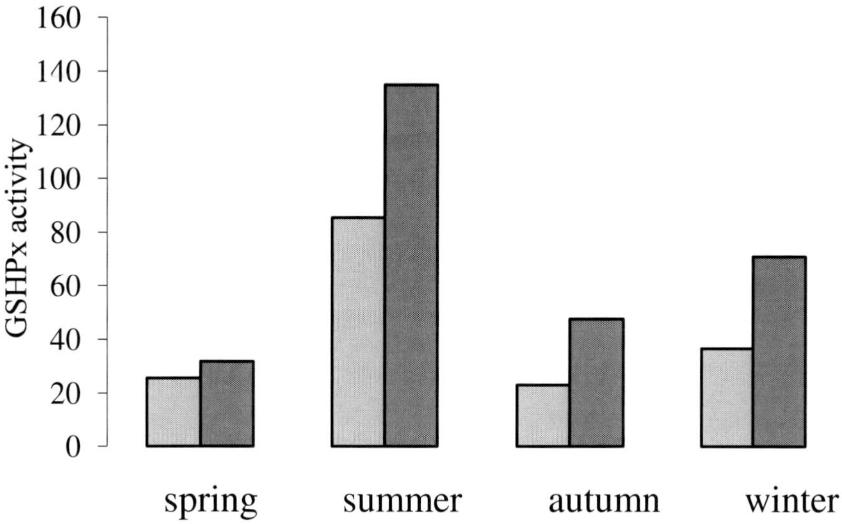

Figure 15. Seasonal fluctuations of GSHPx activity in heart (light columns) and liver (dark columns) of rats (n=6).

As the disease is characterized by a significant decrease of plasma selenium, the authors have concluded that lower incidence of the disease during summer coincides with a higher plasma selenium concentration in summer than in other seasons. It seems important that the selenium antagonist copper exhibits the contra phase dynamics the concentration being much higher in patients with intrahepatic cholestasis than in plasma of healthy women at delivery.

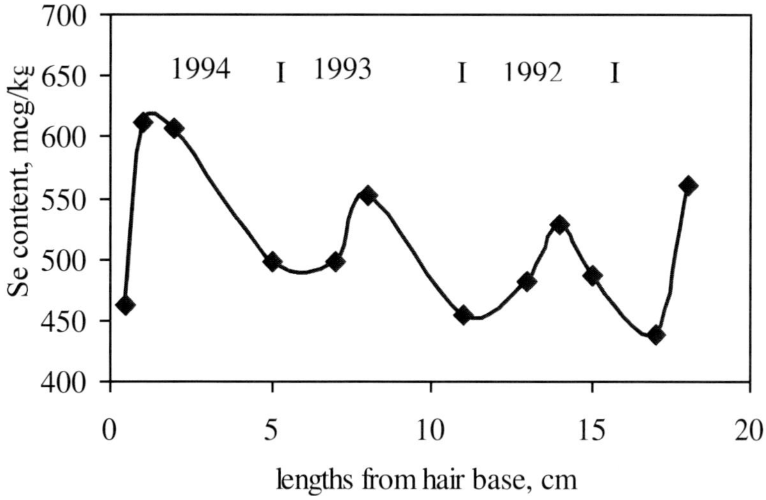

Figure 16. Selenium accumulation by human hair of Moscow residents.

A significant seasonal fluctuation in GSHPx and catalase activity was demonstrated for rats maintained on 12 hours light/dark cycles (Bello-Klein et al, 2000). The authors showed that the highest activity of these enzymes in the heart and liver was typical only in summer (Figure 15).

Trying to achieve a correlation between antioxidant enzyme activities and lipid peroxidation they found the highest lipid peroxidation in the heart in spring with no variations in the liver during the year.

Thus it is obvious that any study of antioxidants, including selenium, or oxidative stress must take into account seasonal variations for a more precise analysis of changes due to any pathological condition.

Another example of seasonal fluctuation of the selenium status was demonstrated in Pakistan, where investigation of seasonal assessment of selenium in lactating and non-lactating ewes and male sheep revealed that although selenium concentration in forage was higher in winter blood plasma selenium contents were higher in summer (Khan et al, 2010).

But the most pronounced seasonal fluctuations of selenium levels are demonstrated in human hair. Hair happens to be highly suitable for selenium biorhythms investigation because it possesses a constant growth rate and we can follow selenium accumulation dynamics using a long hair lock. This approach reveals repeating fluctuations of selenium content characterized by a time period equal to 1 year, an amplitude - about 50μg/Kg and acrophase value of 6 months (Figure 16) (Golubkina et al, 1996a).

The results indicate that the highest hair selenium concentration is registered in summer, whereas the lowest is in winter. This is in positive agreement with the above data.

It seems probable that such fluctuations reflect endogenous biorhythms, as the dynamics character is almost identical to residents living in areas with pronounced seasonal changes (Moscow) and the inhabitants of equatorial countries (Nicaragua) (Somarriba et al, 1998).

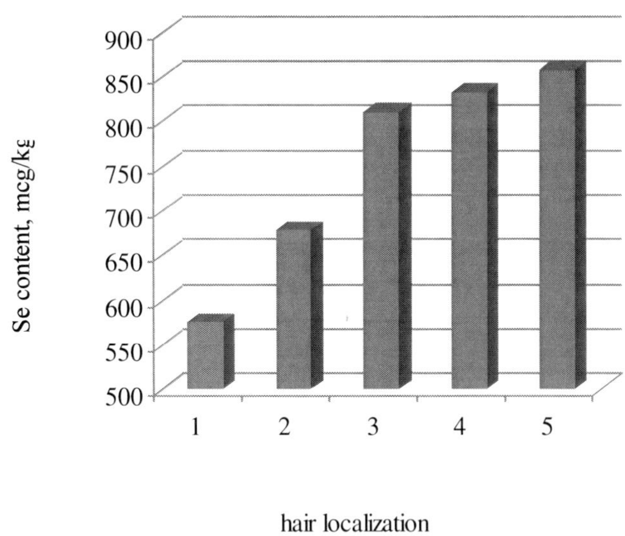

Figure 17. Selenium concentrations in hair of different localizations: 1- head, 2- armpit, 3-beard, 4-celiac plexus, 5-pubis (GolubkinaandPapazian, 2006).

It is significant that the highest mortality in the Northern biosphere is registered in winter (Seasonal mortality, 2003), when the hair selenium level, reflecting the human selenium status, is the lowest. Besides this it is known that training of athletes with identical intensity results in a higher effect in spring/summer and at the beginning of autumn, than in winter (Refinetti, 2006).

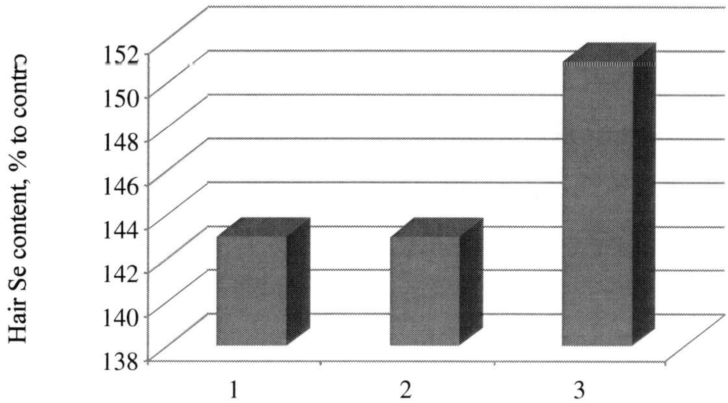

Figure 18. Effect of methandrostenolon (1), stromba (2) and stromba+ GH (3) supplementation by heavy athletes (1 month) on hair selenium accumulation.

The reason for this phenomenon is thought to be influenced by exogenous factors: weather, nutrition, life regime and on periodicity of endocrine system activity (Selye, 1976). Taking into account the significant role of selenium in antioxidant defense of an organism the above data may also be connected with a lower selenium status in winter period. Seasonal changes of hair selenium concentration seem to be a good example of sex hormones' effect on the human selenium status. Seasons are known to influence the system of sex hormones.

The level of gonadotropic hormones in blood plasma is higher in spring (Roenneberg and Aschoff, 1990 Aschoff, 1981). Estrogen concentrations also increase in spring.

Hair growth is known to be androgen-dependant and its highest effect is determined in hair of the pubis, armpit, beard and celiac plexus. Hair in these localizations exhibits the highest selenium concentrations (Figure 17).

Determination of pubis hair selenium concentration for persons with increased sexual activity has shown extremely high values (more than 1700 μg/kg) compared with low concentrations for patients with sexual development delays (up to 550-600 μg/kg).

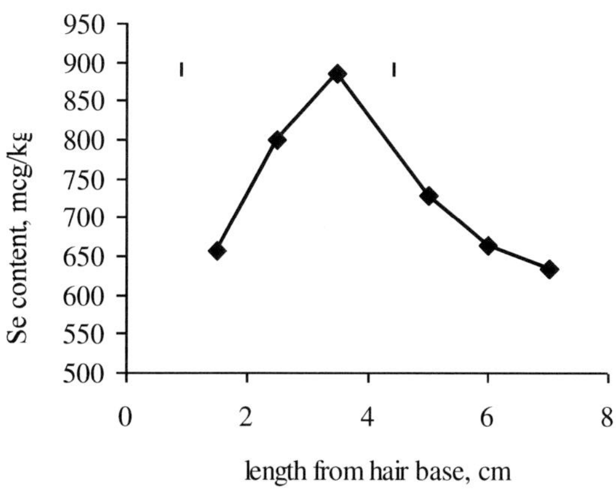

Figure 19. Demonstrates the dynamics of selenium accumulation in hair of an athlete consuming an anabolic (Golubkina et al, 1996b). Though most sportsmen have their hair short cut in several cases it seems possible to use a relatively long lock for detecting anabolic use in the pre competition period (Figure 19).

Thus the increase of hair selenium concentration in spring is in good agreement with the increase of androgen release.

Another proof of hormonal regulation of selenium accumulation by tissues and organs is the effect of doping preparations (many of which are sex hormones) on selenium accumulation in the hair and serum. Investigation of the selenium status and individual questionnaire of athletes on medical treatment in a sports rehabilitation center in Moscow and results received from volunteers have revealed a significant increase of selenium concentration in hair as a result of metandrostenolon, stromba and growth hormone intake. (Figure 18).

Table 6. Plasma and erythrocytes selenium concentration in athlete volunteers as a result of pharmacological loading

Parameter		GH (n=15)	Methandrostenolone (n=15)	Erythropoietin (n=14)	Parlodel (n=15)
Plasma Se	mcg/l	106±5*	109±2*	97.5±5.0	65±5*
	% to control	115	119	106	71
Erythrocyte Se	mcg/l	152±5	127±7	86±8	203±7
	% to control	92	77	52	123
Se plasma/ Se eryth.		0.696*	0.858*	1.134*	0.132*
	% to control	124	153	203	57

*P<0.05 compared to control. Doses and way of administration: GH- 4 Units intramuscularly 2 times per 7 days; MS- 10 mg per os daily during 7 days; erythropoietin- 2000 units intravenously 4 days before blood sampling; parlodel- 2.5 mg daily during 7 days.

Table 7. Effect of methandrostenolone supplementation on concentration of immunoglobulins and selenium in blood of volunteers

Parameter		GH (n=15)	Methandrostenolone (n=15)	Erythropoietin (n=14)	Parlodel (n=15)
Plasma Se	mcg/l	106±5*	109±2*	97.5±5.0	65±5*
	% to control	115	119	106	71
Erythrocyte Se	mcg/l	152±5	127±7	86±8	203±7
	% to control	92	77	52	123
Se plasma/ Se eryth.		0.696*	0.858*	1.134*	0.132*
	% to control	124	153	203	57

Figure 19 portrays the effect of anabolic consumption on hair selenium accumulation of an athlete (Golubkina et al, 1996b). Anabolic steroids also affect the concentration of selenium in serum. Data in Table 6 indicate a significant redistribution of selenium between serum and erythrocytes as a result of methandrostenolone, GH, parlodel and erythropoietin administration (Seifula et al, 1997).

Blood selenium redistribution as a result of administration of sex hormones is also accompanied by changes in immunological parameters (Seifula et al, 1997). Thus methandrostenolone supplementation increases serum selenium level by 18% with an almost twofold decrease of IgG concentration. Decrease of IgA and IgM concentrations in serum is less pronounced and reaches values of 34% and 25% respectively (Table 7).

The data received on experimental animals are in positive agreement with the results received on human beings and prove the existence of GH, stenazol, erythropoietin and parlodel influence on the selenium status (Table.8).

Table 8. Plasma/erythrocytes selenium distribution for mice as a result of pharmacological loading

Parameter		GH (n=16)	Stenazol (n=18)	Erythropoietin (n=15)	Parlodel (n=16)
Plasma Se	μg/l	221±3*	257.0±7.1	216.0±9.5	63.7±5.0*
	% to control	119	138	116	34
Erythro-cyte Se	μg/l	290±15	323.0±8.2	259.0±11.4	277±16
	% to control	99	110	88	94
Se plasma/ Se eryth.		0.762*	0.801*	0.834*	0.230*
	% to control	120	126	132	36

*P<0.05 compared to control. Doses and way of administration: GH- 0.2 Units/mouse intramuscularly 3 times during 8 days; stenazol – 0.0833 mg/mouse intramuscularly 2 times during 8 days; erythropoietin- 0.2 Units/mouse intramuscularly 4 days before blood sampling; parlodel- 0.03 mg/mouse *per os* daily during 7 days (Tutelian et al, 2002).

Dynamics of selenium status changes during pregnancy also coincides with the intensive hormonal regulation of humoral selenium redistribution.

PREGNANCY AND MENSTRUAL CYCLE

A special period in women's life is pregnancy accompanied by significant changes in biochemical characteristics of blood, including trace element levels. A typical curve describing the dynamics of serum selenium concentration during pregnancy is presented in Figure 20.

The fact of serum selenium content declining by the delivery time is often comprehended as a reflection of selenium deficiency development in pregnant women. That opinion is confirmed by a decrease of plasma GSHPx activity from the first to the third trimester (Figure 21, Behne and Wolters, 1979). On the contrary GSHPX activity in erythrocytes remains constant during the pregnancy.

Moreover pregnancy is accompanied by a significant decrease of urine selenium excretion that supposes physiological validity of selenium level changes sufficient for supplementation of the fetus with the element.

Indeed just in several days after the delivery the selenium concentration in serum returns to the initial level. And besides these facts a linear correlation with high statistical significance is demonstrated for serum selenium concentrations of pregnant and non-pregnant women residing in one and the same area (Figure 22).

This supposes that the relatively high selenium status of pregnant women should not correspond to 115-120 μg Se/L serum values, reached for residents of Finland 10 years after the beginning of universal supplementation of fertilizers with sodium selenate (Aro and Alfthan, 1995). Data in Figure 22 indicate that according to a regression equation: y=0.9325x-17.255 (where y- is the serum selenium level at delivery, and x- is the serum selenium concentration for non-pregnant women) real concentrations of serum selenium should be 104-109 μg/L – by the end of the first trimester, 98-103 μg/L – by the end of the second trimester, and 95-100 μg/L – by the moment of delivery (Golubkina, Alfthan, 2002).

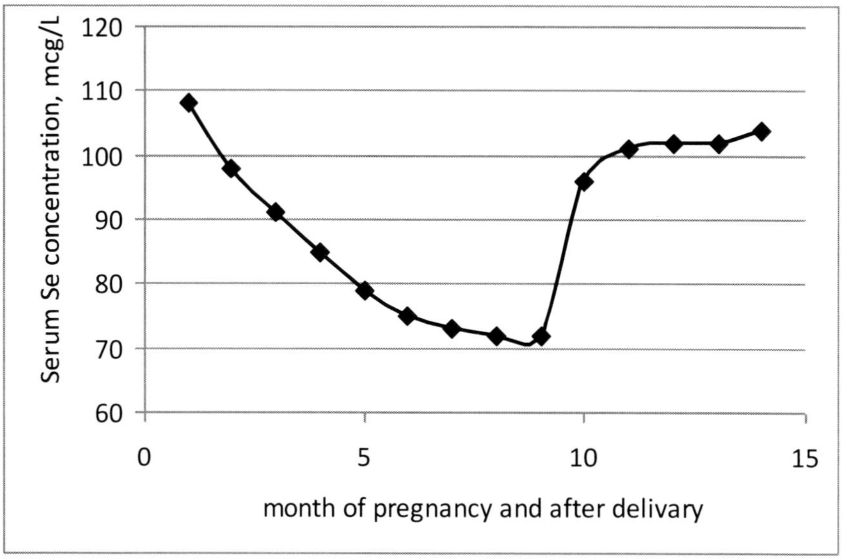

Figure 20. Changes in serum selenium content during pregnancy and after delivary in women of Nigny Novgorod, Russia (Golubkina and Papazian, 2006).

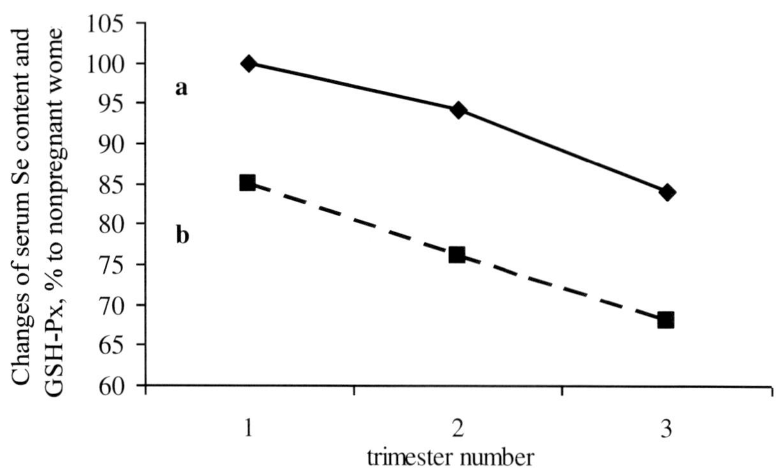

Figure 21. Serum changes in selenium content (a) and glutathione peroxidase activity (b) during pregnancy (Behne and Wolters, 1979).

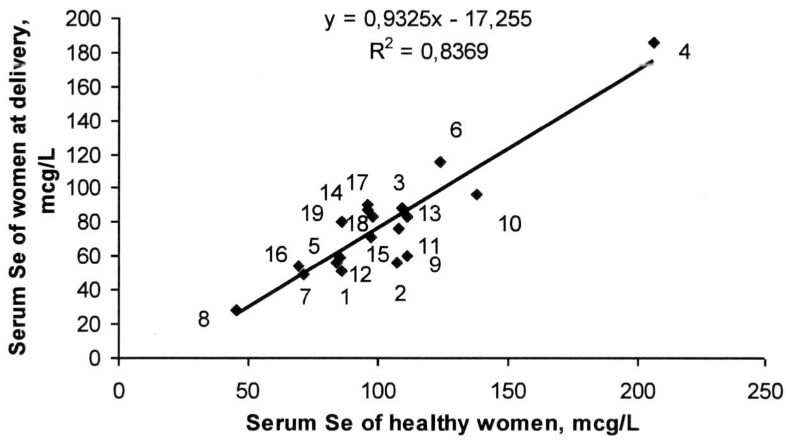

Figure 22. Correlation between serum selenium concentrations in women at delivery and non pregnant women (1-Italy, 2-Germany, 3-Norway, 4-USA, 5-Poland, 6-Danmark, 7,8,9-Finland; Russian Federation: 10-Sakhalin, 11-N.Novgorod, 12-Zelenograd Moscow district, 13-Tashauz Turkmenia, 14-Ioshkar-Ola, Mary El republic, Bashkortostan: 15-Ufa, Bashkortostan, 16-Slavutich, Ukrain, Bashkortostan: 17-Salavat, 18-Sterlitamak, 19-Chishmy).

The significance of adequate provision of pregnant womenwith selenium is obvious as a low selenium status of mothers during the first term increases the risk of premature membrane break that is the main reason for preterm delivery (Rayman et al, 2011), being the significant cause of prenatal mortality and morbidity. Nevertheless at present there is no criterion of optimal values for selenium status during pregnancy.

The most important infradian rhythm of women connected with hormonal regulation is the menstrual cycle.

The menstrual cycle in healthy, ovulating females is regulated as well as defined by changes in the gonadotropic hormones, follicle-stimulating hormone (FSH) and luteinizing hormone (LH), and the sex steroid hormones, estrogen and progesterone.

At the start of the cycle, during the pre-ovulatory follicular phase (FP), FSH stimulates ovarian follicles to grow and develop at which point circulating estrogen levels begin to rise and remain high throughout the FP. This culminates in ovulation at mid-cycle, when LH levels surge and stimulate the release of an oocyte.

The subsequent secretion of sex hormones by the newly formed corpus luteum characterizes this post-ovulatory LP, when progesterone is the dominant hormone. If the egg is not fertilized, sex hormone levels drop at the end of the LP and trigger the shedding of the uteral lining (menstruation) (Farage et al, 2009).

During each cycle other hormones are secreted in different amounts. Changes may undergo other systems, for instance those participating in immune function. A positive correlation between blood selenium parameters and estrogen levels in the menstrual cycle gives another argument of an important role of estrogens in selenium metabolism. Serum selenium concentration (DasandChowdhury, 1997; McAdam et al, 1994) and glutathione peroxidase activity in plasma (McAdam et al, 1994) and erythrocytes (Larsen et al, 1996) are shown to fluctuate during the menstrual cycle however, the selenium concentration in erythrocytes remains constant (Ha and Smith, 2003; Peiker et al, 1991).

A specific direct correlation between estrogen and plasma selenium concentrations and plasma and erythrocytes GSHPx activity has been demonstrated (Ha and Smith, 2003; Massafra et al, 1998; Ha and Smith, 2003) during the menstrual period (Figure 23).

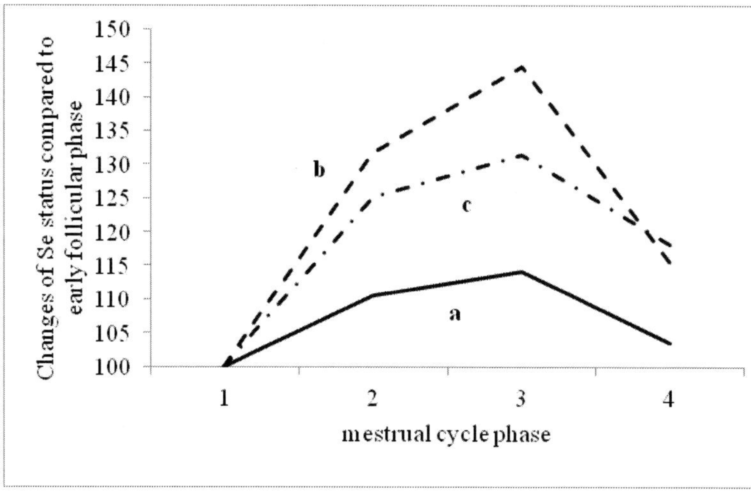

Figure 23. Changes in total serum Se (a), plasma GSHPx (b) and erythrocytes GSHPx (c) during menstrual cycle: 1) early follicular 2) periovulatory, one day before estrogen peak, 3) periovulatory, day of estrogen peak, 4) mid-luteal.

Table 9. Effect of hormones on the selenium status of mammals

Hormone	Effect on the selenium status	References
Estrogens	Increase of Se concentration in serum, erythrocytes, and Uterine GSH-Px , elevated GSHPx activity in plasma, liver and brain, elevated levels of GSHPx1 mRNA and SelP mRNA in liver	Xiaodong Zhou, 2007Ohwada et al, 1996
Progesterone	Increase of serum GSH-Px activity Decrease of uterine GSH-Px activity	Zagrodzki and Ratajczak, 2008Ohwada et al, 1996
Testosterone	Decrease of serum Se in males during puberty and accumulation of the element in testis	Marano et al, 1991
Hormone	Effect on the selenium status	References
Thyroid hormones	Complex interaction between thyroid hormones, progesterone and plasma GSHPx activity	Zagrodzki and Ratajczak, 2008
Prolactin	Low levels of PL in patients with high serum Se	Golubkina et al, 1997
Growth hormone	Increase of serum Se content in mice and human serum	Golubkina et al, 1997
Metandro stenolon	Increase of serum Se concentration in volunteers	Seifula et al, 1997
Erythropoietin	Increase of serum Se concentration and decrease of Se content in erythrocytes in mice	Seifula et al, 1997
Melatonin	Circadian fluctuation of GSHPx and glutathione reductase activity in chick brains Increase of GSHPx activity in blood of sheep during gestation period in selenium deficient region	Pablos et al, 1998 Santiago et al. 2009.

All the above parameters (except the erythrocyte selenium content) are shown to possess a positive correlation with plasma estrogen concentration changes, the phenomenon being typical both for human beings (Ha and Smith, 2003) and rats (Xiaodong Zhou, 2007).

Figure 23 demonstrates that the maximal selenium level in plasma corresponds to the period of the highest estrogen activity - the pre-ovulatory follicular phase. But growth hormone activity is also elevated during this period, which may be another reason for high serum selenium content (Golubkina and Papazyan, 2003). As a whole, one can distinguish several liaisons between the selenium status of an organism and hormones' secretion (Table 9).

PLANTS

Important if not the leading role of hormones in selenium metabolism of human beings and animals supposes that this analogous mechanism may also be possible for plants, though up to the present time the essentiality of selenium to plants has not been estimated. The question of selenium biorhythms existence in plants is of great importance due to the known protective effect of the element against UV-radiation (Germ et al, 2005; Smrkolj et al, 2006), draught (Germ et al, 2007) and herbivore (Eisner et al, 2004; Eisner and Meinwald, 1995; Hartle and Baldwin, 2006; Golubkina and Skryabin, 2010) and the practice of several plants species fortification with the element (Zhu et al, 2009). It is well known that circadian clocks of plants control growth and development via regulation of the activity of key hormones. The circadian effect on genome is widely distributed and clocks modulate genes, participating in the synthesis of phytohormones and signal systems besides other ways determining growth and development. Circadian oscillator affects growth and development of a plant in such a way that specific phytohormone action takes place in a definite day time, often decreasing the rhythmic expression of genes participating in conjugation, signal system and degradation of phytohormones (Thines and Harmon, 2011). Separate investigations prove the interaction of phytohormones with the level of selenium accumulation by plants (Table.10).

Thus widely used in agricultural practice the Epin preparation, being a complex of steroid hormones epibrassinolides, increases the selenium content in leaves of *Brassica chinensese* L. after foliar application during vegetation period (Figure 24) (Golubkina et al., 2005).

Single application of epibrassinolides (Epin) with a concentration of 2 ml/L results in a 1.3 time increase of selenium concentration in perennial onion leaves (Golubkina et al, 2009). Dependence of selenium accumulation levels in plants on the phytohormones activity is confirmed also by the results of indolyl acetic acid (auxin) and gibberellin utilization in vegetable growing (Khrikina et al, 2007).

The data presented in Figure 25 indicate that the effect of gibberellin, heteroauxin and epibrassinolides not only stimulates selenium accumulation by plants but also results in selenium redistribution between roots and leaves.

It is well known that transport of nutritious compounds is achieved predominantly to organs containing a larger amount of hormones (auxins, for instance). Thus indolyl acetic acid (auxin) is synthesized first of all in apical meristem and adjacent to it young leaves, developing germs, seed buds and seed lobes. Meristems of root tops secrete little auxin, while stock tissues of seeds and pollen are rich with this hormone.

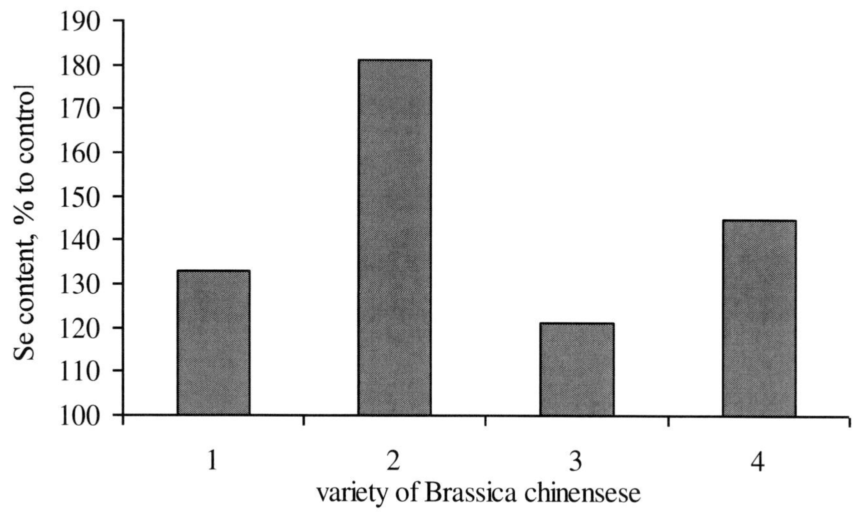

Figure 24. Effect of epibrassinolides on selenium accumulation by several *Brassica chinensese* varieties 1) Bansei Mana, 2) Chrysanthenum heart, 3) Hiroshima Haruna, 4) Kashin Hakusai (single foliar application of epibrassinolide solution, 2 ml/10 L of water at the end of the third week of vegetation (Golubkina et al, 2005).

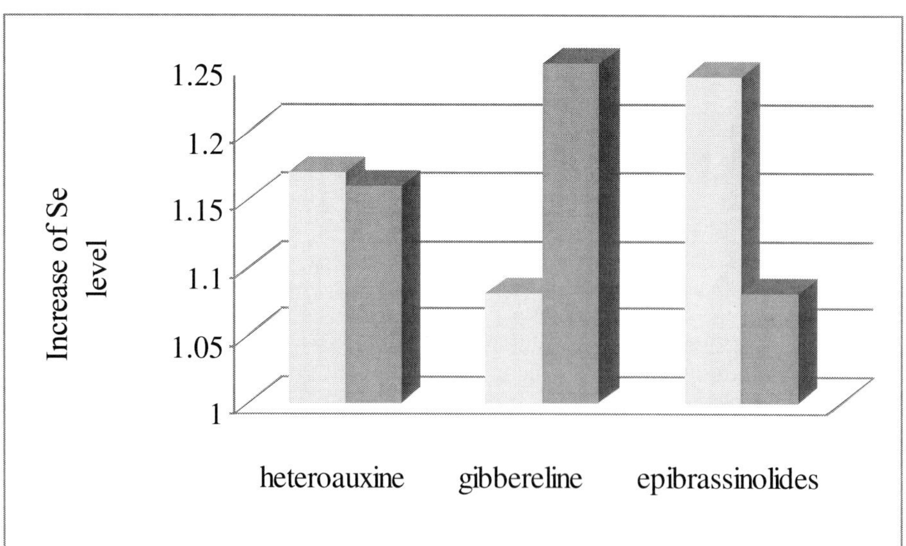

Figure 25. Effect of phytohormones on selenium accumulation by Allium sativum L. (2006-2007 годы). (dark columns-cloves, light columns-leaves): heteroauxine 125 ml/L, epibrassinolides-0.025 ml/L, gibbereline- 125 mg/L- single foliar application during vegetation period.

Plant aging results in the increase of an enzyme destroying indolyl acetic acid. Studies on selenium accumulation in plants show that the highest selenium levels are typical to young leaves and reproductive organs, while old leaves contain only a small amount of selenium (Galeas et al, 2007).

Table 10. Effect of phytohormones on the selenium status of plants

Phytohormone	Biochemical effect	References
Epibrassinolydes	Increase of Se accumulation in leaves of *Allium* species and *Brassica chineneses* varieties	Khrikina et al, 2007 Golubkina et al, 2005
Heteroauxine	High Se levels in young leaves and reproductive organs Increase of Se accumulation in garlic leaves	Galeas et al,-2007 Khrikina et al, 2007
Gybereline	Increase of garlic Se content exclusively in cloves	Khrikina et al, 2007
Methyl jasminic acid, ethylene, salicylic acid	Hyperaccumulators with extremely high Se content possess higher levels of these hormones compared to non-accumulators	Freeman et al, 2006

Seasonal fluctuations of selenium concentration in plants also seem to be connected with the effect of phytohormones. In spring the increase of insulation and temperature leads to the activation of auxin secretion in plants (including aquatic plants) that stimulates growth. Plant-hyperaccumulators of selenium demonstrate the highest selenium concentration in young leaves in spring when an intensive selenium flow is achieved from roots to leaves. After that selenium is transported from the leaves to the reproductive organs (flowers, seeds), while in autumn selenium returns back from old leaves to the roots (Galeas et al,-2007). Pasturable plants (non-accumulators) demonstrate the highest levels of selenium in leaves in March and the content of the element decreases by June-August and increases by October (Gissel-Nielsen, 1975). An increase in the amount of phytohormones or treatment of plants with phytohormones (methyljasmonic acid, salicylic acid, ethylene) results in the increase of selenium accumulation. The concentration of these hormones are higher in hyper-accumulators of selenium (*S.pinnata*) than in non-accumulators (Freeman et al, 2006).

A good example of a leading role of auxin in seasonal peculiarities of growth and development of plants and selenium metabolism are seasonal fluctuations of selenium accumulation by marine grass zostera. It was shown that spring, when plants are rich with auxin, corresponds to the highest levels of selenium (Golubkina et al, 2011).

Table 11. Characteristics of Karakol lake

Parameter	Values
Coordinates of the center	43°32'N 51°18'E
Length	15.44 km
Maximal width	2.9 km
Length of border line	About 37.72 km
Maximal depth	2.0 m
Square	50.7 sq.km
Mean depth	1.0 m
Mean width	2.74 km
Height above the sea level	-28 m abs
Salinity	13.28-13.69°/oo (12.81 in Kaspy)

Table 12. Seasonal fluctuations of selenium content in zostera of Karakol Lake and Eastern Kaspy

Place	April	August	The end of October
Karakol	427±30	203±30	165±6
Eastern Kaspy	301±74	245±32	161±11

The investigation has been carried out on plants from the shoal of the Eastern Kaspy and Karakol Lake (Kazakhstan) (Table 11). Karakol is situated on the East of middle Kaspy and is separated from the sea by a narrow isthmus of 1 km width.

The reservoir is formed as a result of waste water dumping of a heat electropower station and is connected with Kaspy by two channels. Its outstanding characteristic is extremely high salinity exceeding that of neighboring Kaspy by 3.7-6.9%. This provides an opportunity to study the effect of salinity on the selenium metabolism in zostera. Unexpectedly not salinity but the change of seasons happen to be the leading factor affecting selenium accumulation by zostera. Indeed seasonal selenium fluctuations from the highest level in spring to the lowest at the end of autumn are typical both for zostera gathered on the shoal of the Eastern Kaspy and in Karakol Lake (Table 12, Golubkina et al, 2011).

Nevertheless it should be indicated that hormonal regulation of selenium metabolism in plants is still poorly studied and there are still no data about the interaction of citocinines, ethylene and abscisic acid with plant selenium homeostasis.

Interpretation of circadian fluctuations of selenium accumulation in plants is much more difficult though circadian biorhythms of physiological processes in plants are studied with great detailed elaboration.

Thus, it is shown that the circadian biorhythms of plants are regulated by labile proteins which biosynthesis is determined by light (David et al, 2006). Chromosomes regulating the mobility of leaves and flowers have been detected in *Arabidopsis.* The starting points of *Arabidopsis* biorhythms are registered on all levels of gene expression: transcription, translation and protein synthesis (Feldman, 2004). At present circadian biorhythms are known to participate in stem elongation, root pressure, stomatal apeture, cell membrane potential, plastid migration and gas exchange (Sweeney, 1987).

In 1969 circadian biorhythm was registered for the [^{3}H] uridin incorporation in RNA of *Acetabularia mediterranea* (Driessche and Bonotto, 1969). In 1975 a biochemical model for circadian clocks was proposed, including participation of ATP, adenylcyclase, c-ATP, phosphodiesterase and AMP (Cummings, 1975).

Studies of chlorophyll biosynthesis circadian oscillation in leaves of *Phaseolus vulgaris* has led to a conclusion that a new endogenous rhythm is set by imbibition of seeds, light and harvest/cutting (Prombona and Argyroudi-Aloyunoglou, 2004). Circadian clocks help organisms adapt to changes in the environment. Thus circadian biorhythms of specific proteins helps plants to adapt to changes in temperature (Kusakina et al, 2008), crassulacean acid biorhythms determine the stability of orchids to water stress (Motomura et al, 2008).

To study circadian biorhythms of selenium two groups of plants were chosen, the selenium accumulators: garlic (*Allium sativum* L.) and schnitt onion (*Allium schoenoprasum*) and non-accumulators of selenium of different tier: apple tree (*Malus domestica Borkh*), blackberry (*Ribes nigrum* L) and strawberry (*Frigaria viridis* L) (Golubkina, 2010; Golubkina et al, 2008). All plants were grown on sod-podzol heavy loamy soil with 2.5-3.2%

humus content (other characteristics of soil: P-175 mg/kg, Se-240-255 μg/kg, NO_3-10.18 mg/kg, pH-6.0). Garlic and onion leaves were gathered during the days of the Full Moon and New Moon (June). Leaves of the apple tree, blackberry and strawberry were gathered in May during wane. The results of selenium determination in leaves of selenium accumulators (Table 13 and Figure 26, 27) indicate similar characters of fluctuation as for garlic and schnitt-onion.

Table 13. Dynamics of selenium and nitrates accumulation by leaves of selenium accumulators *Allium sativum L.* and *Allium schronoprasum L*

Time, Hours	*Allium sativum L.*				*Allium schronoprasum L.*			
	Full Moon		New Moon		Full Moon		New Moon	
	Se, μg/ Kg d.w.	Nitrates, mg/Kg fresh w.	Se, μg/ Kg d.w.	Nitrates, mg/Kg fresh w.	Se, μg/Kg d.w.	Nitrates, mg/Kg fresh w.	Se, μg/Kg d. w.	Nitrates, mg/Kg fresh w.
7.00	186±8	56±2	101±6	50±2	92±3	49±1	92±3	50±1
9.45	228±10 1)	54±2 7)	120±5 3)	48±2 7)	96±4 7)	45±1 1)	105±4 2)	40±1 1)
12.00	178±10 7)	67±3 1)	112±5 5)	56±3 4)	84±3 4)	53±2 5)	91±3 8)	49±1 7)
14.00	220±10 2)	63±2 3)	146±6 1)	51±2 8)	95±3 7)	44±1 1)	83±3 3)	50±2 8)
16.00	178±10 7)	73±2 1)	132±6 1)	56±2 4)	77±3 1)	47±2 7)	100±4 5)	46±3 6)
18.00	157±9 3)	74±2 1)	139±6 1)	54±2 5)	75±3 1)	50±2 8)	79±3 1)	46±2 5)
20.00	173±10 6)	65±2 1)	143±6 1)	54±1 5)	84±4 4)	47±2 7)	117±5 1)	44±1 1)
22.00	190±10 8)	61±3 5)	165±5 1)	51±3 8)	91±4 8)	43±1 1)	117±4 1)	48±2 7)
24.00	220±	57±	179±	46±	99±4 6)	41±2 1)	122±4 1)	40±1 1)

Significance of differences with the starting level: 1) P<0.001, 2) P<0.002, 3) P<0.005, 4) P<0.05, 5) P<0.1, 6) P<0.2, 7) P<0.5, 8) non significant.

Thus, the *Allium* species demonstrate equal time of oscillations with maximums corresponding to 10 and 14 hours with an increase of selenium concentration by the end of the day. At New Moon selenium content in leaves increases by the end of the day more intensively than at Full Moon. Besides this at New Moon garlic demonstrates lower concentrations of selenium that at Full Moon (Figure 26 a,b).

The dynamics of diurnal selenium fluctuations in plant non-accumulators (apple tree, blackberry and strawberry) is characterized by four maximums of selenium accumulation corresponding to 7, 11, 17 and 21 o'clock and an increase of selenium concentration by 2 o'clock at night (Table 14, Figure 28).

The revealed biorhythms of selenium accumulation seem to be the first example of circadian biorhythms of trace elements in plants reflecting biosymmetry of physiological processes and the mechanism of selenium status endogen regulation via accumulation/emission of the element.

Selenium emission by organisms is well known, but has been considered earlier only in connection with environmental toxic concentration of the element in a whole, particularly for plants, as a possibility to achieve effective phytoremediation of polluted soils and reservoirs (LeDuc and Terry, 2005). Intensive selenium loading is known to cause the formation of volatile methyl selenides by living organisms (Biggar and Jayaweera, 1993; Duckart et al, 1992).

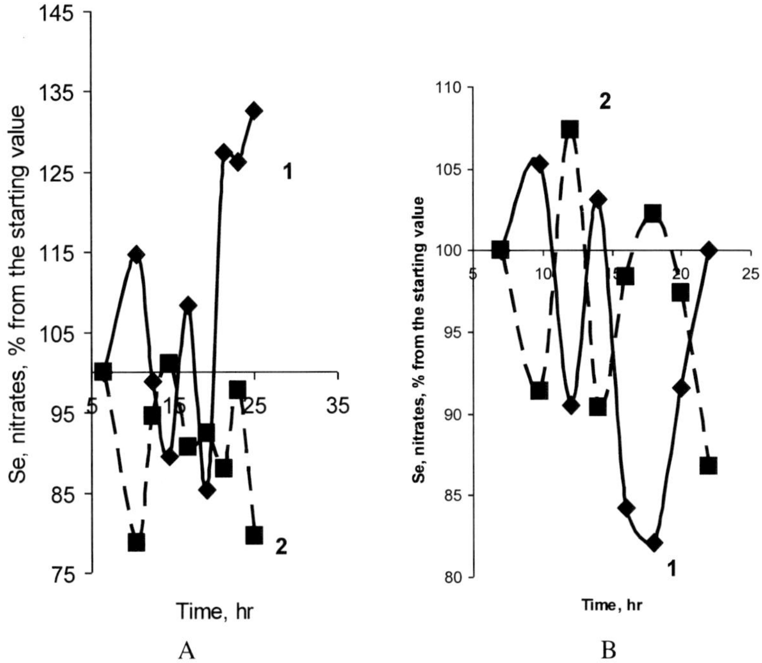

Figure 26. Normalized values of selenium (1) and nitrates (2) accumulation by *Allium sativum* L (a) Full Moon, (b) New Moon.

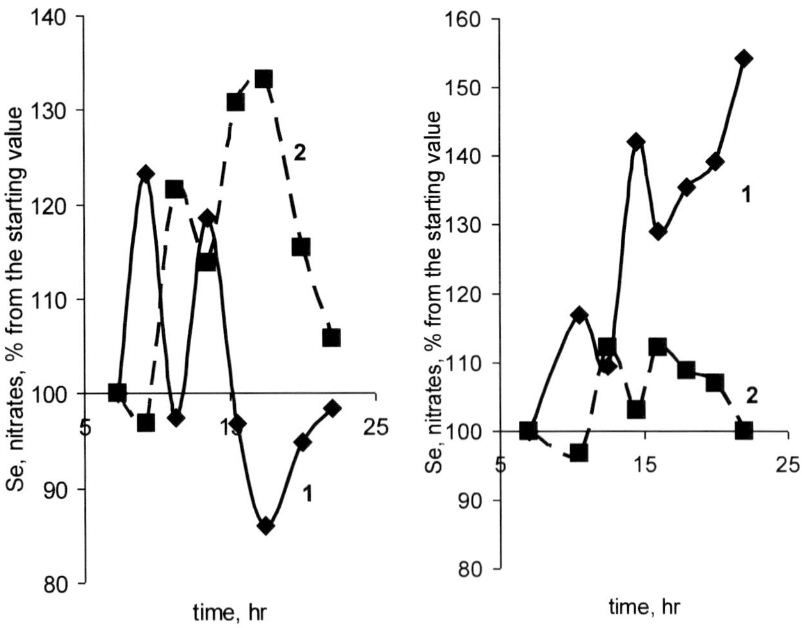

Figure 27. Normalized levels of selenium (1) and nitrates (2) accumulation by leaves of *Allium schoenoprosum* L. a) Full Moon b) New Moon.

Table 14. Dynamics of selenium accumulation by leaves of apple-tree, blackberry and strawberry

Time/hour	Strawberry (n=10)		Blackberry (n=5)		Apple tree (n=3)	
	Se content, µg/kg d.w.	Nitrates content, mg/kg d.w.	Se content, µg/kg d.w.	Nitrates content, mg/kg d.w.	Se content, µg/kg d.w.	Nitrates content, mg/kg d.w.
5	83±4	32±2	121±4	31±2	74±4	21±2
7	117±5[1)]	20±1[1)]	111±4[1)]	32±2[6)]	94±4[1)]	20±2[5)]
9	111±4[1)]	23±1[1)]	72±3[1)]	27±1[1)]	93±5[1)]	24±1[1)]
11	123±5[1)]	24±1[1)]	64±3[1)]	24±1[1)]	107±4[1)]	18±1[1)]
13	86±3[2)]	21±1[1)]	81±3[1)]	30±2[3)]	82±3[1)]	19±2[4)]
15	85±4[5)]	25±2[1)]	79±3[1)]	28±1[1)]	75±5[5)]	24±1[1)]
17	105±3[1)]	20±1[1)]	128±5[1)]	29±2[5)]	111±7[1)]	18±3[1)]
19	102±3[1)]	29±2[1)]	108±4[1)]	30±2[3)]	104±6[1)]	24±1[1)]
21	118±5[1)]	26±1[1)]	126±5[3)]	24±1[1)]	107±4[1)]	21±2[5)]
23	80±3[3)]	34±2[1)]	95±4[1)]	33±2[6)]	86±5[1)]	21±3[5)]
26*	86±3[3)]	28±2[1)]	125±3[2)]	30±1[5)]	98±4[1)]	20±1[5)]

* corresponds to 2 o'clock at night.

Statistically significant differences compared to starting values [1)] P<0.001, [2)] P<0.002, [3)] P<0.005, [4)] P<0.05 [5)] not significant, [6)] P<0.01.

Releasing directly through plant tissues (Lewis et al, 1966), selenium emission by Se-accumulators occur via dimethyl diselenide formation while non-accumulators of selenium synthesize predominantly monomethylated form of selenide (Evans et al, 1968). The precursor of volatile derivative in the first case is selenomethyl selenocystein, and in the second case - selenomethionine (Terry et al, 2000; De Souza et al 2000).

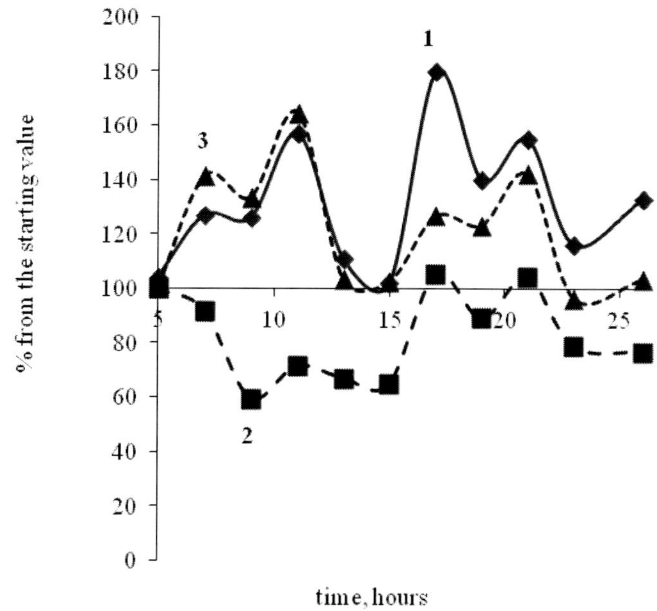

Figure 28. Dynamics of selenium accumulation by leaves of apple tree (1), blackberry (2) and strawberry (3).

The data presented above indicate different dynamics of selenium oscillations for Se-accumulators and non-accumulators. Thus, fluctuations of selenium accumulation for the *Allium* species are similar and possess on average a twice larger period of oscillations than for typical non-accumulators species: apple tree, blackberry, and strawberry. In addition the dynamics of selenium concentration changes for non-accumulators does not differ for plants of a different tier (Figure 28).

The results do not allow affirmation that small and large periods of selenium concentration fluctuations for accumulators and non-accumulators are determined by the dynamics of synthesis of mono- and dimethylated forms and elucidation of this question needs special investigations.

On the other hand it should be indicated that an important peculiarity of selenium biorhythms is the synchronism of selenium and nitrates accumulation. In all cases diurnal oscillations of nitrate accumulation happen to be in contra-phase with the selenium content (Figure 28, Table 15), the correlation coefficients being equal to 0.6-0.65 (Table 15).

High coefficients of correlation between these parameters confirm the known data about the stimulating effect of selenium on the activity of nitrate reductase.

Indeed, administration of selenium salts into soils for potatoes (Munshi and Mindy, 1992), sunflower (Ruiz et al, 2007), and wheat (Nowak et al, 2004) results in the increase of nitrate reductase activity and accordingly to a decrease of nitrates amount in plants.

Table 15. Correlation between nitrates and selenium concentration in plants leaves

Plant	Correlation coefficient	Statistical significance (P<)
Allium sativum L.	-0,651	0,01
Allium schoenoprasum L.	-0,643	0,01
Frigaria viridis L.	-0,604	0,01
Ribes nigrum L.	-0,645	0,01
Malus domestica Borkh	-0,595	0,01

The demonstrated fluctuations of selenium/nitrates as a dual pair are typical for ordinary conditions of vegetation without exogenous selenium loading. Thus the rhythmicity of accumulation is distinctive not only for selenium but also nitrates.

In this context it should be noted that nitrate reductase activation in plants is realized via auxin action (Hayat et al, 2006), that also supposes the possibility of hormonal regulation of selenium concentration. And besides this, it is shown that transcription of the nitrate reductase gene depends not only on nitrate concentration but also on light (Strater and Hachtel, 2000).

At present little is known about the effect of the Luna phase on the chemical composition of plants. The most interesting work in this respect is of Vogt (Vogt et al, 2002), who has shown significant differences in the hemicellulose and calcium content in leaves of the palm tree in the Full and New Moon. The author supposes that the suitability of leave gathering by native people in the Full Moon is determined by less leave damage by herbivores at Full Moon because of the high concentrations of calcium and hemicellulose during this period of vegetation. Though high concentrations of selenium are known to protect plants from herbivores (Eisner et al, 2004; Eisner and Meinwald, 1995), considerably higher concentrations are necessary for such protection than detected at present. Thus physiological significance of synchronic changes in selenium and nitrates concentration remains unfinished.

From a practical point of view works devoted to selenium accumulation in plants should include data of the sampling time. Indeed taking into account circadian oscillation differences may reach 24.2-38.8% for *Allium* species and 30.4-43.6% for plants that are non-accumulators of selenium. Variations of nitrates concentration is 14.3-27.0% and 20.9-41.2% accordingly.

CONCLUSION

The above examples of short and long term selenium status fluctuations in mammals and plants suppose that hormonal regulation of selenium status is a unique mechanism provided by Nature and is accompanied by synchronous fluctuations of selenium content characteristics and hormone activity. The general demand to selenium status investigations in healthy subjects and during pathology is the necessity of time-sampling data to achieve proper comparisons and conclusions.

REFERENCES

Alfthan G, Bogye G, Aro A, Feher J. The human selenium status in Hungary//*J.Trace Elem. Electrolytes Health Dis.*-1992-Vol.6-P.233-238.

Anneren G, Gebre-Medhin M, Gustavson KH. Increased plasma and erythrocyte selenium concentrations but decreased erythrocyte glutathione peroxidase activity after selenium supplementation in children with Down's's syndrome//*Acta. Paediatr.Scand.*-1989-Vol.78-P.879-884.

Aro A, Alfthan G. Effect of supplementation of fertilizers on human selenium status in Finland//*Analyst*-1995-Vol.120-P.841-843.

Aschoff J. Biological Rhythms //in *Handbook of behavioral Neurobiology*-1981-Vol.4-Okenum Press, N.Y., London.

Barlow-Walden L. R., Reiter R. J., Abe M., Pablos M., Menendez-Pelaez A., Chen L-D., Poeggeler B. Melatonin stimulates brain glutathione peroxidase activity//*Neurochem. Int.*-1995-Vol.26 (Iss. 5)- P. 497-502.

Beckett GJ, Arthur JR. Selenium and endocrine systems// *J. Endocrinol.*-2005-Vol.184(3)-P.455-465.

Behne D, Wolters W. Selenium content and glutathione peroxidase activity in the plasma and erythrocytes of non-pregnant and pregnant women//*J.Clin.Chem.Clin.Biochem.*-1979-Vol.17-P.133-135.

Bello-Klein A, Morgan-Martins MI, Barp J, Liesny S, Bello AA, Singai PK. Circa-annual changes in antioxidants and oxidative stress in the heart and liver in rats//*Comp.Biochem.Physiol. Part C. Pharmacology, toxicology and endocrinology*-2000-Vol.126 (Iss.2)-P.203-208.

Biggar JW, Jayaweera GR. Measurement of selenium volatilization in the field//*Soil Sci.*-1993-Vol.155-P.31-35.

Bortoli A, Fazzin G, Marchiori M, Mello F, Brugiolo R, Martelli F. Selenium status and effect of selenium supplementation in a group of elderly women//*J. Trace Elem.Electrolytes Health Dis.*-1991-Vol.5-P.19-21.

Bowrey DJ, Morris-Stiff GJ, Puntis MC Selenium deficiency and chronic pancreatitis: disease mechanism and potential for therapy//*HPB Surg.*-1999-Vol.11 (4)-P.407-415.

Capel ID, Smallwood AE Sex differences in the glutathione peroxidase activity of various tissues of the rat//*Res.Comm.Chem.Pathol.Pharmacol.*-1983-Vol.140-P.367-378.

Cummings F.W. A biochemical model of the circadian clock// *J. Theor. Biol.*-1975-Vol.55-P. 455-470.

Dallaspezia S, Benedetti F Melatonin, circadian rhythms and the clock genes in bipolar disorder//*Curr.Psychiatry*-2009-Vol.11-P.488-493.

Das K, Chowdhury AR. Metallic ion concentration during menstrual cycle in normally menstruating women//*Indian J.Med.Sci.*-1997-Vol.51-P.52-54.

David K.M., Armbrusterm U., Tama N., Putterrill J., Arabidopsis GIGANTEA protein is post-transcriptionally regulated by light and dark//*FEBS Lett.*-2006-Vol.580-P.1193-1197.

De Souza MP, Lytle CM, Mulholland MM, Otter MZ, Terry N, Selenium assimilation and volatilization from dimethylselenon by Indian mustard//*Plant Physiol.*-2000-Vol.122-P.1281-1288.

Driessche T.V., S. Bonotto, The circadian rhythm in RNA synthesis in *Acetabularia mediterranea*//*Biochim.Biophys.Acta* - nucleic Acids and protein synthesis-1969-Vol.179-P. 58-66.

Duckart EC, Waldron LJ, Doner HE Selenium uptake and volatilization from plants growing in soil//*Soil.Sci.*-1992-Vol.153-P.94-99.

Ducros V, Ferry M, Faure P, Belin N, Renversez J-Ch, Ruffieuz D, Favier A. Distribution of selenium in plasma of French Women: relation to age and Se status//*Clin.Chem.*-2000-Vol.46-P.731-733.

Eisner T., Meinwald J. The chemistry of sexual selection//*Proc. Metl. Acad. Sci. USA*-1995-Vol.92-P. 50-55.

Eisner T, Eisner M, Rossini C, Iyengar VK, Roach BL, Benedikt E., Meinwald J. Chemical defense against predation in an insect egg//*Proc .Matl. Acad. Sci. USA* -2004-Vol.101-P. 1634-1639.

Evans CG, Asher CJ, Hohnson CM. Isolation of dimethyl diselenide and other volatile selenium compounds from *Astragalus racemosus* (Pursh.)//*J.Biol.Sci.*-1968-Vol.21-P.13-20.

Farage MA, Neill S, MacLean AB. Physiological changes associated with the menstrual cycle: a review//*Obstetrical and Gynecological Survey*- 2009-Vol.64(1)-P.58–72.

Feldman H. Circadian clocks and natural antisense RNA//*FEBS Lett.*-2004-Vol.567-P.49-54.

Ferri T., Favero G., Frasconi M. Selenium speciation in foods: Preliminary results on potatoes//*Microchem. J.*-2007-Vol. 85 (2)-P.222-227.

Finley JW, Kincaid RL Effect of sex and time of sampling on selenium and glutathione peroxidase activity in tissues of mature rats//*Biol.Trace Elem.Res*-1991-Vol.3-P.181-191.

Freeman JL, Zhang LH, Marcus MA, Fakra S, Pilon-Smits EAH Spatial imaging, speciation and quantification of selenium in the hyperaccumulator plants Astragalus bisulcatus and Stanleya pinnata//*Plant Physiol.*-2006-Vol.142-P.124-134.

Galeas ML, Zhang LH, Freeman JL, Wegner M, Pilon-Smiths EA. Seasonal fluctuations of selenium and sulfur accumulation in selenium hyperaccumulators and related nonaccumulators//*New Phytol*.-2007-Vol.173 (3)-P.517-525.

Germ M., Kreft I., Osvald J. Influence of UV0B exclusion and selenium treatment on photochemical efficiency of photosystem II, yield and respiratory potential in pumpkins (Cucuebita pepo L.)//*Plant Physiol.Biochem*.-2005-Vol.43 (Iss 5)-P. 445-448.

Germ M., Kreft I., Stibilj V., Urbanc-Bersis O. Combined effects of selenium and drought on photosynthesis and mitochondrial respiration in potato//*Plant Physiol. Biochem*.-2007-Vol.45 (Iss.2)-P. 162-167.

Gissel-Nielsen G. Selenium concentration in Danish forage crops//*Acta Agric.Res.Scand*-1975-Vol.25-P.216-220.

Golubkina NA. Circadian biorhythms of selenium in plants//*Proc.of the VI-XII biogeochemical readings in memory of V.V.Kovalsky-V.V.Ermakov ed.-Moscow-GEOKHI of RAS*-2010-P.145-151(Russ).

Golubkina NA; Alfthan G. Selenium status of pregnant women and newborns in the former Soviet Union//*Biol. Trace Elem. Res*.-2002-Vol.89 (1)-P. 13-23.

Golubkina NA., Agafonov AF., Dudchenko NS. Trace elements content in perennian onion//*Gavrish*-209-No5-2009-P.18-21 (Russ).

Golubkina NA, Chigenkiva OA, Karapun MJ, Melyakina EI Selenium content in hydrobionts of Kaspy sea//Proc. Biogeochemical school "*Fundamental and innovational aspects of biogeochemistry*-2011 Astrakhan, Russia Sept 202-206 (Russ).

Golubkina NA, Dibirova A. Selenium status of Dagestan residents//*Trace elements in medicine*-2012 in press.

Golubkina NA., Golubev FV, Dobrutskaya EG Circadian biorhythms of selenium in several *Allium* species//*Ecologica* -2008-Vol.15-P.9-1.

Golubkina NA., Papazyan T.T., *Selenium in Nutrition. Plants, animals, human beings*, Pechatny Gorod, Moscow, 2006 (Russ).

Golubkina NA, Skriabin K. Anomalous accumulation of selenium by genetically modified potato, stable to Colorado beetle// *J. Food Comp. Anal*.- 2010 –Vol.23 (2)-P.190-193.

Golubkina NA., Sokolov YA., Yemelyanov BN. Modulation of infectious process with bromocryptine, somatotropin and selenium in mice; analogous correlation between severity of infectional disease (viral hepatitis) and levels of selenium, somatotropin, prolactine in blood of patients//*Immunology*-1997-No6-P.27-29 (Russ).

Golubkina NA., Sokolov YA., Yemelyanov BN. In vitro dynamics of selenium distribution in blood//*Trace Elem. Med*. -2003-Vol.4-P. 17-20.

Golubkina NA., Sokolov JA, Emelianov BN, Martianov SS, Portugalov SN, Vedenin AN, Somarriba OJ. Method of doping control//RU pat. № 212971-1996a.

Golubkina NA, Somarriba O.J., Sokolov JA, Hair selenium in the human selenium status estimation//*Vopr. pitania*-1996b-Vol.4-№3-P.14-18 (Russ).

Golubkina NA, Temichev AV, Startsev VI, Effect of growth stimulator Epin and sodium selenate on trace elements content in *Brassica chinensese* L.//*Agricultural Biol*.-2005-No1-P.88-91 (Russ).

Goodwin WJ., Lane HW., Bradford K., Marshall MV., Griffin AC., Geopfert H., Jesse RH. Selenium and glutathione peroxidase levels in patients with epidermoid carcinoma in the oral cavity and oropharunx//*Cancer*-1983-Vol.51-P.110-115.

Guemouri L, Arthur Y, Herbeth B, Jeandel C, Cuny G, Siest G. Biological variability of superoxide dismutase, glutathione peroxidase and catalase in blood//*Clin.Chem.*-1991-Vol.37-P.1932-1937.

Ha EJ, Smith AM. Selenium-dependent glutathione peroxidase activity is increased in healthy post-menopausal women//*Biol.Trace Elem.Res.*-2009-Vol.131 (1)-P.90-95.

Ha EJ, Smith AM. Plasma selenium and glutathione peroxidase activity increase during the pre-ovulatory phase of the menstrual cycle//*J.Am.Coll.Nutr.*-2003-Vol.22-P.43-51.

Halberg F., Chronobiology//*Ann. rev. physiol.*-1969-Vol. 31-P. 675—725

Hartle M., Baldwin I.T. Evolution: The Ecological Reverberations of Toxic trace Elements// *Current Biol.*-2006-Vol. 16-P. R958-R960.

Hayat S, Fariduddin Q, Barket Ali, Ahmad A. Effect of chloroindole auxins on the growth and nitrate reductase activity in Solanum melongene//*Asian J.Plant Sci.*-2006-Vol.5(1)-P.14-16.

Hill KE, Xia Y, Akesson B, Boeglin ME, Burk RF. Selenoprotein P concentration in plasma is an index of selenium status in selenium deficient and selenium supplemented Chinese subjects//*J.Nutr.*-1996-Vol.126-P.138-145.

Ho KY, Veldhuis JD, Johnson ML, Furianetto R, Evans WS, Alberti KGMM, Thorner MO. Fasting enhances growth hormone secretion and amplifies the complex rhythms of growth hormone secretion in man//*J.Clin.Invest*-1988-Vol.81(4)-P.968-975.

Ito Y, Kajkenova O, Feuers RJ, Udupa KB, Desai VG, Epstein J. Impaired glutathione peroxidase activity accounts for the age-related accumulation of hydrogen peroxide in activated human neutrophils//*J.Gerontol. A Biol.Sci.Med.Sci.*-1998-Vol.53-P.M169-M175.

Jasmin G, Panisset B, Selye H. Polyarthritis produced in rats by treatment with prolactin and growth hormone//*Endocrinology*-1959-Vol.65 (3)-P.494-499.

Karita K, Yamanouchi Y, Takano T, Oku J, Kisaki T, Yano E. Association of blood selenium and serum lipid levels in Japanese premenopausal and postmenopausal women//*Menopause*-2008-Vol.15(1)-P.119-124.

Khan ZI, Ashraf M, Ahmad K, Al-Qurainy F. Seasonal assessment of selenium as a hazardous element in pasture and animal systems: A case study of Kajli sheep in Sargodha, Pakistan//*J. Hazardous Mater.*-2010-Vol.179 (Iss. 1-3)- P.1111-1114.

Khrikina JA., Golubkina NA, Nikulshin VP, Grigoriants IK, Bogachev VN. Selenium accumulation by Allium sativum L.//*Vestnik RAAS*- 2007. №5 C 32-33 (Russ)

Kimball H, Lipsett MB, Odell WD. Comparison of the effect of the pyrogens, etiocholanolone and bacterial endotoxin on plasma cortisol and growth hormone in man//*J.Clin.Endocrinol.*-1968-Vol.28-P.337-342.

Klevecz RR, Shymko RM, Blumen feld D, Braly PS. Circadian gating of S phase in human ovarian cancer//*Cancer Res.*-1987-Vol.47-P.6267-6271.

Kondratov RV, Gorbacheva VY, Antoch MP The role of mammalian circadian proteins in normal physiology and genotoxic stress responses//*Curr.Top.Dev.Bio.*-2007-Vol.78-P.173-216.

Kukes V.G., Aslanian N.V., Golubkina N.A., Khotimchenko S.A., Shikh E.V. The concentration of selenium in serum of volunteers after administration of different selenium preparations//*Trace Elem. Med.* -2002-Vol.3-P. 13-16.

Kusakina J, Hartwell J, Hall A Investigating effects of temperature on the circadian clock in plants//*Comp. Biochem. Physiol.*-P.A-2008-Vol.150 (Iss3)-Suppl.1-P.154-54.

Larsen JF, Andersen HR, Hansen AB, Andersen O. Variation in risk indicators of cardiovascular disease during the menstrual cycle: an investigation of within-subject variations in glutathione peroxidase, haemostatic variables, lipids and lipoproteins in healthy young women //*Scand.J.Clin.Lab.Invest*.-1996-Vol.56-P.241-249.

Lazarev AF, Schneeweiss U, Fabricius EA. Inhibition of subcutaneous Ehrlich carcinoma by antihuman-somatotropin hormone serum of the horse as determined by the tumor tetanus assay of the mouse//*Arch.Geschwulstforsch*-1976-Bd.46-No4-P.261-270.

LeDuc DL., Terry N. Phytoremediation of toxic trace elements in soil and water//*J.Ind. Microbial. Biotech*.-2005-Vol. 32 (11-12)-P.514-520.

Levanver O, Burk RF. Selenium//EE Ziegler, LJ Filer (eds) *Present knowledge in nutrition*-1998- 7th ed.-P.320-328.

Lewis BG., Johnson CM., Delwiche CC. Release of volatile selenium compounds by plants: Collection procedures and preliminary observations//*J. Agric. Food Chem*.-1966-Vol. 14-P. 638-640.

Li Y, Yang L, Wang W, Li H, Zou X. Trace element concentrations in hair of healthy Chinese centenarians//*Sci.Total Env*.-2011-Vol.409 (8)-P.1385-1390.

Lin K., Xu X., Jin X., Shao Z., Xiang Y., Eco-toxicological effects of selenium and its critical value on Oryza sativa//*Ying Yong Sheng Tai Xue Bao* -2005-Vol.16-P.678-682 (Chinese).

Livio M. The Golden Ratio. *Broadway Books*, 2003.

Lowrey PL, Takahashi JS. Mammalian circadian biology: elucidating genome-wide levels of temporal organization//*Annu.Rev.Genomics Hum.Genet*.-2004-Vol.5-P.407-441.

Marano G, Spagnolo A, Morisi G, Menotti A. Changes of serum selenium and serum cholesterol in children during sexual maturation//*J Trace Elem Electrolytes Health Dis*.-1991-Vol.5(1)-P.59-61.

Marchaluk E, Persson-Moschos M, Thorling EB, Akesson B. Variations in selenoprotein P concentration in serum from different European regions//*Eur. J. Clin. Nutr*.-1995-Vol.49-P.42-48.

Martin BJ, Lyon TDB, Fell GS. Comparison of inorganic elements from autopsy tissue of young and elderly subjects//*J. Trace Elem. Electrolytes Health Dis*.-1991-Vol.5-P.203-211.

Massafra C, Buonocore G, Berni S, Gioia D, Giuliani A, Vezzosi P Antioxidant erythrocyte enzyme activities during oral contraception//*Contraception*-1993-Vol.47-P.591-596.

Massafra C, Felice C, Gioia D, Buonocore G. Variations in erythrocyte antioxidant glutathione peroxidase activity during the menstrual cycle//*Clin.Endocrinol*-1998-Vol.49-P.63-67.

Massafra C, Gioia D, Fellice C, Musceola M, Longini M, Buonocore G. Gender-related differences in erythrocyte glutathione peroxidase activity in healthy subjects//*Clin.Endocrinol*-2002-Vol.57-P.663-667.

McAdam PA., Morris VC, Forman MR, Lanza E, Campbell W, Beecher GR, Levander OA. Differences in plasma selenium and glutathione peroxidase activity during three phases of the menstrual cycle in normal healthy women//*FAEIB*-1994-J8:A436.

McClung CA. Circadian genes, rhythms and the biology of mood disorders//*Pharmacol. Ther*.-2007-Vol.114-P.222-232.

McMurray CH, Davidson WB. In vitro metabolism of selenite in sheep blood: factors controlling the distribution of selenium and the labeling of plasma protein//*Int.J.Biochem. Biophysics*- 1979-Vol.583-P.332-343.

Miller L, Mills BJ, Blotsky AJ, Licleman RD. Red blood cell and serum selenium concentrations as influenced by age and selected diseases//*Am.J. Coll.Nutr.*-1983-Vol.4-P.331-341.

Motomura H, Yukawa T, Ueno O, Kagawa A. The occurrence of crassulacean acid metabolism in Cymbidium (Orchidaceae) and its ecological and evolutionary implications//*J.Plant Res.*-2008-Vol.121 (2)-P.163-177.

Mulero-Navatto S, Esreller M. Epigenetic biomarkers for human cancer: the time is now//*Crit.Rev.Oncol.Hematol.*-2008-Vol.68-P.1-11.

Munshi C.B., Mindy N.I. Glycoalkaloid and nitrate content of potatoes as affected by method of selenium application//*Biol.Trace Elem.Res.* -1992-Vol.33-P. 121-127.

Muntau AC, Streiter M, Kappler M, Roschinger W, Schmid I, Rehnert A, Schramel P, Roscher AA. Age-related reference values for serum selenium concentrations in infants and children//*Clin.Chem.*-2002-Vol.48-P.555-560.

Navarro-Alarcon M, Lopez G, de la Serrana H., Perez-Valero V, Lopez-Martinez C. Serum and urine selenium concentrations as indicators of body status in patients with diabetes mellitus//*Sci.Total Environ.*-1999-Vol.228 (1)-P.79-85.

Nowak J., Kaldewski K., Ligocki M. Influence of selenium on oxidoreductive enzymes activity in soil and plants//*Soil Biol. Biochem.*-2004-Vol.36-P. 1553-1558.

Ohwada M, Suzuki M, Sato I, Tsukamoto H, Watanabe K Glutathione peroxidase activity in endometrium: effect of sex hormones and cancer//*Gynecol.Oncol*-1996-Vol.60 (2)-P.277-282.

Ozgen IT, Dagdemir A, Elli M, Saraymen R, Pinarli FG, Fisgin T, Albayrak D, Acar SJ. Hair selenium status in children with leukemia and lymphoma//*Pediatr Hematol Oncol.*-2007-Vol.29(8)-P.519-22.

Pablos MI., Reiter RJ., Ortiz GG., Guerrero JM.,. Agapito MT, Chuang J-I., Sewerynek E. Rhythms of glutathione peroxidase and glutathione reductase in brain of chick and their inhibition by light//*Neurochem. Int.*-1998-Vol. 32 (Iss. 1)- P. 69-75.

Peiker G, Muller B, Dawczynski H, Winnefeld K. Behavior of selenium in serum and erythrocytes in females with normal and risk pregnancy (pregnancy induced hypertension, fetal retardation and hepatosis)//*Zentralbl Gynakol*-1991-Vol.113-P.45-48.

Prombona A., Argyroudi-Aloyunoglou J. Diverse signals synchronize the circadian clock controlling the oscillations in chlorophyll content of etiolated Phaseolus Vulgaris leaves//*Plant Sci.*-2004-Vol.167-P. 117-127.

Rayman MP, Wijnen H, Vader H, Kooistra L, Pop V. Maternal selenium status during early gestation and risk for preterm birth//*CMAJ*-2011-Vol.183(5)-P.549-555.

Refinetti R Circadian Physiology-2006- 2d ed., *Taylor and Francis Group*, N.Y

Reyes H, Báez ME., González MC., Hernández I, Palma J., Ribalta J., Sandoval L., Zapata R. Selenium, zinc and copper plasma levels in intrahepatic cholestasis of pregnancy, in normal pregnancies and in healthy individuals, in Chile//*J. Hepatol.*-2000-Vol. 32 (4)-P.542-549.

Richter HG, Torres-Farfan C, Rojas-Garcia PP, Campino C, Torrealba F, Seron-F-Perre M. The circadian timing system: making sense of day/night gene expression//*Biol.Res.*-2004-Vol.37 (1)-P.11-28.

Robinson MF, Godfrey PJ, Thomson GD, Rea HM, van Rij AM, Blood selenium and glutathione peroxidase activity in normal subjects and in surgical patients with and without cancer in New Zealand//*Am.J.Clin.Nutr.*-1979-Vol.32-P.1477-1985.

Roenneberg T, Aschoff J: Annual rhythm of human reproduction: II. Environmental correlations. //*J. Biol. Rhythms* -1990-Vol. 5-P.217-239.

Ruiz JM., Rivero RM, Romero L. Comparative effect of Al, Se and Mo toxicity on NO_3^- assimilation in sunflower (Helianthus annuus L.) plants//*J.Environ. Manage*.-2007-Vol.83 (2)-P.207-212.

Sahar S, Sassone-Corsi P. Metabolism and cancer: the circadian clock connection //*Nat.Rev.Cancer*-2009-Vol.9-P.886-896.

Sammalkorpi K, Valtonen V, Alfthan G. Serum selenium in acute infection//*Infection*-1988-Vol.16(4)-P.222-224.

Sandholm M. Function of erythrocytes in attaching selenite onto specific plasma proteins//*Acta Pharmac. Toxicol.* (Copenhagen)-1975-Vol.36-P.321-327.

Santiago A., Joaquín S., Antonio J. Evaluation of the influence of melatonin implants during the gestation period in sheep from a selenium-deficient region.// *Am. J. Vet. Res*.-2009-Vol. 70 (3)- P.404-408.

Schernhammer ES, Laden F, Speizer FE, Willett WC, Hunter DJ, Kawachi I, Fuchs CS, Colditz GA: Night-Shift Work and Risk of Colorectal Cancer in the Nurses' Health Study//*J. Natl. Cancer Inst*.-2003-Vol.95-P.825-828.

Schibler U, Ripperger J, Brown SA, Peripheral circadian oscillators in mammals: time and food// *J.Biol.Rhythms*-2003-Vol.18-P.250-260.

Schrauzer GN, The nutritional significance, metabolism and toxicity of selenomethionine //*Adv. Food Nutr.Res*.-2003-Vol.47-P.73-112.

Schulz P, Steimer T, Neurobiology of circadian systems//*CNS Drugs*-2009-Vol.23 (Supl.2)-P.3-13.

Seasonal mortality in Denmark//*Demographic Research*-2003-Vol.9, art.9, P.197-222.

Seifula RD, Golubkina NA, Sokolov JA. Ecological problems of spermatogenesis disturbances in sportsmen//*Sci.Works of Russian Institute of physical training*-Moscow-1997-P.322-333 (Russ).

Selye, H. *Stress in Health and Disease*. Boston: Butterworths Inc. 1976.

Seppanen M., Turakainen M., Hartikainen H. Selenium effects on oxidative stress in potato// *Plant Sci*.-2003-Vol.165-P. 311-319.

Smith AM, Chang MP-H, Medeiros LC, Generational differences in selenium status of women//*Biol.Trace Elem.Res*.-2000-Vol.75-P.157-165.

Smrkolj P., Stibilj V., Kreft I., Germ M. Selenium species in buckwheat cultivated with foliar addition of selenium (VI) and various levels of UV-B radiation//*Food Chem*.-2006-Vol.96 (Iss.4)-P.675-681.

Sobia Rana, Saqib Mahmood Circadian rhythm and its role in malignancy//*J. Circadian Rhythms*- 2010-Vol. 8 (1)-P.3-49.

Somarriba OJ., Golubkina NA., Sokolov JA. Estimation of the human selenium status of Managua (Nikaragua) residents using hair selenium parameter//*Vopr.pitania*-1998-№2-P.22-24.

Strater T, Hachtel W. Identification of light- and nitrate-responsive region of the nitrate reductase promoter from birch// *Plant Sci*.-2000-Vol.150 (Iss.2)-P.153-161.

Sweeney B.M. Rhythmic Phenomena in Plants, Academic Press, San Diego, Calif. -1987.

Terry N., Zayed AM., de Souza MP., Tarun AS. Selenium in higher plants// *Ann.Rev.Plant Physiol. Plant Mol. Biol*. – 2000 - Vol. 51 - P. 401 - 432.

Thines B, Harmon FG. Four easy pieces: mechanisms underlying circadian regulation of growth and development//*Current Opinion in Plant Biology*-2011-Vol.14 (Iss.1)-P.31-37.

Tokunaga H, Takebayashi Y, Utsunomiya H, Akahari J, Higashimoto M, Mashiko M, Ito K, Niikura H, Takenoshita S, Yaegashi N. Clinocopathological significance of circadian rhythms-related gene expression levels in patients with epithelial ovarian cancer//*Acta Obstetricia et Gynecologica*-2008-Vol.87-P.1060-1070.

Tsuneyoshi T, Yoshida J, Sasaoka T. Hydroponic cultivation offers a practical means of producing selenium-enriched garlic// *J. Nutr.*- 2006-Vol.136(3 Suppl)-P.870S-872S.

Tutelian VA, Kniagev VA, Khotimchenko SA, Golubkina NA, Kushlinsky NE, Sokolov JA. Selenium in human organism. *Metabolism, antioxidant properties*, role in carcinogenesis-Moscow RAMS publ.-2002-P.150-152 (Russ).

Valkama E., Kivimaenpaa M., Hartikainen H., Wulff A. The combined effects of enhances UV-B radiation and selenium on growth, chlorophyll fluorescence and ultrastructure in strawberry (*Fragaria Xananassa*) and barley (*Hordeum vulgare*) treated in the field//*Agr. Forest Meteor.*-2003-Vol.120-P. 267-278.

Vernie LN, De Vries M, Benckhuijsen C, De Goeij JJ, Zegers C. Selenium levels in blood and plasma, and glutathione peroxidase activity in blood of breast cancer patients during adjuvant treatment with cyclophosphamide, methotrexate and 5-fluorouracil// *Cancer letters*, 1983-Vol. 18 (3)-P.283-289.

Vogt KA., Beard KH., Hammann S., Palmiotto JO., Vogt DJ., Scatena FN., Hecht BP., Indigenous knowledge informing management of tropical forests: the link between rhythms in plant secondary chemistry and lunar cycles//*Ambio*-2002-Vol.31-P.485-490.

Walker JJ, Terry JR, Tsaneva-Atanasova K, Armstrong SP, McArdle CA, Lightman SL. Encoding and decoding mechanisms of pulsatile hormone secretion//*J.Neuroendocrinol.*-2010-Vol.22-P.1226-1238.

Xiaodong Zhou The effect of estrogen status on selenium metabolism in female rats-2007, Thesis Ohio State University.

Yonghua Li, Linsheng Yang, Wuyi Wang, Hairong Li, Jinmei LV, Xiaoyan Zou Trace element concentrations in hair of healthy Chinese centenarians //*Sci. Total Environ.* - 2011-Vol. 409 (Iss.8)- P. 1385-1390.

Zagrodzki P, Ratajczak R. Selenium status, sex hormones and thyroid function in young women//*J.Trace Elem.Med.Biol.*-2008-Vol.22-P.296-304.

Zhu Y, Brown HN, Zhang Y, Stevens RG, Zheng T. Period 3 structural variation: a circadian biomarker associates with breast cancer in young women//*Cancer Epidemiol. Biomarkers Prev.*-2005-Vol.14-P.268-270.

Zhu Yong-Guan, Pilon-Smits, E A.H., Zhao Fang-Jie, Williams P.N., Meharg AA. Selenium in higher plants: understanding mechanisms for biofortification and phytoremediation //*Trends in Plant Science*-2009-Vol.14 (8)-P.436-442.

In: Selenium: Sources, Functions and Health Effects
Editors: Chinatsu Aomori and Megumi Hokkaido
ISBN: 978-1-61942-061-8

Chapter 3

SELENIUM IN SEAFOOD AND AQUACULTURE PRODUCTS

***Cláudia Afonso*[1], *Carlos Cardoso*[1], *Narcisa M. Bandarra*[1] and *Maria Leonor Nunes*[1]**
[1]National Institute of Biological Resources,
Avenida Brasília, Lisboa, Portugal

1. INTRODUCTION

Seafood and aquaculture products offer a wide diversity of species, a broad range of foods—from whole fresh fish to canned products—and large amounts of several nutrients, invaluable to human health. Consequently, they are considered to have an important role in the human diet. The consumption of these products has been increasing over recent decades. Moreover, the health benefits of a seafood rich diet have been extensively acknowledged in the past decade. In fact, these foods contain various nutritionally beneficial components, such as, readily digested proteins, ω3 polyunsaturated fatty acids (ω3-PUFA), namely, eicosapentaenoic acid (EPA, 20:5ω3) and docosahexaenoic acid (DHA, 22:6ω3) —which are associated with decreased morbidity and mortality from cardiovascular and other diseases as well as with foetal development (Simopoulos, 2002; EFSA, 2005)—, vitamins, and minerals. Regarding the latter nutrients, seafood is a rich source of selenium (Se) (Klapec et al., 2004; Sirichakwal et al., 2005; Alexander et al., 2007). Norwegian studies have shown that the quantity of Se in seafood, in average, varies widely from 10 to 290 μg per 100 grams of edible part (Alexander et al., 2007). The majority of seafood species in the Norwegian diet contains from 20 to 40 μg per 100 grams of edible part. However, some seafood reach higher values, for instance, crab meat is one of the richest sources of this element (84 ± 28 μg/100 g) according to a survey of over 700 food samples representing 100 different types of food product consumed in the United Kingdom (Barclay et al., 1995). Some estimates based only on Northern European consumers, from Alexander et al., 2007, have shown that seafood account for an average of 31 % of the total Se intake for adults, 24 % for 13-year-olds, 23 % for 9-year-olds, 25 % for 4-year-olds, and 14 % for 2-year-olds. In Japan, which has a very high level of seafood consumption, it was found that the major Se contribution comes from fish (up to 60 % of daily total intake) rather than the staple foods, such as rice and vegetables (Haratake et al., 2007).

On the other hand, for farmed fish, the amount of Se depends on the content and the form in which Se is present in the feed (Alexander et al., 2007). According to the literature (Kölbl, 1995), Se can be in inorganic form (elemental Se, metal selenides, selenite, selenate) or in organic forms with Se-C bonds (Figure 1). Elemental Se is water-insoluble and, as such, accumulates in anaerobic aquatic sediments (Maier et al., 1987). The metal selenides (CdSe, HgSe, PbSe, etc.) are also water-insoluble and display low bioavailability (Figure 2). Moreover, only some organic selenium compounds, such as selenocystine, selenocysteine, selenomethionine or selenoproteins like glutathione peroxidase, have been identified and quantified in seafood and aquaculture products (Wrench, 1978; Luten et al., 1987; Braddon-Galloway and Balthrop, 1985). Up to 91 % of total Se found in aquatic organisms is in the organic form (Maier et al., 1987).

In addition, Se content can be reduced by food processing such as cooking (boiling, baking or grilling) due to the volatilization phenomena (Dumont et al., 2006; Sager, 2006). Particularly, Se loss has been reported for roasted fish (Thomson and Robinson, 1990).

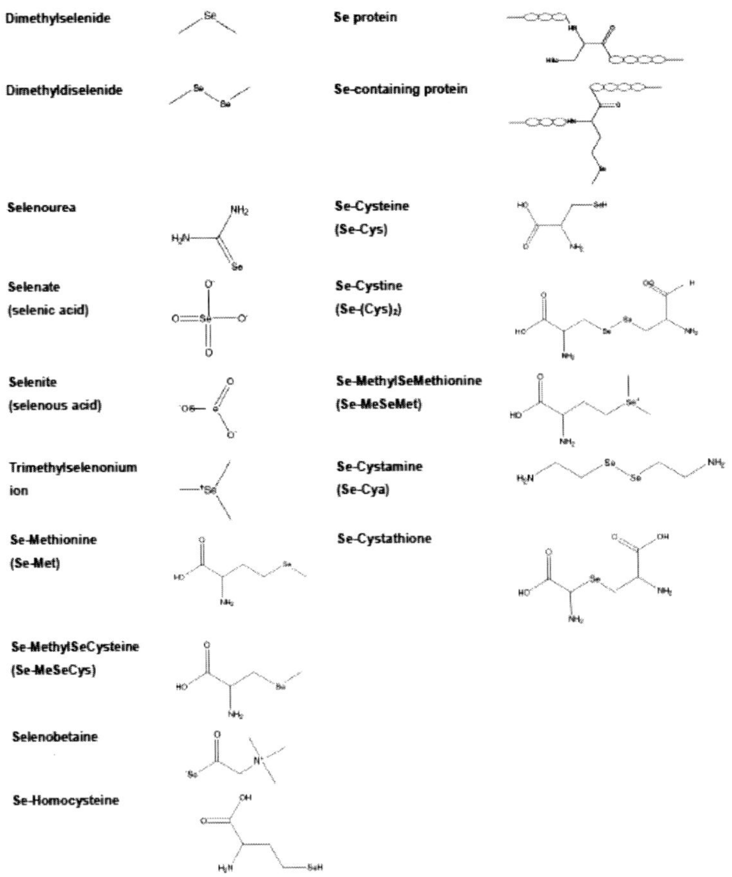

Figure 1. Se compounds found in food and their metabolites (from: Dumont et al., 2006).

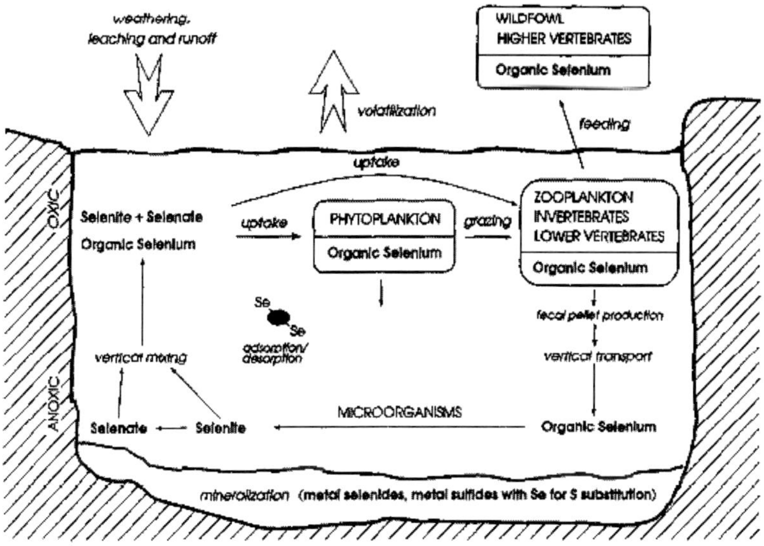

Figure 2. The biogeochemical cycle of Se in the aquatic environment (from: Cutter and Bruland, 1984; McNeal and Balistrieri, 1989; Fishbein, 1991; and Kölbl, 1995).

For wild and aquaculture fish products, there is a specific advantage of high Se contents. In fact, it is well established that mercury (Hg) and Se bind to form Hg selenides with extremely low solubility, which are thought to be metabolically inert. Moreover, scientific works have found that Se supplementation thwarts the negative effects of Hg exposure in all investigated marine species (Beijer and Jernelov, 1987). Being Hg content high in some marine species, this fact accrues a special importance to Se in seafood and aquaculture products.

2. Analysis of Selenium in Seafood and Aquaculture Products

The content variability of Se in seafood and aquaculture products as well as the aforementioned phenomena makes the identification and determination of all Se species a fundamental requirement for the achievement of a complete understanding of the biochemical changes, actions, and functions of Se.

Of course, if the total Se content is to be determined, technical difficulties are minor with respect to a Se speciation analysis. Total Se determination can be achieved through a drastic mineralization (involving addition of concentrated acids, such as sulfuric and nitric, and high temperatures) of the sample and Se detection in the mineralized extract through an atomic spectrometry technique. Pappa et al. (2006) applied this atomic spectrometry technique for the quantification of total Se in several fish species, such as bogue, sardine or trout. Other authors (Miklavčič et al., 2011) used also hydride generation atomic fluorescence spectrometry for Se determination in fresh and canned fish. This methodology operates via formation of hydrogen selenide. Besides this detection technique, other used methods comprise fluorometry (after reaction of selenite with an aromatic *orto*-diamine to yield a piazselenole), polarography and voltametry, gas chromatography (after conversion of selenite to a volatile piazselenole), graphite furnace atomic absorption spectrometry, and neutron activation analysis (Kölbl et al., 1993; Kölbl, 1995). The reaction of selenite with an aromatic *orto*-diamine to form fluorescent piazselenoles is one of the simplest analytical procedures. Alternatively, selenite can be selectively reduced to hydrogen selenide through electrochemical methods. Selenate does not react. Only after being reduced to selenite with hydrochloric or hydrobromic acid, the selenate fraction can be quantified. The difference between total Se (obviously after thorough sample mineralization) and the sum of selenite plus selenate is ascribed to organic and elemental Se. For instance, this route was used with hydride generation atomic absorption spectrometry for the determination of selenite, selenate, organic, elemental, and total Se in biogenic samples from the marine environment (Cutter and Bruland, 1984; Cutter, 1985).

Over the past decade, it has been perfected a different route combining ultrasound-assisted extraction with detection by electrothermal atomic absorption spectrometry with Zeeman background correction (Lavilla et al., 2008). This enabled a fast and reliable quantification of total Se in fish and shellfish, such as hake, sole, clam, cuttlefish, and shrimp. Indeed, for Se determination in fish, matrix usually needs to be removed in order to convert the sample to the liquid state before determination. Wet and dry oxidation procedures have been extensively applied involving drastic reaction conditions such as the use of strong acids, oxidizing agents, and high temperature and pressure conditions (Plessi et al., 2001, Kwockek et al., 2006). In recent years, ultrasound-assisted Se extraction has emerged as an efficient approach for sample preparation. Regarding this matter, ultrasonic probes have some advantages compared to ultrasonic baths, such as, short extraction time, low acidic conditions, and an efficiency of almost 100 % (Bendicho and Lavilla, 2000).

Concerning interferents and most continuous flow systems, no significant problem has been found for Se determination in fish and shellfish samples, at least, within the concentration ranges of the potential interferents (Na^+, Ca^{2+}, K^+, Cl^-, SO_4^{2-}, $H_2PO_4^-$, CO_3^{2-}, Fe^{3+}, Mg^{2+}) typically found in seafoods (Yebra-Biurrun and García-Garrido, 2001).

However, if an accurate and complete Se speciation is aimed at, there are several problems concerning such a thorough Se analysis, namely, its low concentration levels and the shear wide variety of Se species (Figure 1). Hence, Se species need to be quantitatively extracted from the matrix without changing the original species form and correctly identified and quantified. This excluded the more aggressive mineralization procedures mentioned above for the total Se determination. Moreover, the assessment of the methodological accuracy is troublesome, given the rarity of available Se-compounds standards and certified reference materials. In order to overcome these difficulties, the use of isotopic dilution analysis for species transformation during the sample treatment phase, the application of different chromatographic separation mechanisms for a correct determination of Se-peaks, and the participation in intercomparison exercises are advised (Pedrero and Madrid, 2009). More advanced analytical techniques, such as electrospray ionization-mass spectrometry (ESI-MS) and matrix-assisted laser desorption ionization-time of flight (MALDI-TOF) could offer further accuracy.

The Se speciation, that is, the set of analytical procedures for the identification and quantification of the various Se species, encompasses the following phases: sample treatment (extraction, pre-concentration, derivatization); species separation; and identification. The accuracy and quality of the determination is affected by all these phases.

Se in seafood and aquaculture products is found mostly in organic form (Maier et al., 1987), being commonly presented as Se-amino acids either incorporated into proteins or free. The extraction of Se-amino acids from samples can be performed by alternative routes, such as aqueous leaching, acid hydrolysis (with HCl), alkali hydrolysis (using tetramethylammonium hydroxide), and enzymatic extraction (Pedrero and Madrid, 2009). Whereas the first alternative leads to poor recovery levels, the second and third are prone to the degradation and/or transformation of the Se species. Accordingly, the enzymatic hydrolysis is the most used extraction method. It uses proteolytic enzymes and allows the extraction of Se species under mild conditions of pH (7.0) and temperature (37 ºC), thereby avoiding degradation and transformation phenomena. Regarding this subject, an interesting method was developed for the determination of Se bioaccessibility and its speciation in fishery products (Luten et al., 1987). It involved an in-vitro gastrointestinal digestion and comprised two phases: gastric (addition of gastric juice containing 6 % pepsin and incubation during 4 h at 37 ºC) and gastrointestinal digestion (addition of intestinal juice containing 1.5 % pancreatine and 0.5 % amylase to the solid residue of the 1st phase). Since this groundbreaking work in fish products, several other authors have followed this same approach; for instance, Cabañero et al. (2004) have studied the bioaccessibility of Se in fish, simulating the stomachal and intestinal phases of the digestion process in vitro.

Following isolation of Se species from the matrix, separation, identification, and quantification are necessary. Concerning this matter, a great leap forward was achieved recently with the use of the so called hyphenated techniques, based on coupling chromatographic separation with inductively coupled plasma-mass spectrometry (ICP-MS) and its conjugation with ESI-MS, ESI-MS-MS, and MALDI-TOF. These techniques permit the identification and quantification of Se species in real time. They combine an effective separation technique —not only gas and liquid chromatography, but also electrophoretic methodologies— with an extremely sensitive detector. ICP-MS is deemed as the most accurate and reliable analytical tool for Se speciation (Pedrero and Madrid, 2009). Of course, there are some problems, namely, spectral (caused by atomic or molecular species presenting the same mass of the analyte) and non-spectral (caused by sample matrix) interferences. Regarding the former, the most abundant Se isotope, ^{80}Se, overlaps with the $^{40}Ar^{2+}$ dimer in the conventional quadrupole ICP-MS. This problem may be overcome by the ^{82}Se isotope. Nevertheless, the advantages of this detector make the tandem liquid chromatography-ICP-MS the most used separation, identification, and quantification method (Pedrero and Madrid, 2009). For instance, this tandem (comprising an ion exchange chromatography column) has been used for the speciation of Se in fish and oysters (Moreno et al., 2004; Cabañero et al., 2005).

3. The Importance of Selenium in Seafood and Aquaculture Products

Meat and fish products are Se sources. These products contain Se in its functional form, selenoproteins (selenomethionine and selenocysteine) (IOM, 2000; Goyer and Clarkson, 2001). Virtually all animal proteins contain selenomethionine, which is obtained by the consumption of vegetables (IOM, 2000). However, compared with red meat and poultry, fish, especially marine species (Burger and Gochfeld, 2011), contain higher Se contents, being considered the best dietary source of this element (Lall, 1995, Shen et al., 1997; Oehlenschläger, 1997).

Se is absorbed from fish and other aquatic organisms from the surrounding water and diet (Lall, 1995; Watanabe et al., 1997; Lall, 2002), either by gills, epidermis or gut (Hamilton, 2004). According to the National Research Council (NRC), Se is widely distributed at low concentrations in freshwater (0.2–10 μg/l) and seawater (around 0.09 μg/l) (Lall, 2002). Yet, most Se accumulated comes from the marine organisms (phytoplankton and zooplankton) and not directly from the water (Navarro-Alarcon and Cabrera-Vique, 2008).

In what concerns the relationship between the trophic level and Se content in marine organisms, some studies have not found any correlation (Miklavčič et al., 2011). However, Burger et al. (2001) have reported that, differently from Hg, Se levels were higher in species that were lowest in the food chain. On the other hand, Phibbs et al. (2011) have suggested that the proportion of organic Se compounds (predominantly selenomethionine) is likely to increase at higher trophic levels, even though the trophic transfer of Se has been observed to be relatively low at top trophic levels (Stewart et al., 2010). Nevertheless, in general, Se levels did not differ greatly between fish species (Oehlenschläger, 1997, Shen et al. 1997; Plessi et al., 2001, Carvalho et al. 2005; Fabris et al., 2006; Afonso et al., 2008; Burger and Gochfeld, 2011). So far, within the same fish species, Se content can vary widely (Oehlenschläger, 1997; Plessi et al., 2001, Carvalho et al., 2005; Afonso et al., 2008), probably not only due to environmental factors but also to certain characteristics of the species (size, age, food availability, physiological status, etc.) and, in the case of farmed fish, of the production system and the composition of the feeds used (Lall, 1997). Se concentration in fish and other seafood products typically ranges between 10 μg/100 g and 100 μg/100 g (Table 1). In addition, several authors indicate that Se levels in the liver are higher than those found in fish muscle (Agusa et al., 2005; Capelli et al., 2008; Afonso et al., 2008).

Se essentiality in the diet of several animal species and its metabolic relations is well known (Lall, 2002). Whereas, Se level in most plants and terrestrial animals reflects the amount of Se present in the geographical area on which they are grown (Lall, 1995; McDonald et al., 2002), in fish this does not necessarily occur (Lall, 1995).

The Se level of feedstuffs of plant origin differs according to the level and biological availability of Se in the soil at the various geographical locations (Lall, 2002). Regarding the common fish feedstuffs, fish meals and marine by-products offer the best natural Se sources. But, certain fish meals, whenever they contain high levels of heavy metals, present a poor biological availability because of heavy metal complexing with Se (Lall, 2002).

Se occurs naturally in foods and feedstuffs in organic complexes, primarily in the form of selenomethionine, selenium–methylselenomethionine, selenocystine, and selenocysteine. Common Se supplements include selenite, selenate, selenomethionine, selenium–methylselenomethionine, selenocystine, and selenocysteine (Lall, 2002).

It should be stressed that the biochemical distinctions between the amino acids selenocysteine (actively synthesized in animal tissues) and selenomsethionine (predominant form present in plants) are particularly significant (Kohrle et al., 2005). However, upon eventual degradation, the Se freed from selenomethionine becomes available for *de novo* selenocysteine synthesis in animal cells (Figure 3), thus making vegetable sources viable feeds for farmed fish.

Table 1. Selenium content in seafood and aquaculture products μg/100 g of edible part)

Seafood product	n	Selenium	
		Average	Range
Wild seawater fish			
Angler[a]	10	43 ± 22	21-90
Angler[b]	3	41.4 ± 16.9	22.0-52.9
Atlantic bluefin tuna[c]	23	43.0 ± 3.8	---
Black scabbardfish[d]	10	46 ± 13	33-81
Black sea bass[c]	19	20.0 ± 1.8	---
Blackbelly rosefish[a]	9	36 ± 7	28-48
Bogue[c]	2	50.67 ± 1.32	49.36-51.99
Cod[d]	≥3	30	---
Cod[b]	4	30.5 ± 10.0	23.4-45.3
Haddock[d]	≥3	30	---
Halibut[d]	≥3	40	---
Herring[d]	≥3	50	---
Ling[c]	39	18.0 ± 1.5	---
Mackerel[d]	≥3	30	---
Megrim[a]	12	41 ± 9	24-57
Redfish[d]	≥3	50	---
Saithe[d]	≥3	30	---
Salmon[d]	≥3	50	---
Sardine[d]	≥3	57	---
Sea trout[d]	≥3	26	---
Sprat[a]	≥3	10	---
Striped bass[c]	178	30.0 ± 1.0	---
Tuna[b]	3	74.7 ± 27.0	46.1-99.7
Yellowfin tuna[c]	45	47.0 ± 2.7	---
Wild freshwater fish			
Perch[d]	≥3	28	---
Pike[d]	≥3	22	---
Farmed fatty fish			
Brown trout[d]	≥3	30	---
Gilthead seabream[b]	4	18.5 ± 2.1	16.7-20.7
Salmon[d]	≥3	30	---
Crustaceans			
Crab[f]	20	84 ± 28.0	28.0-126.0
Cooked crab[g]	≥3	200	---
Lobster[f]	16	54 ± 8.0	41.0-76.0
Prawn[f]	18	19 ± 1.8	16.0-23.0
Cooked prawn[f]	≥3	30	---
Tiger prawn[f]	18	26 ± 9.0	8.0-42.0
Molluscs			
Raw blue mussel[g]	≥3	51	---
Innards and fish oils			
Angler liver[a]	10	142 ± 64	71-269
Black scabbardfish liver[a]	9	386 ± 96	277-578
Cod liver[g]	≥3	30	---
Cod liver oil[g]	≥3	n.d.	---
Cod roe[g]	≥3	20	---
Processed fish products			

Table 1. (Continued)

Seafood product	n	Selenium	
		Average	Range
Canned mackerel in tomato sauce[d]	≥3	28	---
Canned sardine in oil[b]	8	13.8 ± 28.5	24.2-118.0
Canned tuna in own juice[b]	8	54.8 ± 20.4	28.9-88.9
Canned tuna in oil[b]	9	74.7 ± 13.7	50.1-99.5
Canned tuna in oil[h]	≥3	81	---
Canned tuna in olive oil[b]	15	68.9 ± 19.1	46.2-108.0
Cod roe pâté[d]	≥3	41	---

[n]number of samples. Data are expressed as mean±standard deviation. n.d. not detected.
Data from: [a]Afonso et al., 2008; [b]**Miklavčič et al.**, 2011; [c]Burger and Gochfeld, 2011; [d]Alexander et al., 2007; [e]Pappa et al., 2006; [f]Barclay et al., 1995; [g]Fardy et al., 1994; [h]Akl et al., 2006.

A new approach to the development of functional seafood products may include dietary modulation of farmed fish. The content of Se may increase via dietary modulation of feed with organic Se-compounds attaining the level of 120 □g/100 g in African catfish as was reported by Mierke-Klemeyer et al. (2008).

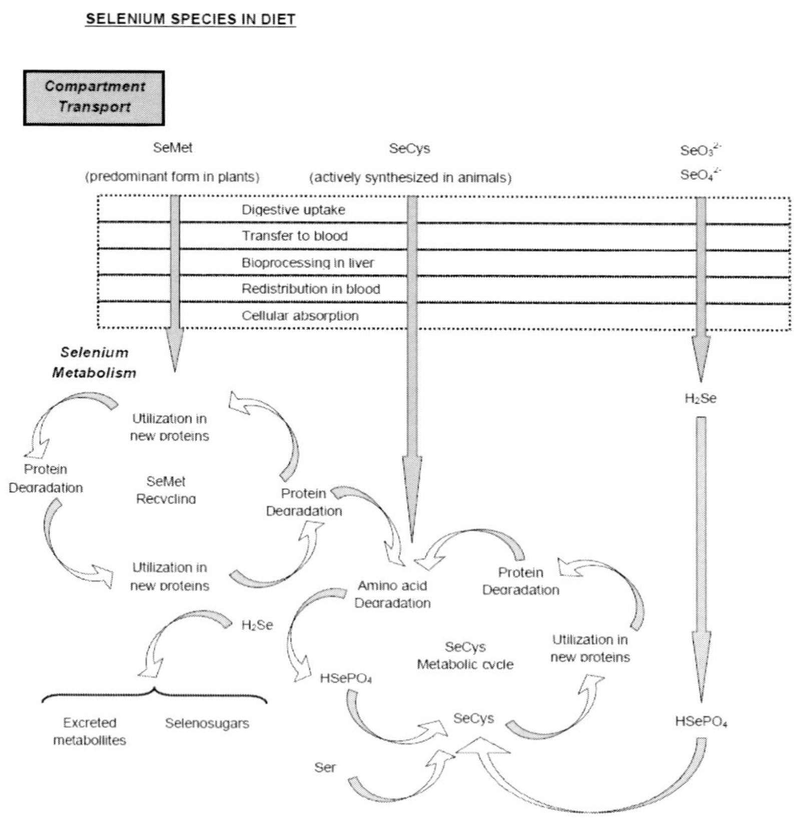

Figure 3. Metabolic cycles of selenomethionine, selenocysteine, and inorganic selenium (adapted from: Ralston et al., 2008).

Concerning the fish diet, Lall (2002) concluded that the minimum required Se is dependent on the level of vitamin E and concentration of Se in water and generally ranges from 0.2 to 0.5 mg/kg (Lovell, 1987; Watanabe et al., 1997). In general, in several countries the legislated limit for Se supplementation in fish and animal feeds is about 0.1 mg Se/kg (Lall, 2002). However, the U.S. Food and Drug Administration established a Se supplementation limit of 0.3 mg/kg from sodium selenate or sodium selenite in feeds for animals including those for aquaculture species (NRC, 2011).

Se toxicity in animals was described well before its essentiality and can vary with several factors, including source and duration of the exposition (NRC, 2011). Based on laboratory and field studies, threshold concentration of adverse effects of Se on fish have been estimated on 3-4 mg/kg on diet and 2-5 μg/l in water (NRC, 2011).

Lemly (2002) explained the biochemical basis of Se toxicity as a simple but important flaw in the process of protein synthesis due to physical and chemical similarities between Se and sulphur (S). Studies have shown that mammalian cells do not distinguish between Se and sulphur during the protein synthesis. So, when in high concentrations, Se replaces the sulphur and form a mistaken connection tri-selenium (Se-Se-Se) or a seleno disulfide bond (S-Se-S) and both prevent the chemical formation of essential disulfide bonds (SS). As a result, dysfunctional proteins are formed and disturbances of the normal cellular biochemistry occur. Chronic intoxication is characterized by, among other symptoms, anemia (Lemly, 2002; NRC, 2011), pathological changes in liver, kidney, heart, and ovaries, increased lymphocytes, abnormal reproduction, teratogenic deformities (Lemly, 2002), and high mortality (Lall, 2002, NRC, 2011). Diets high in S or protein provide some protection against Se toxicity (Lall, 2002).

Yet, Se is an essential element and the physiological role of a deficiency of Se is well recognized (Lall, 1995; 2002). In fish, Se deficiency is associated with reduced growth (Watanabe et al., 1997; Lall, 2002), lethargy, loss of appetite, and other symptoms (Watanabe et al., 1997). Like vitamin E, Se deficiency causes muscular dystrophy (Lovell, 1987; Lall, 2002, NRC, 2011).

4. The Biochemical Role of Selenium

The physiological and/or pathological mechanisms occurring in living organisms produce reactive oxygen species that cause damage to the structure of the cell membrane or even within the nucleus itself. The most relevant of these reactive oxygen species are the superoxide, hydroxyl, and hydrogen peroxide radicals. The action of hydroxyl radical upon the phospholipid chains within the cell membranes triggers a chain of reactions usually referred to as lipid peroxidation, which, in turn, yield degradation products such as the peroxyl radical (ROO •) (Halliwell and Gutteridge, 2007). Furthermore, there are endogenous antioxidant systems such as the enzymes superoxide dismutase, catalase, and glutathione peroxidase (containing Se), which are the first line of defense against any formed radicals (McDonald et al., 2002, Halliwell and Gutteridge, 2007). In fact, Se-containing glutathione peroxidase is a well known antioxidant, which is fundamental for preventing oxidative damage at the cellular level (Gao et al., 2011). The enzyme catalyses the reduction of hydrogen peroxide (see Equation 1) as well as various lipid hydroperoxides, thereby protecting cell membranes from oxidative degradation (Rayman, 2000). Moreover, glutathione peroxidase and the selenoprotein P are involved in the regulation of the inflammatory response (Van Cauwenbergh et al., 2004).

$ROOH + 2GSH \rightarrow ROH + GSSG + H_2O$ (Equation 1)
(from: Belitz et al., 2004; Halliwell and Gutteridge, 2007)

It should be remarked that while at low concentration, Se can decompose reactive oxygen radical and hydroxyl radical, and therefore prevent the oxidative damage, at high concentration, Se can catalyse the production of reactive oxygen radical resulting in the oxidative damage (Xi et al., 2003).

The antioxidative action of Se can also contribute to countervail the damage induced by the ultraviolet-β radiation in humans. Particularly, the selenoprotein P acts as an extra cellular antioxidant conjugated with the vascular endothelium, reducing the peroxinitrile ($ONOO^-$) content, a reactive nitrogen species (Li et al., 2007). Furthermore, it has been established an important role for Se-cysteine, which is the main Se-compound present in the selenoproteins of the body tissues (Sunde, 2000). On the other hand, on the basis of the antioxidant role of selenium and vitamin E, it has been reported that at low ingestion levels of either Se or vitamin E, higher dietary supplementation of Se spares the metabolic requirements of vitamin E (Lin and Shiau, 2009).

Se interacts with several trace elements, and these interactions can be additive, antagonistic, or synergistic, and in some cases they reverse the interaction, i.e. antagonism changed to synergism (Hamilton, 2004; Akl et al., 2006). Particularly, the antagonistic interaction between Hg and Se is much important, given the high Hg contents in some fish species, such as black scabbard fish (Afonso et al., 2007; 2008). It is reported that Se reduces Hg toxicity whenever both elements are simultaneously administrated (Caurant et al., 1996; Cabañero et al., 2007). It has been shown that the chemical forms of Hg and Se are important in the toxicology of both elements (Cuvin-Aralar and Furness, 1991). For instance, regarding neurological problems associated to Hg toxicity, selenite (SeO_3^{2-}) proved to be more effective than elemental Se in tuna (Ohi et al., 1976). The presence of other elements and compounds can also affect the interaction between Hg and Se. Namely, arsenic as sodium arsenite, Na_2AsO_3, modified the ability of SeO_3^{2-} to thwart the toxicity of methyl-Hg in animals (El Begearmi et al., 1982). In fact, it enhanced the effectiveness of SeO_3^{2-} to prolong the survival of Hg intoxicated quails. Moreover, the association of SeO_3^{2-} and cystine presented an additive effect in reducing Hg toxicity (Stillings et al., 1974). Among the possible mechanisms for the protective Se effect against Hg toxicity, some deserve mention: Hg redistribution in the presence of Se; competition for binding sites between Se and Hg; formation of a Hg-Se complex; conversion of Hg toxic forms to other forms; and Se prevention of oxidative damage (Cuvin-Aralar and Furness, 1991). In fact, studies in rats fed with fish containing methylmercury do not show the characteristic neurotoxic effects of poisoning by this compound. Raymond and Ralston (2004) suggested that this is due to the high levels of Se in fish, at least 2 times, higher than those of mercury on a molar basis. Thus, due to the interaction between these two elements, Se seems to present an effect on the Hg bioavailability (Raymond and Ralston, 2004; Eisler, 2006). So, it can be assumed that Se has effects not only on the Hg bioavailability, but also on its availability for the synthesis and normal activity of Se-dependent enzymes. This confirms the importance of maintaining sufficient amounts of Se in the body (Raymond and Ralston, 2004).

In seafood and aquaculture products, Se may present a special protective role against the oxidation of ω3 PUFA. Indeed, it is recognized that lipid peroxidation is thwarted by antioxidant defense systems, comprising antioxidant nutrients such as vitamin A, vitamin C, alpha-tocopherol, and Se (Sies, 1986; Halliwell, 1994). The stabilization of ω3 PUFA such as DHA (22:6 ω3) is thought to be ensured by Se as well as tocopherols and tocotrienols, albeit by quite different pathways (Girotti, 1998). Moreover, the consumption of fish rich in ω3 PUFA only exerts a beneficial health effect if antioxidant protection against oxidative stress is adequate (Song et al., 2000). It has also been shown that in rats with adequate vitamin E and enhanced fish oil intake, Se deficiency modifies the lipid concentration and fatty acid composition in the liver (Schäfer et al., 2004). Such alterations were related to a reduction of the levels of selenoenzymes (not only glutathione peroxidase, but also deiodinases and the thioredoxin reductases) with antioxidative functions. Additionally, Se was found to indirectly affect the desaturation and the absorption of fatty acids (Schäfer et al., 2004). This is clearly different of assuming for Se only an antioxidant defense function. If Se were involved in fatty acid metabolism merely as antioxidant, Se-deficient rats would not exhibit an increase of the hepatic content of 18:2 ω6 (Schäfer et al., 2004). This fatty acid is converted by elongation and desaturation to long chain PUFA and Δ-6-desaturase is deemed to be the rate-limiting enzyme in the conversion process. Hence, the increased concentrations of 18:2 ω6 and EPA (20:5 ω3) in Se deficient rats may be related to a decreased conversion of 18:2 ω6 to 20:4 ω6 and of EPA to DHA. In fact, it has been reported that the enzymatic activity of Δ-6-desaturase may be affected by Se (Infante, 1986).

The experimental results mentioned above help not only to understand the role of Se during seafood storage and processing, but also to support the health benefits ascribed to Se discussed below.

5. The Stability of Selenium during Seafood Storage and Processing

The seafood and aquaculture products comprise a broad range of foods—such as whole fresh fish, ready-to-cook and ready-to-eat fillets, canned fish, salted fish, smoked fish, or surimi-derived products—, which are subjected to a great variety of technological processes, some of them quite particular and without equivalence in other food groups. Besides, the group of seafood and aquaculture products is much more prone to spoilage than other food groups due to their intrinsic characteristics and to their natural habitats. Among these aspects, deserve special mention the high moisture level, the presence of many nitrogen-containing compounds with a low molecular weight and, as such, very volatile, the low importance of the connective tissues, the high levels of ω3-PUFA in fatty fish, and the importance of psychrophilic bacterial flora. All these aspects lead to several alterations that rapidly generate significant quality losses and product rejection. These changes occur even at low refrigeration temperatures (0-2 °C) in unprocessed whole fish. In particular, ω3-PUFA have a greater tendency to oxidize due to their very high unsaturation levels, thereby causing the formation of off-flavours and substantial reduction of the products' shelf life (Pérez-Mateos et al.,

2004). In fact, hydrolytic and oxidative reactions during processing and storage are among the basic processes leading to the formation of hydroperoxides (Osman et al., 2001), free fatty acids (Chaijan et al., 2006), and rancidity (Donelli and Robinson, 1995). Hydroperoxides are unstable and decompose, yielding a complex mixture of secondary oxidation products, mainly aldehydes and ketones, which are directly responsible for the changes in aroma and flavour (Frankel, 1991). The basic mechanism of free radical induced lipid oxidation may be described by three different steps: initiation, propagation, and termination. This phenomenon is affected by both intrinsic and extrinsic factors, such as fatty acid composition, concentration of pro-oxidants, endogenous ferrous iron, myoglobin, enzymes, pH, temperature, ionic strength, and oxygen consumption (Andreo et al., 2003).

Concerning Se itself, it has been reported that food processing such as cooking (boiling, baking or grilling) may reduce Se food content by volatilization (Dumont et al., 2006; Sager, 2006). Namely, a decrease of Se levels has been found in roasted fish, such as trout (Thomson and Robinson, 1990). Lamand et al. (1997) found that whatever cooking method was used, Se was weakly retained in fish. On the other hand, some studies have shown that fish cooking did not affect Se bioavailability (Fox et al., 2004). A large integrated study to produce farmed fish enriched with Se, evaluated whether ice storage, cold-smoking and hot-smoking influenced Se concentrations in African catfish and was concluded that Se was well retained and did not changed during process (Larsen et al. 2010a, 2010b). Moreover, other studies did not report any Se loss and even concluded that processes such as cooking or lyophilization significantly increase Se content in seafood (Zhang et al., 1993). For instance, the Se content of shrimp augmented from 58.5 ± 2.1 □g/100 g in raw to 73.5 ± 0.7 □g/100 g in the cooked meal. This variation was ascribed to moisture loss and it was hypothesized a close association between Se and protein (Zhang et al., 1993). Cabañero et al., (2004) concluded that cooking did not affect total Se concentration and did not modify the selenomethionine in tuna samples. Moreover, Mierke-Klemeyer et al. (2008) reported that cooking did not significantly reduce Se content in African catfish fillets, being attained retention factors between 91 and 104 %. The contradictory findings of these different scientific works warrant more research into this scientific field in order to assess the effect of different cooking processes upon the Se content of food (Navarro-Alarcon and Cabrera-Vique, 2008). This is particularly important, since any reduction of Se content, besides representing a loss of nutritional value for the food, may expose seafood and aquaculture products to oxidative damage.

6. Health Benefits and General Dietary Recommendations

Ironically, approximately half a century ago, Se was only known as a toxic trace element. Today, it is perfectly established that Se is essential for the normal functioning of many systems in the body and that its deficiency can bring about adverse consequences (Lall, 1995; Raymond and Ralston, 2004).

Two of the twenty-two primary amino acids are distinguished by having selenium: they are the selenomethionine and the selenocysteine (Raymond and Ralston, 2004; O'Brien et al., 2006). The selenomethionine, which can not be synthesized by animals, including humans, is initially synthesized by plants (IOM, 2000). It is biochemically equivalent to methionine and is considered an unregulated storage compartment of Se. In contrast, selenocysteine is highly regulated and it is incorporated specifically in many proteins in order to perform essential biological functions (IOM, 2000; Raymond and Ralston, 2004).

Thus, within a specific range of Se intake in the diet, the concentration of selenoproteins in the body depends on the amount of Se intake (IOM, 2000).

According to Belitz et al. (2004) Se levels in the human body vary between 10 and 15 mg and the intake of this micronutrient is about 50 to 100 μg/day. Nevertheless, depending on the region, it can change significantly (Navarro-Alarcon and Cabrera-Vique, 2008) because of the variation of Se level in the soil (Belitz et al., 2004).

Se can enhance tocopherol activity (Watanabe et al. 1997; IOM, 2000, Lall, 2002; Belitz et al., 2004; Halliwell and Gutteridge, 2007, NRC, 2011). Animal studies have shown that induction of a deficiency in vitamin E and Se causes lipid peroxidation, necrosis in the liver, and heart damage (IOM, 2000). Particularly, the hepatocyte damage and necrosis usually occur concurrently with low vitamin E contents in the body, being endoplasmic reticule changes at the origin of these liver problems (Simonoff and Simonoff, 1991). Se can reduce also the toxicity in fish of high dietary or waterborne copper (NRC, 2011). On the other hand, Se reduces the toxicity of Hg (Lall, 1995; Lall, 2002; Raymond and Ralston, 2004; Gaile, 2007; Sivaperumal et al., 2007), arsenic (Peraza et al. 1998; Kotsonis et al. 2001; ATSDR, 2007), silver (Lall, 2002), and cadmium (Watanabe et al. 1997; Sivaperumal et al., 2007), and thus a deficiency of this element may contribute to increase the toxicity caused by these metals.

Several studies suggest that disruption of the methyl-Hg molecule leads to the formation of free radicals, which affect the lipid membranes of neuronal cells, causing damage to them (Miura et al. 1995). So, to this situation should not be strange that the toxic effects of methyl-Hg can be reduced by antioxidants such as Se (Chang and Cockerham, 1994; Lall, 1995; Raymond and Ralston, 2004; Eisler, 2006; Gaile, 2007; Sivaperumal et al., 2007). The deficiency of Se is also associated with Keshan's disease in children (the occurrence of cardiomyopathy) (IOM, 2000; Goyer and Clarkson, 2001; Halliwell and Gutteridge, 2007) and with Kashin-Beck disease (endemic disease that occurs in cartilage in pre-adolescence and adolescence) (IOM, 2000; Halliwell and Gutteridge, 2007). Se may act as growth factor, contribute to the normal functioning of thyroid hormone homeostasis, immunity and fertility as well as have anticancer properties (IOM, 2000; Raymond and Ralston, 2004). Regarding this latter subject, evidence based on epidemiological studies carried out during the past half century supports an inverse relationship between Se intake and cancer mortality. The anticancer properties of Se against leukaemia and cancers of the colon, rectum, pancreas, breast, ovaries, prostate, bladder, lung, and skin look unequivocal, at least, under some conditions (Sunde, 2000). A FDA review of the evidence for Se (Trumbo, 2005) showed that there is some basis for a qualified health claim concerning Se and certain cancers. Namely, of the 36 observational studies reviewed, approximately half supported an association with all cancers, being the evidence stronger for breast and prostate cancers (Trumbo, 2005).

What is more, an effect of Se upon different cardiopathies has been sustained by various studies (Navarro-Alarcon and Cabrera-Vique, 2008). In particular, it has been reported that a serum Se level <55 μg/l is associated with an enhanced risk of coronary heart disease. Helmersson et al. (2005) found that low Se levels predict mortality and cardiovascular disease (CVD) in a study on 50 year old Swedish men. These authors concluded that high concentrations of serum Se reduced not only levels of oxidative stress, but also subclinical cyclooxigenase-mediated (but no cytokinemediated) inflammation. Accordingly, the association between Se, oxidative stress and inflammation may be at the root of the cardiovascular protective properties ascribed to Se (Helmersson et al., 2005). To the contrary, other studies (Navas-Acien et al., 2008) have concluded that current evidence is not enough to support using Se as an agent of CVD prevention. Hence, despite some recent studies providing promising results, there is no basis to conclude that higher Se intake decreases risk for CVD. Additional long term intervention trials are warranted. In accordance, indiscriminate Se supplementation still cannot be advised for the prevention of CVDs (Navarro-Alarcon and Cabrera-Vique, 2008).

Although toxicity from chronic excessive consumption of Se has been found in humans, its occurrence is rare (IOM, 2000). Clinical signs include dental caries (Lall, 1995; Raymond and Ralston, 2004; Sivaperumal et al., 2007), loss of hair and nails, skin lesions, tooth decay, and abnormalities in the nervous system (IOM, 2000; Goyer and Clarkson, 2001; Raymond and Ralston, 2004).

6.1. Mercury-Selenium Interaction in Fish Products

Recently, based on the importance of the benefits of fish as an essential part of a healthy diet, there was a great promotion of the fish consumption. However, the balance between risks and benefits of fish consumption due to chemical contaminants present in the edible part of fish is still not well characterized (Domingo et al., 2007).

The consumption of contaminated fish is responsible for an important route of human exposure to toxic elements (Gašpić et al. 2002; Martí-Cid et al. 2007; Lavilla et al., 2008). Diversity and quantity of fish consumed by a population determine the different Hg intakes. Thus, the risk associated with consumption of this contaminant varies considerably with the target population. This observation has particular importance in certain sectors of the population, such as fishermen and their families, who, having greater access to fishery products than other people constitute a group with a potentially high risk (Gaggi et al. 1996; Renzoni et al.; 1998; Storelli et al. 2003; Storelli et al., 2005).

However, according to Raymond and Ralston (2004), after the episodes of Minamata and Nigata, no cases of children with symptoms of fetal poisoning caused by consumption of Hg were observed. Nevertheless, according to these authors, this is still a cause of much controversy regarding fish consumption and its risks linked to Hg intake. This controversy may be addressed taking into account that, for instance, in the cases of Faroe and the Seychelles Islands, different outcomes were reported. Indeed, in the latter, no adverse effects were detected, differently from the Faroe Islands. Whereas, in the Faroe Islands, whales were a part of the diet —containing these sea mammals large amounts Hg (around 2000 mg/kg), PCBs as well as other toxins not found in fish—, in Seychelles Islands are not. Effectively, in the Seychelles Islands, studies of MeHg child exposure from consumption of fish with typical Hg levels have reported beneficial effects and no danger of fish consumption (Raymond and Ralston, 2004). Moreover, in eastern Finland, there were also adverse effects reported in a male population with high freshwater fish consumption levels. However, their diet is deficient in Se as a result of the poor Se levels in the country soil (Raymond and Ralston, 2004). As mentioned by these authors, the exhaustion of Se reserves in the formation of a metal complex with Hg reduces Se bioavailability for glutathione peroxidase, thereby promoting lipid peroxidation. Hence, it can be sustained that consumption of fish with typical Hg levels within a diet containing enough Se does not entail any risk.

Several studies have reported Se-dependent protective effects against Hg toxicity through fish consumption (Raymond and Ralston, 2009), namely, of yellowfin tuna (Ohi et al., 1976; Ganther et al., 1972), menhaden (Stillings et al., 1974), swordfish (Freidman et al., 1978), and rockfish (Ohi et al., 1980). Moreover, Ralston et al. (2008) have found out that methyl-Hg toxicity was not predicted by tissue Hg, but was inversely related to tissue Se ($P<0.001$) and directly related to Hg:Se ratios ($P<0.001$).

Although Se content in fish is high, sometimes fish is a poor source of available Se, precisely as a result of its high Hg content and other heavy metals, which bind to Se leading to the formation of insoluble inorganic complexes (Van der Torre et al., 1991; Pappa et al., 2006). Nonetheless, the bioavailability of Se in fish is affected by source and species. For instance, it has been reported a high level of Se availability from salmon (Ornsrud and Lorentzen, 2002). Absorption of Se from fish by humans has been considered similar to that from plants (Navarro-Alarcon and Cabrera-Vique, 2008). Moreover, Fox et al. (2004) defended that fish is a highly bioavailable source of dietary Se. These opposite conclusions may be due to differences in the study populations as well as in the chemical speciation of Se.

As it is well established, Hg toxicity is related to its high affinity for sulphydryl groups (-SH) of proteins (Raymond and Ralston, 2004; Baeyens et al., 2003; Balshaw et al., 2008; Hajeb et al., 2009). Although physiologically sulfur (S) is more abundant than Se, due to its high affinity to Se, mercury binds selectively to the latter. It has therefore been assumed that Se effectively reduces the adverse effects caused by Hg (Lall, 1995; Raymond and Ralston, 2004; Eisler, 2006; Sivaperumal et al., 2007).

Se is known to bind and sequester Hg on a 1:1 molecular basis, forming the biological inactive complex, mercuryselenide. Studies on rats fed with fish containing methyl-Hg confirm this fact, since the animals do not exhibit the characteristic neurotoxic effects of poisoning by this compound. Raymond and Ralston (2004) suggested that this is due to high levels of Se in fish —at least 2 times higher than those of Hg on a molar basis.

It is reasonable then to assume that Se has effects on the bioavailability of Hg but also interferes on its availability for the synthesis and normal activity of selenium-dependent enzymes. This reflects the importance of maintaining sufficient amounts of selenium in the body (Raymond and Ralston, 2004).

According to several authors (Plessi et al. 2001; Cabañero et al., 2004), the molar ratio between Se and Hg content (Table 2) can help to anticipate which seafood is more favourable for human consumption. Indeed, in several fish species, Se:Hg molar ratio is above 1:1 in several tissues (Dietz et al., 2000; Plessi et al., 2001; Cabañero et al., 2004; Afonso et al., 2008; Burger and Gochfeld, 2011). This molar ratio can be much higher in the liver tissue, reaching values above 20, for instance, 26.6 for black scabbardfish. Therefore, instead of focusing on methyl-Hg alone, assessing the Se abundance in fish is fundamental for an adequate evaluation of seafood safety (Raymond and Ralston, 2009). For this reason, the Se health benefit value (Se-HBV) (Equation 2) has been proposed as a more comprehensive indicator of the benefits and potential risks of Hg exposure through seafood consumption (Kaneko and Ralston, 2007).

$$Se - HBV = \left(\frac{[Se]}{[Hg]} \text{ molar } \times [Se] \right) - \left(\frac{[Hg]}{[Se]} \text{ molar } \times [Hg] \right)$$ (Equation 2)

The large majority of ocean fish species yield very high Se-HBVs, given their richness of dietary Se, which is commonly present in far greater molar quantities than methyl-Hg. This fact helps to explain the reduction of the Hg toxicity by supplementing diets with fish, even though fish contributes additional methyl-Hg to the diets (Raymond and Ralston, 2009).

The correlation between [Hg] and [Se] on a molar basis is quite poor in fish muscle tissues, but is more clear in liver tissues. This is supported by several studies cited by Ruiter (1995).

Table 2. Molar ratio between selenium and mercury contents

Seafood product	n	Se-Hg ratio	
		Average	Range
Wild seawater fish			
Alaskan pollack[a]	3	6.67	---
Anchovy[a]	3	8.25	---
Angler fish[a]	3	5.00	---
Angler fish[b]	12	4.60 ± 2.1	2.60-8.40
Atlantic croaker[c]	63	10.13	---
Atlantic menhaden[c]	5	60.78	---
Black scabbardfish[b]	50	3.1 ± 2.3	1.00-7.90
Black seabass[c]	19	3.20	---
Cod[a]	3	5.00	---
Grayfish[a]	3	1.25	---
Hake[a]	3	14.29	---
Halibut[a]	3	5.00	---
Ling[c]	39	11.39	---
Mackerel[a]	3	7.15	---
Porgy[a]	3	3.58	---
Redfish[a]	3	6.25	---
Sardine[a]	3	11.11	---
Sardine[d]	5	13.00	---
Scorpion fish[a]	3	7.69	---
Sole[a]	3	12.50	---
Striped bass[c]	178	1.95	---
Swordfish[a]	3	1.23	---
Tuna[a]	3	7.69	---
Tuna[d]	5	1.50	---
White seabream[e]	55	0.40	0.05-33.33
Whiting[a]	3	3.85	---
Yellowfin tuna[c]	45	5.95	---
Wild freshwater fish			
Perch[a]	3	1.33	---

Seafood product	n	Se-Hg ratio	
		Average	Range
Farmed fatty fish			
Salmon[a]	3	5.88	---
Crustaceans			---
Blue crab[a]	3	10.00	---
Mediterranean prawn[a]	3	20.00	---
Spiny lobster[a]	3	6.25	---
Molluscs			
Clam[a]	3	16.67	---
Mussel[a]	3	33.33	---
Octopus[a]	3	11.11	---
Squid[a]	3	12.50	---
Innards			
Angler liver[b]	12	21.40	6.80-28.60
Black scabbardfish liver[b]	50	26.60	3.10-64.00

[n]number of samples. Data are expressed as mean±standard deviation.

Data from: [a]Plessi et al., 2001; [b]Afonso et al., 2008; [c]Burger and Gochfeld, 2011; [d]Cabañero et al., 2004; [e]Andersen and Depledge, 1997.

7. General Recommendations and Dietary Advice for Selenium Intake

The Se intake depends obviously on its concentration in food and amount of food consumed (Navarro-Alarcon et al., 2005). Combs (2001) pointed at least, 40 µg/day of Se to an adequate adult intake, in order to support the maximum expression of Se enzymes. The German Nutrition Society recommends a daily Se intake of 30–70µg (Mierke-Klemeyer et al., 2008). Furthermore, it was hypothesized that as much as 300 µg/day would be required for cancer risk reduction. Hamilton (2004) considered three different Se levels of biological activity: (1) trace concentrations are needed for normal growth and development; (2) moderate concentrations can be stored and homeostatic functions maintained; and (3) high concentrations can lead to toxic effects.

Accordingly, intake recommendations for Se have been set by the Food and Nutrition Board (FNB) at the Institute of Medicine (IOM) of The National Academies (IOM, 2000) and are provided as Dietary Reference Intakes (DRIs). The DRIs are reference values that are used to plan and measure nutrient intakes for apparently healthy people. These values vary by life stage and gender group and include four different reference values: recommended dietary allowances (RDA – dietary intake level sufficient to meet the nutrient requirements of nearly all (97 %–98 %) healthy people); adequate intake (AI – used when the RDA can not be established and set at a level to ensure the nutritional adequacy); estimated average requirement (EAR - estimated nutrient needs of half of the individuals in a life stage and group); and tolerable upper intake level (UL - highest intake improbable to cause adverse health effects) (IOM, 2000). The complete information provided by IOM (2000) about Se DRIs is shown in Table 3. This information together with knowledge of the content of Se in fish and fish products can be used to estimate the proportion of Se in the human diet provided by this food.

The RDA, EAR, and UL for individuals, male or female (excluding the states of pregnancy and lactation), aged over 14 years have been set respectively at 55, 45 and 400 μg (IOM, 2000).

Se content in raw and cooked fish products has to be combined with the bioavailability data in order to estimate, with caution, the advisable consumptions of seafood and aquaculture products that guarantee the DRI.

Table 3. Dietary reference intake (DRIs) values for the selenium (μg selenium/day)

	DRIs			
	RDA[a]	AI[b]	EAR[c]	UL[d]
Life Stage group				
0 -6 mo	-	15	-	45
6-12 mo	-	20	-	60
1-3 y	20	-	17	90
4-8 y	30	-	23	150
9-13 y	40	-	35	280
14-18y	55	-	45	400
19- >70 y	55	-	45	400
Pregnancy				
14-18 y	60	-	49	400
19-50 y	60	-	49	400
Lactation				
14-18 y	70	-	59	400
19-50 y	70	-	59	400

[a]RDA, recommended dietary allowance the intake that meets the nutrient needs of almost all (97%–98%) of individuals in a group.

[b]AI, adequate intake: the observed average or experimentally determined intake by a defined population or subgroup that appears to sustain a defined nutritional status, such as a growth rate, normal circulating nutrient values, or other functional indicators of health. The AI is used if sufficient scientific evidence is not available to derive an EAR. The AI is not equivalent to an RDA.

[c]EAR, estimated average requirement: the intake that meets the estimated nutrient needs of half of the individuals in a group.

[d]UL, tolerable upper intake level. (source: IOM, 2000).

REFERENCES

Afonso, C., Lourenço, H. M., Dias, A., Nunes, M. L. and Castro, M. (2007). Contaminant metals in black scabbard fish (Aphanopus carbo) caught off Madeira and the Azores. *Food Chemistry*, 101, 120-125.

Afonso, C., Lourenço, H. M., Pereira, C., Martins, M. F., Carvalho, M. L., Castro, M. and Nunes, M. L. (2008). Total and organic mercury, selenium and α-tocopherol in some deep-water fish species. *Journal of the Science of Food and Agriculture,* 88, 2543-2550.

Agusa, T., Kunito, T., Yasunaga, G., Iwata, H., Subramanian, A., Ismail, A. and Tanabe, S. (2005). Concentrations of trace elements in marine fish and its risk assessment in Malaysia. *Marine Pollution Bulletin*, 51, 896 911.

Akl, M. A., Ismael, D. S., El-Asmi, A. A. (2006). Precipitate flotation-separation, speciation and hydride generation atomic absorption spectrometric determination of selenium (IV) in foodstuffs. *Microchemical Journal*, 83, 61-69.

Alexander, J., Froyland, L., Hemre, G. I., Jacobsen, B. K., Lund, E., Meltzer, H. M. and Skåre, J. U. (2007). A comprehensive assessment of fish and other seafood in the Norwegian diet. *Norwegian Scientific Committee for Food Safety*. 173 p.

Andersen, J. L. and Depledge, M. H. (1997). A survey of total mercury and methylmercury in edible fish and invertebrates from Azorean waters. *Marine Environmental Research*, 44, 331-350.

Andreo, A. I., Doval, M. M., Romero, A. M. and Judis, M. A. (2003). Influence of heating time and oxygen availability on lipid oxidation in meat emulsions. *European Journal of Lipid Science*, 105, 207-213.

ATSDR (2007). Toxicological profile for arsenic. Atlanta, USA: *Agency for toxic substances and disease registry,* 500 p.

Baeyens, W., Leermakers, M., Papina, T., Saprykin, A., Brion, N., Noyen, J., Gieter, M. D., Elskens, M. and Goeyens, L. (2003). Bioconcentration and biomagnification of mercury and methylmercury in North Sea and Scheldt estuary fish. *Archives of Environmental Contamination and Toxicology*, 45, 498-508.

Balshaw, S., Edwards, J. W., Ross, K. E. and Daughtry, B. J. (2008). Mercury distribution in the muscular tissue of farmed southern bluefin tuna (Thunnus maccoyii) is inversely related to the lipid content tissues. *Food Chemistry*, 111, 616-621.

Barclay, M. N. I., MacPherson, A. and Dixon, J. (1995). Selenium content of a range of UK foods. *Journal of Food Composition and Analysis*, 8, 307-318.

Barghigiani, C., Ristori, T., Biagi, F. and De Ranieri, S. (2000). Size related mercury accumulations in edible marine species from an area of the northern Tyrrhenian Sea. *Water, Air, and Soil pollution*, 124, 169-176.

Beijer, K. and Jernelov, A. (1987). Ecological aspects of mercury-selenium interactions in the marine environment. *Environmental Health Perspectives*, 25, 43-45.

Belitz, H. D., Grosch, W. and Schieberle, P. (2004). *Food chemistry.* Berlin, Germany: Springer-Verlag, 1070 p.

Braddon-Galloway, S. and Balthrop, J. E. (1985). Se-dependent GSH-peroxidase isolated from Black Sea bass (Centropristis striata). *Comparative Biochemistry and Physiology* C, 83, 13-17.

Burger, J., Gaines, K. F., Boring, C. S., Stephens, W. L., Snodgrass, J. and Gochfeld, M. (2001). Mercury and selenium in fish from the Savannah river: species, trophic level, and locational differences. *Environmental Research* A, 87, 108-118.

Burger, J. and Gochfeld, M. (2011). Mercury and selenium levels in 19 species of saltwater fish from New Jersey as a function of species, size, and season. *Science of the Total Environment*, 409, 1418-1429.

Cabañero, A. I., Madrid, Y. and Cámara, C. (2004). Selenium and mercury bioaccessibility in fish simples: an in vitro digestion method. *Analytica Chimica Acta,* 526, 51-61.

Cabañero, A. I., Carvalho, C., Madrid, Y., Batoréu C. and Cámara, C. (2005). Quantification and speciation of mercury and selenium in fish samples of high consumption in Spain and Portugal. *Biological Trace Element Research,* 103, 17-35.

Cabañero, A. I., Madrid, Y. and Camara, C. (2007). Mercury-selenium species ratio in representative fish samples and their bioaccessibility by an in vitro digestion methods. *Biological Trace Element Research*, 119, 195-211.

Capelli, R., Das, K., Pellegrini, R. D., Drava, G., Lepoint, G., Miglio, C., Minganti, V. and Poggi, R. (2008). Distribution of trace elements in organs of six species of cetaceans from the Ligurian Sea (Mediterranean), and the relationship with stable carbon and nitrogen ratios. *Science of the Total Environment,* 390, 569-578.

Carvalho, M. L., Santiago, S. and Nunes, M. L. (2005). Assessment of the essential element and heavy metal content f edible fish muscle. *Analytical and Bioanalytical Chemistry*, 382, 426-432.

Caurant, F., Navarro, M. and Amiard, J. C. (1996). Marcury in pilot whales: possible limits to the detoxification process. *Science of the Total Environment,* 186, 95-104.

Chaijan, M., Benjakul, S., Visessanguan, W. and Faustman, C. (2006). Changes of lipids in sardine (Sardinella gibbosa) muscle during iced storage. *Food Chemistry*, 99, 83-91.

Chang, L. W. and Cockerham, L. (1994). Toxic metals in the environment. In: *Basic environmental toxicology*. L. G. Cockerham and B. S. Shane (Eds). Boca Raton, USA: CRC Press, pp. 109-129.

Combs, G. F. (2001). Selenium in global food systems. *British Journal of Nutrition*, 85, 517-547.

Cutter, G. A. and Bruland, K. W. (1984). The marine biochemistry of selenium: A re-evaluation. *Limnology and Oceanography*, 29, 1179-1192.

Cutter, G. A. (1985). Determination of selenium speciation in biogenic particles and sediments. *Analytical Chemistry*, 57, 2951-2955.

Cuvin-Aralar, M. L. A. and Furness, R. W. (1991). Mercury and selenium interaction: A review. *Ecotoxicology and Environmental Safety*, 21, 348-364.

Dietz, R., Riget, F. and Born, E. W. (2000). *The Science of the Total Environment,* 245, 15-24.

Domingo, J. L., Bocio, A., Falcó, G., Llobet, J. M. (2007). Benefits and risks of fish consumption Part I. A quantitative analysis of the intake of omega-3 fatty acids and chemical contaminants. *Toxicology,* 230, 219-226.

Donelli, J. K. and Robinson, D. S. (1995). Free radicals in foods. *Free Radical Research,* 22, 147-176.

Dumont, E., Vanhaecke, F., Cornelis, R. (2006). Selenium speciation from food source to metabolites: a critical review. *Analytical and Bioanalytical Chemistry,* 385, 1304-1323.

EFSA (2005). Opinion of the scientific panel on contaminants in the food chain on a request from the European Parliament related to the safety assessment of wild and farmed fish. *The EFSA Journal*, 236, 118 p.(URL: http://dx.doi.org/doi:10.2903/j.efsa.2005.236)

Eisler, R. (2006). *Mercury hazards to living organisms*. Boca Raton, USA: Taylor and Francis group, CCR press, 312 p.

El Begearmi, M. M., Ganther, H. E. and Sunde, M. L. (1982). Dietary interaction between methyl mercury, selenium, arsenic and sulfur amino acids in Japanese quail. *Poultry Science*, 61, 272-279.

Fabris, G., Turoczy, N. J. and Stagnitti, F. (2006). Trace metal concentration in edible tissue of snapper, flathead, lobster, and abalone from coastal waters of Victoria, Australia. *Ecotoxicology and Environmental Safety*, 63, 286-292.

Fardy, J. J., McOrist, G. D., Farrar, Y. J., Bowles, C. J., Warner, I. M. and Mingguang, T. (1994). Application of neutron activation analysis and inductively coupled plasma mass spectrometry to the determination of toxic and essential elements in Australian foods. Nuclear techniques for toxic elements in foodstuffs. Report on an IAEA Co-ordination Research Programme. Vienna, Austria: *International Atomic Energy Agency*. 70 p.

Fishbein, L. (1991). Selenium. In: *Metals and their compounds in the environment*. E. Merian (Ed). Weinheim, Germany: VCH Verlag.

Fox, T. E., Van den Heuvel, E. G., Atherton, C. A., Dainty, J. R. and Lewis, D. J. (2004). Bioavailability of selenium from fish, yeast and selenate: a comparative study in humans using stable isotopes. *European Journal of Clinical Nutrition,* 58, 343–349.

Frankel, E. N. (1991). Review: Recent advances in lipid oxidation. *Journal of the Science of Food and Agriculture*, 54, 495-511.

Freidman, M. A., Eaton, L. R. and Carter, W. H. (1978). Protective effects of freeze-dried swordfish on methylmercury content. *Journal of Environmental Contamination and Toxicology,* 19, 436–443.

Gaggi, C., Zino, F., Duccini, M. and Renzoni, A. (1996). Levels of mercury in scalp hair of fishermen and their families from Camara de Lobos-Madeira (Portugal): Preliminary study. *Bulletin of Environmental Contamination and Toxicology*, 56, 860-865.

Gailer, J. (2007). Arsenic-selenium and mercury-selenium bonds in biology. *Coordination Chemistry Reviews*, 251, 234-254.

Ganther, H. E., Goudie, C., Sunde, M. L., Kopecky, M. J., Wagner, P., Oh, S. H. and Hoekstra, W. G. (1972). Selenium: relation to decreased toxicity of methylmercury added to diets containing tuna. *Science*, 175, 1122-1124.

Gao, J., Liu, Y., Huang, Y., Lin, Z. Q., Bañuelos, G. S., Lam, M. H. W. and Yin, X. (2011). Daily selenium intake in a moderate selenium deficiency area of Suzhou, China. *Food Chemistry*, 126, 1088-1093.

Gašpić, Z. K., Zvonarić, T., Vrgoč, N., Odžak, N. and Barić, A. (2002). Cadmium and lead in selected tissues of two commercially important fish species from Adriatic Sea. *Water Research*, 36, 5023-5028.

Girotti, A. W. (1998). Lipid hydroperoxide generation, turnover, and effector action in biological systems. *Journal of Lipid Research*, 39, 1529-1542.

Goyer, R. A. and Clarkson, T. W. (2001). Toxic effects of metals. In: Casarett and Doull's Toxicology: *The basic science of poisons*, C. D. Klaassen (Ed). New York, USA: McGraw-Hill, pp. 811 -867.

Hajeb, P., Jinap, S., Ismail, A., Fatimah, A. B., Jamilah, B. and Rahim, M. A. (2009). Assessment of mercury levels in commonly consumed marine fishes in Malaysia. *Food Control*, 20, 79-84.

Halliwell, B. (1994). Free radicals and antioxidants: a personal view. *Nutrition Reviews,* 52, 253-265.

Halliwell, B. and Gutteridge, J. M. C. (2007). *Free radicals in biology and medicine.* New York, U.S.A.: Oxford University press Inc., 851 p.

Hamilton, S. J. (2004). Review of selenium toxicity in the aquatic food chain. *Science of the Total Environment*, 326, 1-31.

Haratake, M., Takahashi, J., Ono, M. and Nakayama, M. (2007). An assessment of Niboshi (a processed Japanese anchovy) as an effective food source of selenium. *Journal of Health Science*, 53, 457-463.

Helmersson, J., Arnlov, J., Vessby, B., Larsson, A., Alfthan, G. and Basu, S. (2005). Serum selenium predicts levels of F2-isoprostanes and prostaglandin F2α in a 27 year follow-up study of Swedish men. *Free Radical Research*, 39, 763–70.

Infante, J. P. (1986). Vitamin E and Se participation in fatty acid desaturation: a proposal to an enzymatic function of these nutrients. *Molecular and Cell Biochemistry*, 69, 93-108.

IOM (2000*). Dietary reference intakes for vitamin C, vitamin E, selenium, and carotenoides.* Washington (DC), U.S.A.: National Academic Press, 509 p.

Kaneko, J. J., Ralston, N. V. C. (2007). Selenium and Mercury in Pelagic Fish in the Central North Pacific near Hawaii. *Biological Trace Element Research*, 19, 242–254.

Klapec, T., Mandic, M. L., Grgic, J., Primorac, L. G., Perl, A. and Krstanovic, V. (2004). Selenium in selected foods grown or purchased in eastern Croatia. *Food Chemistry*, 85, 445-452.

Kohrle, J., Jakob, F., Contempré, B., Dumont, J. E. (2005). Selenium, the thyroid, and the endocrine system. *Endocrine Reviews*, 26, 944–984.

Kölbl, G., Kalcher, K., Irgolic, K. J. and Magee, R. J. (1993). Identification and quantification of inorganic and organic selenium compounds with high performance liquid chromatography: A review. *Applied Organometal Chemistry,* 7, 443-466.

Kölbl, G. (1995). Concepts for the identification and determination of selenium compounds in the aquatic environment. *Marine Chemistry*, 48, 185-197.

Kotsonis, F. N., Burdock, G. A. and Flamm, W. G. (2001). Food toxicology. In: Casarett and Doull's Toxicology: *The basic science of poisons*, C. D. Klaassen (Ed). New York, USA: McGraw-Hill, pp. 1049 -1088.

Kwockek, M., Szefer, P., Hac, E. and Grembecka, M. (2006). Essential and toxic elements in seafood available in Poland from different geographical regions. *Journal of Agriculture and Food Chemistry*, 54, 3015-3024.

Lall, S. P. (1995). Macro and trace elements in fish and shellfish. In: Fish and fishery products. Composition, nutritive properties and stability, A. Ruiter (Ed.). Wallingford, USA: *Cab International*, pp. 187-213.

Lall, S. P. (2002). *The Minerals.* In: Fish nutrition, J. E. Halver and R. W. Hardy (Eds). San Diego, USA: Academic Press.

Lamand, M., Tressol, J. C. and Villart, S. (1997). Évaluation des pertes minérales à la cuisson dans quelques aliments. *Cahiers de Nutrition et de Diététique,* 32, 28-30.

Larsen, R., Mierke-Klemeyer, S, Maehre, H., Elvevoll, E.O., Bandarra, N. M., Cordeiro, A. R., Nunes, M. L., Schram, E., Luten, J. and Oehlenschläger, J. (2010a). Retention of health beneficial components during hot and cold-smoking of African Catfish (Clarias gariepinus) fillets. *Journal of Food Safety and Food Quality,* 61, 31-35.

Larsen, R., Mierke-Klemeyer, S, Maehre, H., Elvevoll, E. O., Bandarra, N. M., Cordeiro, A. R., Nunes, M.L., Schram, E., Luten, J. and Oehlenschläger, J. (2010b). Changes in health beneficial components during ice storage of African catfish (Clarias gariepinus). *Archive fur Lebensmittelhygiene*, vol.61 (4): 139-144.

Lavilla, I., Vilas, P. and Bendicho, C. (2008). Fast determination of arsenic, selenium, nickel and vanadium in fish and shellfish by electrothermal atomic absorption spectrometry following ultrasound-assisted extraction. *Food Chemistry,* 106, 403-409.

Lemly, A. D. (2002). Symptoms and implications of selenium toxicity in fish: the Belews lake case exemple. *Aquatic Toxicology*, 57, 39-49.

Li, N., Gao, Z., Luo, D., Tang, X., Chen, D. and Hu, Y. (2007). Selenium level in the environment and the population of Zhoukoudian area, Beijing, China. *Science of the Total Environment*, 381, 105-111.

Lin, Y. H. and Shiau, S. Y. (2009). Mutual sparing of dietary requirements for alpha-tocopherol and selenium in grouper, Epinephelus malabaricus. *Aquaculture*, 294, 242-245.

Lovell, R. T. (1987). Requerimientos minerales de los peces. In: Nutrición en Acuicultura II, J. E. de los Monteros and U. Labarta (Eds). Madrid, Spain: *Plan de Formación de Técnicos Superiores en Acuicultura*, pp. 281-289

Luten, J. B., Bouquet, W., Burggraf, M. M. and Rus, J. (1987). The in-vitro availability of copper, zinc and selenium and the speciation of selenium in fishery products. In: *Trace element-analytical chemistry in medicine and biology.* P. Brätter and P. Schramel (Eds). Berlin, Germany: De Gruyter.

Maier, K. J., Foe, C., Ogle, R. S., Williams, M. J., Knight, A. W., Kiffney, P. and Melton, L. A. (1987). The dynamics of selenium in aquatic ecosystems. *Trace Substances in Environmental Health*, 21, 361-408.

Martí-Cid, R., Bocio, A., Llobet, J. M. and Domingo, J. L. (2007). Intake of chemical contaminants through fish and seafood consumption by children of Catalonia, Spain: health risks. *Food and Chemical Toxicology*, 45, 1968-1974.

McDonald, P., Edwards, R. A., Greenhalgh, J. F. D. and Morgan, C. A. (2002). *Animal nutrition*. Harlow, England: Pearson Education Ltd., Prentice Hall, 693 p.

McNeal, J. M. and Balistrieri, L. S. (1989). Geochemistry and occurrence of selenium: An overview. In: Selenium in agriculture and the environment. L. W. Jacobs (Ed). Madison, USA: *Soil Science Society of America Special Publication.*

Mierke-Klemeyer, S., Larsen, R., Oehlenschläger, J., Maehre, H., Elvevoll, E. O., Bandarra, N. M., Parreira, R., Andrade, A. M., Nunes, M. L., Schram, E. and Luten, J. (2008). Retention of health-related beneficial components during household preparation of selenium-enriched African catfish (Clarias glariepinus) fillets. *European Food Research and Technology*, 227, 827-833.

Miklavčič, A., Stibilj, V., Heath, E., Polak, T., Tratnik, J. S., Klavž, J., Mazej, D. and Horvat, M. (2011). Mercury, selenium, PCBs and fatty acids in fresh and canned fish available on the Slovenian market. *Food Chemistry*, 124, 711-720.

Miura, K., Naganuma, A., Himeno, S. and Imura, N. (1995). Mercury toxicology. In: Toxicology of metals: *Handbook of Experimental Pharmacology,* R. Goyer and G. Cherian (Eds). Berlin, Germany: Springer-Verlag, pp. 163-180.

Moreno, P., Quijano, M. A., Gutiérrez, A. M., Pérez-Conde, M. C. and Cámara, C. (2004). .Study of selenium species distribution in biological tissues by size exclusion and ion exchange chromatography inductively coupled plasma-mass spectrometry. *Analytica Chimica Acta*, 524, 315-327.

Navarro-Alarcon, M., Gil-Hernández, F. and Gil-Hernandez, A. (2005). Selenio, manganeso, cromo, molibdeno, yodio y otros oligoelementos minoritarios. In: Tratado de Nutrición Tomo I: Bases fisiológicas y Bioquímicas de la Nutrición, A. Gil-Hernández, A. (Ed). Madrid, Spain: *Acción Medica*, p. 997–1036.

Navarro-Alarcon, M. and Cabrera-Vique, C. (2008). Selenium in food and the human body: A review. *Science of the Total Environment*, 400, 115-141.

Navas-Acien, A., Bleys, J. and Guallar, E. (2008). Selenium intake and cardiovascular risk: what is new? *Current Opinion in Lipidology*, 19, 43–9.

NRC (National Research Council) (2011). *Nutrient requirements of fish and shrimp.* Washington DC, USA: The National Academic Press.

O'Brien, G., Jolley, D. and Morrison, J. (2006). Evolution of chemical contaminant and toxicology studies, part 2–case studies of selenium and arsenic. *South Pacific Journal of Natural Science*, 21, 6-14.

Oehlenschläger, J. (1997). Marine fish – A source for essential elements?! In: Seafood from producer to consumer, integrated approach to quality, J. B. Luten, T. Børresen and J. Oehlenschläger (Eds), Amsterdam, The Netherlands: *Elsevier Science* B.V., pp. 641-651.

Ohi, G., Nishigaki, S., Seki, H., Tamura, Y., Maki, T., Konno, H., Ochiai, S., Yamada, H., Shimamura, Y., Mizoguchi, I. and Yagyu, H. (1976). Efficacy of selenium in tuna and selenite in modifying methyl mercury intoxication. *Environmental Research*, 12, 49-58.

Ohi, G., Nishigaki, S., Seki, H., Tamura, Y., Maki, T., Minowa, K., Shimamura, Y., Mizoguchi, I., Inaba, Y., Takizawa, Y. and Kawanishi, Y. (1980). The protective potency of marine animal meat against the neurotoxicity of methylmercury: its relationship with the organ distribution of mercury and selenium in the rat. *Food and Cosmetics Toxicology*, 18, 139–145.

Ornsrud, R. and Lorentzen, M. (2002). Bioavailability of selenium from raw or cured selenomethionine-enriched fillets of Atlantic salmon (Salmo salar) assessed in selenium-deficient rats. *British Journal of Nutrition*, 87, 13-20.

Osman, N. H., Suriah, A. R. and Law, E. C. (2001). Fatty acid composition and cholesterol content of selected marine fish in Malaysian waters. *Food Chemistry*, 73, 55-60.

Pappa, E. C., Pappas, A. C. and Surai, P. F. (2006). Selenium content in selected foods from the Greek market and estimation of the daily intake. *Science of the Total Environment*, 372, 100-108.

Pedrero, Z. and Madrid, Y. (2009). Novel approaches for selenium speciation in foodstuffs and biological specimens: a review. *Analytica Chimica Acta*, 634, 135-152.

Peraza, A. M., Ayala-Fierro, F., Barber, D. S., Casarez, E. and Rael, L. (1998). Effects of micronutrients on metal toxicity. *Environmental Health Perspectives,* 106(Supp.1), 1-27.

Pérez-Mateos, M., Boyd, L. and Lanier, T. (2004). Stability of omega-3 fatty acids in fortified surimi seafoods during chilled storage. *Journal of Agricultural and Food Chemistry,* 52, 7944-7944.

Phibbs, J., Franz, E., Hauck, D., Gallego, M., Tse, J. J., Pickering, I. J., Liber, K. and Janz, D. M. (2011). Evaluating the trophic transfer of selenium un aquatic ecosystems using caged fish, X-ray absorption spectroscopy and stable isotope analysis. *Ecotoxicology and Environmental Safety*, to be printed.

Plessi, M., Bertelli, D. and Monzani, A. (2001). Mercury and selenium content in selected seafood. *Journal of Food Composition and Analysis*, 14, 461-467.

Ralston, N. V. C., Ralston, C. R., Blackwell, J. L. and Raymond, L. J. (2008). Dietary and tissue selenium in relation to methylmercury toxicity. *Neurotoxicology*, 29, 802-811.

Rao, M. V. and Sharma, P. S. N., (2001). Protective effect of vitamin E against mercuric chloride reproductive toxicity in male mice. *Reproductive Toxicology,* 15, 705-712.

Rayman, M. P. (2000). The importance of selenium to human health. *The Lancet,* 356, 233-241.

Raymond, L. J. and Ralston, N. V. C. (2004). Mercury: selenium interactions and health implications. *Seychelles Medical and Dental Journal*, 7(1), 72-77.

Raymond, L. J. and Ralston, N. V. C. (2009). Selenium's importance in regulatory issues regarding mercury. *Fuel Processing Technology*, 90, 1333-1338.

Renzoni, A., Zino, F. and Franchi, E. (1998). Mercury levels along the food chain and risk for exposed populations. *Environmental Research*, 77, 68-72.

Ruiter, A. (1995). Contaminants in fish. In: Fish and fishery products – composition, nutritive properties and stability, A. Ruiter (Ed), Wallingford, USA*: CAB International,* pp. 261-285.

Sager, M. (2006). Selenium in agriculture, food, and nutrition. *Pure and Applied Chemistry,* 78, 111-133.

SCF (2000). Opinion of the Scientific Committee on Food on the Tolerable Upper Intake Level of Selenium. *SCF/CS/NUT/UPPLEV/25*. Brussels, Belgium: European Commission, Health and Consumer Protection Directorate-General.

Schäfer, K., Kyriakopoulos, A., Gessner, H., Grune, T. and Behne, D. (2004). Effects of selenium deficiency on fatty acid metabolism in rats fed fish oil-enriched diets. *Journal of Trace Elements in Medicine and Biology*, 18, 89-97.

Shen, L. H., Nieuwenhuizen, H. and Luten, J.B. (1997). Speciation and in vitro bioavailability of selenium in fishery products. In: Seafood from producer to consumer, integrated approach to quality, J. B. Luten, T. Borresen and J. Oehlenschläger (Ed). Amsterdam, The Netherlands: *Elsevier Science* B. V, pp. 653-663.

Sies, H. (1986). *Oxidative stress*. London, UK: Academic Press.

Simonoff, M. and Simonoff, G. (1991). *Le selenium et la Vie*. Paris, France: Masson.

Simopoulos, A. P. (2002). Omega-3 fatty acids and cardiovascular disease: the epidemiological evidence. *Environmental Health and Preventive Medicine*, 6, 203-209.

Sirichakwal, P. P., Puwastein, P., Polngam, J. and Kongkachuichai, R. (2005). Selenium content of Thai foods. *Journal of Food Composition and Analysis*, 18, 47-59.

Sivaperumal, P., Sankar, T. V. and Viswanathan, N. (2007). Heavy metal concentrations in fish, shellfish and fish products from internal markets of India vis-a-vis international standards. *Food Chemistry*, 102, 612-629.

Song, J. H., Fujimoto, K., Miyazawa, T. (2000). Polyunsaturated (n-3) fatty acids susceptible to peroxidation are increased in plasma and tissue lipids of rats fed docosahexenoic acid-containing oils. *Journal of Nutrition*, 130, 3028-3033.

Stewart, R., Grosell, M., Buchwalter, D., Fisher, N., Luoma, S., Mathews, T., Orr, P. and Wang, W. (2010). Bioaccumulation and trophic transfer of selenium. In: *Ecological Assessment of Selenium in the Aquatic Environment,* Chapman, P. M., Adams, W. J., Brooks, M. L., Delos, C. G., Luoma, S. N., Maher, W. A., Ohlendorf, H. M., Presser, T. S. and Shaw, D. P. (Eds.). Boca Raton, USA: CRC Press, pp. 93–139.

Stillings, B., Lagally, H., Bauersfeld, P. and Soares, J. (1974). Effect of cystine, selenium and fish protein on the toxicity and metabolism of methyl mercury in rats. *Toxicology and Applied Pharmacology*, 30, 243-254.

Storelli, M. M., Giacominelli Stuffler, R., Storelli, A. and Marcotrigiano, G. O. (2003). Total mercury and methylmercury content in edible fish from the Mediterranean Sea. *Journal of Food Protection*, 60, 300-303.

Storelli, M. M., Storelli, A, Giacominelli-Stuffler, R. and Marcotrigiano, G. O. (2005). mercury speciation in the muscle of two commercially important fish, hake (Merluccius merluccius) and striped mullet (Mullus barbatus) from the Mediterranean sea: estimated weekly intake. *Food Chemistry*, 89, 295-300.

Sunde, R. A. (2000). Selenium. In M. H. Stipanuk (Ed.), *Biochemical and Physiological aspects of human nutrition* (pp. 782-809). New York, NY: W.B. Saunders Company.

Thomson, C. D. and Robinson, M. F. (1990). Selenium content of foods consumed in Otago, New Zealand. *The New Zealand Medical Journal*, 28, 103-105.

Trumbo, P. R. (2005). The level of evidence for permitting a qualified health claim: FDA's review of the evidence for selenium and cancer and vitamin E and heart disease. *Journal of Nutrition*, 135, 354–6.

Van Cauwenbergh, R. V., Robberecht, H. and Van Vlaslaer, V. (2004). Comparison of the serum selenium content of healthy adults living in the Antwerp region (Belgium) with recent literature data. *Journal of Trace Elements in Medicine and Biology*, 18, 99-112.

Van der Torre, H. W., Van Dokkum, W., Schaafsma, G., Wedel, M. and Ockhuizen, T. (1991). Effects of various levels in wheat and meat on blood Se status indices and on Se balance in Dutch men. *British Journal of Nutrition,* 65, 69-80.

Watanabe, T., Kiron, V. and Satoh, S. (1997). Trace minerals in fish nutrition. *Aquaculture,* 151, 187-207.

Wrench, J. J. (1978). Selenium metabolism in the marine phytoplankters Tetraselmis tetrathele and Dunaliella minuta. *Marine Biology*, 49, 231-236.

Xi, L., Yi, L., Feng-Jiao, D. and Song-Sheng, Q. (2003). Study on the action of sodium selenite on the mytochondria metabolism of *Carassius auratus* hybrid carps liver by microcalorimetry. *Wuhan University Journal of Natural Sciences,* 8, 95-98.

Yebra-Biurrun, M. C. and García-Garrido, A. (2001). Continuous flow systems for the determination of trace elements and metals in seafood. *Food Chemistry,* 72, 279-287.

Zhang, X., Shi, B. and Spallholz, J. E. (1993). The selenium content of selected meats, seafoods, and vegetables from Lubbock, Texas. *Biological Trace Element Research,* 39, 161-169.

In: Selenium: Sources, Functions and Health Effects
Editors: Chinatsu Aomori and Megumi Hokkaido
ISBN: 978-1-61942-061-8

Chapter 4

SELENIUM AND PLANT HEALTH: THE PHYSIOLOGICAL ROLE OF SELENIUM

Mirza Hasanuzzaman[1,2] and Masayuki Fujita[1*]

[1]Laboratory of Plant Stress Responses, Department of Applied Biological Science, Faculty of Agriculture, Kagawa University, Miki-cho, Kita-gun, Kagawa, Japan

[2]Department of Agronomy, Faculty of Agriculture, Sher-e-Bangla Agricultural University, Bangladesh

ABSTRACT

Selenium (Se) is a widely studied trace element in human and animal due to its role in antioxidant defense system which is needed for the maintenance of health and hormone balance. During last two decades the physiological role of Se in plants has been explored by researchers. Plant roots take up Se from soil water in either the selenate or the selenite ionic forms. In higher plants metabolism of Se is closely related to that of sulfur due to their chemical similarity. Although, Se is not yet confirmed to be required by higher plants, but several studies demonstrate that at low concentrations it may exert diverse beneficial effects, including growth-promoting activities. Moreover, some plant species grown in Se-enriched media have shown enhanced resistance to certain abiotic stresses, e.g. drought, salinity, extreme temperature, metal toxicity and UV-irradiation. Se exerts its beneficial effects on growth and stress tolerance of plants by enhancing their antioxidative capacity. It enhances plants' resistance against oxidative stress caused by reactive oxygen species (ROS). However, agricultural crops plants are sensitive to high Se concentrations which vary among plant species. Although a number of report indicated the protective role of Se in plants, to date the research works conducted on the physiological role of Se is scarce. In this chapter, we attempted to review the recent findings related to the physiological role of Se in plants.

Keywords: abiotic stress, antioxidants, oxidative stress, plant growth, selenium, trace elements

[*] Corresponding Author: Telephone: +81-87-891-3133; Fax: +81-87-891-3021; E-mail: fujita@ag.kagawa-u.ac.jp.

1. INTRODUCTION

Selenium (Se) is a non-metal, whose Greek name (*Selene*) means moon. Selenium was first discovered by J.J. Berzelius in 1817. Selenium is probably the most widely investigated of all the trace elements. It is also considered as an essential micronutrient necessary for antioxidative and hormonal balance in human and animal cells (Ellis and Salt, 2003). During last two decades the physiological roles of Se in plants have been studied by many researchers. In general, Se can be found in plants both in inorganic and organic forms, including selenoamino acids and methylated compounds. Recent studies on Se are focused on the evidence that Se might be an essential micronutrient for accumulator plants species (Terry et al., 2000; Sors et al., 2005). Moreover, it has been observed that Se is required for an optimal growth of the unicellular green alga. In contrast, Se non-accumulator plants, including most species of crops, do not appear to require Se for their growth, and in general, these plants have a low tolerance to this element (Terry et al., 2000). The essentiality of Se to higher plants, however, is still under debate. Although it is harmful for plants in high concentrations, it can exert beneficial effects at low concentrations (Germ et al., 2007).

Although Se has not been confirmed to be an essential micronutrient in higher plants, there is rising evidence that Se acts as an antioxidant in plants. Selenium has been shown to exert a positive effect on plant growth and productivity at low concentrations (Terry et al., 2000; Xue et al., 2001; Turakainen et al., 2004; Hasanuzzaman et al., 2010). However, the specific physiological mechanisms underlying the beneficial role of Se in plants have not been clearly elucidated (Turakainen, 2007). Moreover, some plant species supplemented with Se have shown enhanced resistance to certain abiotic stresses, e.g. drought (Yao et al., 2009; Hasanuzzaman et al., 2010; Hasanuzzaman and Fujita, 2011), salinity (Djanaguiraman et al., 2005; Hawrylak-Nowak, 2009; Hasanuzzaman et al., 2011), extreme temperature (Chu et al., 2010; Hawrylak-Nowak et al., 2010; Djanaguiraman et al., 2010), toxic metal metals (Hawrylak et al., 2007; Pedrero et al., 2008; Filek et al., 2008; Srivastava et al., 2009; Cartes et al., 2010) and UV-radiation (Valkama et al., 2003; Yao et al., 2010a, b). Selenium exerts its beneficial effects on growth and stress tolerance of plants by enhancing their antioxidative capacity (Xue et al., 2001; Djanaguiraman et al., 2005; Hasanuzzaman and Fujita, 2011). Different plant studies showed that Se enhances plant's resistance to oxidative stress caused by reactive oxygen species (ROS). It was reported that under stressful conditions Se upregulated the activities of antioxidant enzymes viz. superoxide dismutase (SOD), catalase (CAT), ascorbate peroxidase (APX), monodehydroascorbate reductase (MDHAR), dehydroascorbate reductase (DHAR), glutathione reductase (GR), glutathione *S*-transferase (GST), glutathione peroxidase (GPX) and peroxidases (POD) as well as elevated the levels of non-enzymatic antioxidants such as ascorbate (AsA), glutathione (GSH), carotenoids and tocopherols (Hasanuzzaman et al., 2010; 2011; Hasanuzzaman and Fujita, 2011). However, there are plenty of reports that agricultural crop plants are sensitive to high Se concentrations depending on the plant species (Hartikainen et al., 2001; Lyons et al., 2005; Rani et al., 2005).

However, the specific physiological and molecular mechanisms that underlie the beneficial effects of Se in plants have not been fully elucidated so far. In this chapter, we discuss the progress in understanding the possible role of Se in plant growth and physiology as well as its protective role in enhancing tolerance to environmental stresses.

2. Role of Selenium in Plant Growth and Physiology

Different plant studies have showed the beneficial role of Se at its lower concentration. It has also been reported that higher plants metabolize Se through the sulfur (S) assimilation pathway (Zayed et al., 2000), synthesizing Se analogues of various S metabolites (Ellis and Salt, 2003). This involves the fact that the non-specific incorporation of Se into selenoamino acids and proteins as well as Se volatilization occurs, when the metalloid is supplied to plants in excess of any potential. However, plants differ in their ability to accumulate Se in their tissues. In fact, some plants are able to hyperaccumulate Se in their shoots when they grow on seleniferous soil. On the other hand, most forage and crop plants, as well as grasses, contain less than 25 mg Se kg^{-1} dry weight and do not accumulate Se much above 100 mg Se kg^{-1} dry weight when grown on seleniferous soils. These plants are referred to as Se non-accumulators (Gregorio, 2008). A third category of plants, known as secondary Se accumulators, grow on soil of low to medium Se content and accumulate up to 1000 mg Se kg^{-1} dry weight. *Brassica juncea* and *Brassica napus*, are also secondary Se accumulator plant species with a typical Se concentration of several hundred μg of Se g^{-1} dry weight in their shoots when grown on soils containing moderate levels of Se (Parker and Page, 1994; Bañuelos et al., 1996; White et al., 2004). Although there are evidences that Se is required for the growth of algae, the question of the essentiality of Se as a micronutrient in higher plants is controversial (Gregorio, 2008). In addition, the rate and form of Se uptake by plants depend on the concentration and chemical form of Se in soil solution, as well as rhizosphere conditions (Dhillon and Dhillon, 2003).

2.1. Selenium and Plant Growth

Se has not yet been classified as an essential element for plants, although its role has been considered to be beneficial for plant which is capable of accumulating large amounts of the element (Shanker, 2006). The role of Se in plant depends mainly on its concentration (Hasanuzzaman et al., 2010). According to Hamilton (2004), Se has three levels of biological activity: (i) trace concentrations are required for normal growth and development; (ii) moderate concentrations can be stored to maintain homeostatic functions; and (iii) elevated concentrations can result in toxic effects. Studies on ryegrass (*Lolium perenne*) and lettuce (*Lactuca sativa*) show that, although Se is harmful for plants at high concentrations >10 mg kg^{-1} and 1.0 mg kg^{-1} respectively (reduction of biomass), it can exert beneficial effects at low concentrations, namely 0.1 mg kg^{-1} soil (Hartikainen et al., 2000; Xue et al., 2001). The positive growth responses of plants to Se added at small concentrations have been attributed to the antioxidative effect of Se counteracting the oxidative stress (Seppänen et al., 2003).

Perhaps, the first positive effect of Se on plant growth was reported by Singh et al. (1980), who showed that the application of 0.5 mg kg^{-1} Se as selenite stimulated growth and dry matter yield of Indian mustard (*Brassica juncea* L.). Later, it was revealed that Se, applied at low concentrations, enhanced growth and antioxidative capacity of both mono- and dicotyledonous plants. The growth-promoting response to Se was demonstrated in *L. sativa* and *L. perenne* (Hartikainen et al., 1997; Hartikainen and Xue, 1999) and in soybean (*Glycine max* L.) (Djanaguiraman et al., 2005), potato (*Solanum tuberosum*) (Turakainen et al., 2004), and green tea leaves (Hu et al., 2003). Selenium affected plant growth promotion might be the result of increased starch accumulation in chloroplasts (Pennanen et al., 2002) and that protected cell content (Xue et al., 2001). Se could also delay senescence and promote the growth of aging seedlings (Xue et al., 2001). Foliar application, performed manually only on plants, is the most appropriate way to add Se, since contamination of the soil is minimal. Foliar application to barley at 10 and 20 g Se ha^{-1}, as sodium selenate (Na_2SeO_4), increased the Se content of barley grain and straw and red clover forage (MacLeod et al., 1998).

Se has also demonstrated its effect on seed germination. Carvalho et al. (2003) reported that at higher supplementation level than 29 mg kg^{-1} soil, Se inhibited the growth and germination of tomato (*Solanum lycopersicum*), *L. sativa* and radish (*Raphanus sativus*) seeds. In contrast, priming of seeds with selenite promoted germination of bitter gourd (*Momordica charantia* L.) seeds at sub-optimal temperatures (Chen and Sung, 2001). The positive effect on germination was linked to antioxidative activity.

In lettuce shoots, Simojoki (2003) found that small Se addition (<4 μg l^{-1}) increased Se contents up to 1–5 mg kg^{-1} dry matter (DM) tend to enhance plant growth. But the yields drop drastically at Se contents above 20 mg kg^{-1} DM. Selenium fertilization changes root morphology, and the effects are diverse in different parts of the root system. Moderate Se additions decrease the specific length and specific surface area of basal and lateral roots, whereas large additions increase the specific volume of roots (Simojoki, 2003). However, Se treatments did not change the morphology of hypocotyls. On the contrary, the specific length and area of basal and lateral roots were smallest at intermediate Se additions, whereas the specific volume was largest at the largest Se addition. Interestingly, these effects of Se on root morphology were, however, not unambiguously related to plant growth (Simojoki, 2003). In another study, Xue et al. (2001) demonstrated that a low Se dosage (0.1 mg kg^{-1} soil) enhanced the growth of senescing lettuce seedlings (14% increases in dry matter) despite a decreased chlorophyll concentration. This growth-promoting function was related to diminished lipid peroxidation. Turakainen (2007) demonstrated that Se concentrations in the upper leaves, roots, stolons and tubers of potato increased with increasing Se supplementation and consequently the highest Se concentration was reached in young upper leaves, roots and stolons, indicating that added selenate was efficiently utilized and taken up at an early stage. During the growing period the Se concentration declined in the aerial parts, roots and stolons of potato plants whereas an intensive accumulation took place in immature and mature tubers. It was also observed the increased carbohydrate accumulation in the young upper leaves and in stolons, roots and tubers at maturity when supplemented with Se. However it could not be explained by increased production of photoassimilates as net photosynthesis did not differ among Se treatments (Turakainen, 2007). The Se treated plants produced higher tuber yields than control plants, while the addition of higher Se concentration (0.3 mg kg^{-1}) resulted in lower numbers of larger tubers.

Increased yield of Se treated plants indicated that Se may enhance the allocation of photoassimilates for tuber growth, acting as a strong sink for both Se and for carbohydrates.

It was also observed that Se improves the processing and storage quality of potato tubers (Turakainen et al., 2004; Turakainen, 2007). The positive impact of Se on the yield of potato plants could be related to its antioxidative effect in delaying senescence. Selenium at a concentration of 1.5 mg l^{-1} increased yield in *Cucurbita pepo* (Germ et al., 2005). Recently, Ramos et al. (2010) supplemented lettuce plant with seven Se concentrations (0, 2, 4, 8, 16,

32, and 64 μmol l^{-1}) and two forms of Se (sodium selenate – Na_2SeO_4 and sodium selenite – Na_2SeO_3).

Figure 1. Phenotypic appearance of *Brassica napus* seedling supplemented with Se. a. Control; b. Pretreated with 50 μM Na_2SeO_4 (24 h) and c. and 100 μM Na_2SeO_4 (24 h). Twelve day-old seedlings were pretreated with Se and then grown for more 48 h.

They observed that application of Se as selenate at low concentrations is more appropriate for lettuce biofortification because it favors shoot biomass growth and Se levels in the shoot biomass. Selenium in both forms had two effects on lettuce plant metabolism: at low doses it acted as an antioxidant and enhanced plant growth, whereas at higher levels it reduced yield. In our laboratory, we observed better phenotypic appearance on *B. napus* seedlings supplemented with exogenous Se pretreatment (24 h) (Figure 1).

2.2. Selenium and Plant Physiology

Although there are limited numbers of reports on the physiological role of Se on plants, different lines of study suggested that selenium regulates the physiology of plants in different ways. In a study, Pennanen et al. (2002) showed that plant growth promoted by Se is the result of increased starch accumulation in chloroplasts. It was shown that Se also has positive effects on potato carbohydrate accumulation (Turakainen et al., 2004). At the highest Se addition (0.3 mg Se kg^{-1}), the highest soluble sugar concentration was observed in the upper leaves 4 weeks after planting (from 75 to 90 mg g^{-1} DW) and in the roots (from 25 to 50 mg g^{-1} DW) and stolons (from 150 to 175 mg g^{-1} DW) at maturity. The addition of Se induced a higher respiratory potential, measured by electron transport system (ETS) activity, in young Se enriched plants of *Pisum sativum* (Smrkolj et al., 2006), and young chicory (*Cichorium intybus*) plants (Germ et al., 2007). The respiratory potential of *Eruca sativa* significantly increased in plants grown from Se-treated seeds (Germ and Osvald, 2005). Possible explanations are: (i) a higher ETS activity reflect increased GPx activity in mitochondria. It was shown (Hartikainen et al., 2000; Xue and Hartikainen, 2000; Xue et al., 2001; Cartes et al., 2005) that Se exposure increased GPx activity in ryegrass and lettuce; (ii) plants need energy to repair damage caused by Se. The latter is consistent with the fact that Se can mimic S, forming Se analogues of S compounds, for example replacing S in amino acids (methionine and cysteine). The conformation of proteins containing seleno amino acids could be perturbed, and their catalytic activity thereby disturbed (Brown and Shrift, 1982).

Ekelund and Danilov (2001) investigated the role of Se in light-enhanced dark respiration (LEDR) and photosynthesis in the flagellate *Euglena gracilis*, after exposure to UV-radiation (UV-A, 320-400 nm, of 1.02 W m^{-2} plus UVB, 280-320 nm, of 0.73 W m^{-2}). Se was added into the growth medium at different concentrations of selenite (10^{-7}, 10^{-8}, 10^{-9} and 10^{-10} M, $Na_2SeO_3 \cdot 5H_2O$). *E. gracilis* were subjected to six different light pulses with a photon fluence rate of 59, 163, 600, 1180, 2080 and 3340 μmol m^{-2} s^{-1} and periods of darkness between the light pulses. Photosynthetic saturation occurred at irradiances higher than 600 μmol m^{-2} s^{-1} and at the highest irradiance the photosynthetic rate decreased due to photoinhibition. Different plant studies also indicated that Se might affect biosynthesis of glycoalkaloids (Munshi and Mondy, 1992), chlorophylls (Padmaja et al., 1989) as well as nitrogen assimilation (Aslam et al., 1990). Konze et al. (1978) showed that selenomethionine enhanced ethylene biosynthesis in plants; it was a better substrate than methionine for *S*-adenosylmethionine decarboxylase (SAMDC). *S*-adenosylmethionine is required in the synthesis of selenomethionine as well as in polyamine biosynthesis (Terry et al., 2000, Martin-Tanguy, 2001). The preliminary results obtained by Germ and Osvald (2005) showed that the addition of Se did not disturb the flow of electrons in photosystem II (PSII), while increased the respiratory potential of *Eruca sativa*. However, the underlying mechanisms of the effect of Se in the respiratory chain are still not clear and deserve continued research.

Bekheta et al. (2008) found significant increases in all the studied vegetative and reproductive growth parameters as well as photosynthetic pigments (chlorophyll a, b and carotenoides) in *Gerbera jasmonii* when supplemented with Se (5–20 ppm). Filek et al. (2010) investigated the effect of Se (2 μM of Na_2SeO_4) influence on photosystem activity in Cd-stressed (400 μM $CdCl_2$) *Brassica napus* seedlings using fluorescence and electron paramagnetic resonance measurements. The seedlings were cultured till the first leaf reached about 1 cm in length. They found that Cd-induced alteration of the activity of both photosystems were partly diminished by Se. Electron microscopy photographs confirmed less degradation in chloroplasts of plants cultured on media supplemented with Se. It is suggested that higher amount of sugar accumulation in Cd stressed plants, may act as traps for free radicals produced under those conditions.

3. Beneficial Role of Selenium in Plant Abiotic Stress Tolerance

The role of Se as a trace element and its beneficial effect on plant growth and physiology has already been reported by many authors. Moreover, some plant species supplemented with

Se have shown enhanced resistance to certain abiotic stresses, e.g. drought, salinity, extreme temperature, toxic metals UV radiation etc.

3.1. Drought

Drought is one of the most devastating environmental stresses that affects the growth and development of plants. One of the earliest responses of plants to drought is the accumulation of ROS. However, there are a number of reports on the protective role of exogenous Se on drought stress in plant. It has been shown that Se has the ability to regulate the water status of plants under conditions of drought (Kuznetsov et al., 2003). They showed that the addition of 0.1 or 0.25 mM selenium caused a 2–6% increase in the content of water in leaves, increasing thereby the drought-resistance of plants. The Se-induced improvement of the water status of leaf tissues was accompanied by a sharp (2- to 4-fold) increase in the extent of inhibition of the stress-induced accumulation of proline and a significant inhibition of peroxidase activity (Kuznetsov et al., 2003). Their results support the suggestion that Se exerts an antioxidant effect directed towards a decreased concentration of intracellular ROS, which induce the biosynthesis of proline and peroxidase production de novo. They also indicated that the Se-induced effect on the content of intracellular water was less pronounced at the optimal rate of water supply.

In common buckwheat (*Fagopyrum esculentum*), Tadina et al. (2007) observed that the water deficit plants exhibited the significantly lower stomatal conductance (*gs*), while Se-supplementation had significantly improved *gs*. A significantly higher actual photochemical efficiency of PSII was obtained in Se-supplimented water-deficit plants, which was possibly due to improvement of plant water management during treatment. Yao et al. (2009) suggested that optimal Se supply is favorable for growth of drought-stressed wheat (*Triticum aestivum*) seedlings. The growth and physiological responses of seedlings were different, depending on the Se concentration. Use of higher (3.0 mg Se kg^{-1}) and lower amount used (0.5 mg Se kg^{-1}) did not significantly affect on biomass accumulation. Treatments with 1.0 and 2.0 mg Se kg^{-1} promoted biomass accumulation of wheat seedlings. Treatments with 1.0, 2.0 and 3.0 mg Se kg^{-1} significantly increased root activity, proline content, POD and CAT activities, carotenoids, chlorophyll, and reduced MDA contents of wheat seedlings. Lower Se treatment did not significantly affect the chlorophyll and MDA contents, although it also increased some antioxidant index (proline and carotene contents, POD and CAT activities. In contrary, Xiaoqin et al. (2009) observed that extra Se supply did not affect the activities of CAT and POD, while the combination of drought and Se significantly increased activities of POD and CAT, but the increase did not counteract the lipid peroxidation.

In an experiment, Valadabadi et al. (2010) observed that drought stress markedly reduced the physiological and growth indices viz. dry weight, leaf area index, relative growth rate and crop growth rate of *Brassica napus* cv. Opera. However, application of Se (30 g l^{-1}) could reduce the water deficit stress damages and resulted in higher values of those indices. Recently, Wang et al. (2011) examined the effect of Se (5 µM Na_2SeO_4) on the AsA–GSH cycle in *Trifolium repens* seedlings subjected to polyethylene glycol (PEG)-induced water deficit. They observed that compared to the control, H_2O_2, thiobarbituric acid reactive substances (TBARS), AsA, dehydroascorbate (DHA), and GSSG contents increased, whereas a constant content of GSH and decreases in AsA/DHA and GSH/GSSG ratios were observed in the presence of PEG. The activities of APX, DHAR and GR were upregulated, except for MDHAR activity during water deficit. On the other hand, Se application decreased the contents of H_2O_2, TBARS, DHA, and GSSG; increased the levels of GSH and AsA; and inhibited the decreases of AsA/DHA and GSH/GSSG ratio. Although it did not affect APX activity significantly, Se addition improved the activities of MDHAR, DHAR, and GR. Furthermore, GR activity showed the highest increase followed by that of DHAR and MDHAR in decreasing order. They concluded that regulation of AsA–GSH metabolism resulting from Se may have a positive effect on drought stress mitigation.

In our laboratory we studied the beneficial role of Se in plants under drought stress condition (Hasanuzzaman and Fujita, 2011). In this study, rapeseed (*Brassica napus*) seedlings were grown in Petri dishes. A set of 10-day-old seedlings was pretreated with 25 μM Se (Na_2SeO_4) for 48 h. Two levels of drought stress (10% and 20% PEG) were imposed separately as well as on Se-pretreated seedlings, which were grown for another 48 h. Drought stress, at any level, caused a significant increase in GSH and GSSG content; however, the AsA content increased only under mild stress (Table 1).

The activity of APX was not affected by drought stress. The MDHAR and GR activity increased only under mild stress (10% PEG). The activity of DHAR, GST and GPX activity significantly increased under any level of drought stress, while CAT activity decreased (Table 2).

A sharp increase in H_2O_2 and MDA content was induced by drought stress (Table 1).

On the other hand, Se-pretreated seedlings exposed to drought stress showed a rise in AsA and GSH content, and evidenced increased activities of CAT, APX, DHAR, MDHAR, GR, GST and GPX as compared with the drought-stressed plants without Se.

Selenium-supplimented seedlings showed a concomitant decrease in H_2O_2, and the level of lipid peroxidation (Table 1, 2). The results indicate that the exogenous application of Se increased the tolerance of the plants to drought-induced oxidative damage by enhancing their antioxidant defense and methylglyoxal detoxification systems.

Table 1. Malondialdehyde (MDA), H_2O_2 content, reduced ascorbate (AsA), reduced glutathione (GSH) and oxidized glutathione (GSSG) in rapeseed seedlings induced by selenium under drought stress conditions (Hasanuzzaman and Fujita, 2011)

Treatment	MDA content (nmol g^{-1} fresh weight)	H_2O_2 content (μmol g^{-1} fresh weight)	AsA content (nmol g^{-1} fresh weight)	GSH content (nmol g^{-1} fresh weight)	GSSG content (nmol g^{-1} fresh weight)
Control	21.16 c	4.44 c	4970.52 d	283.31 d	8.48 d
D_1	30.77 b	5.92 b	5923.34 bc	439.93 bc	18.26 c
D_2	42.06 a	7.36 a	5203.01 d	414.97 c	32.61 a
Se	19.87 c	4.69 c	5476.88 cd	269.73 d	9.84 d
Se+D_1	23.42 c	5.00 c	7006.01 a	514.72 a	17.61 c
Se+D_2	29.46 b	5.82 b	6295.85 b	467.83 b	21.52 b

D_1, D_2, Se, Se+D_1, and Se+D_2 indicates 10% PEG, 20% PEG, 25 μM Na_2SeO_4, 10% PEG+Na_2SeO_4, and 20% PEG+Na_2SeO_4 treatment, respectively. Values with different letters are significantly different at $P<0.05$ applying LSD test.

Table 2. Activities of antioxidant enzymes in rapeseed seedlings induced by selenium under drought stress conditions (Hasanuzzaman and Fujita, 2011)

Treatment	CAT activity (μmol min^{-1} mg^{-1} protein)	APX activity (μmol min^{-1} mg^{-1} protein)	MDHAR activity (nmol min^{-1} mg^{-1} protein)	DHAR activity (nmol min^{-1} mg^{-1} protein)	GR activity (nmol min^{-1} mg^{-1} protein)	GST activity (nmol min^{-1} mg^{-1} protein)	GPX activity (nmol min^{-1} mg^{-1} protein)
Control	24.98 a	0.344 bc	37.40 c	201.73 e	33.67 c	45.17 c	0.1177 d
D_1	18.79 b	0.366 b	41.09 b	233.38 cd	41.54 b	56.52 b	0.1390 c
D_2	16.02 b	0.307 c	34.42 c	261.90 bc	39.25 bc	59.24 b	0.1402 c
Se	25.52 a	0.343 bc	37.48 c	212.83 de	36.52 bc	49.98 c	0.1306 c
Se+D_1	25.32 a	0.432 a	45.30 a	270.46 b	54.48 a	67.39 a	0.1748 a
Se+D_2	23.68 a	0.453 a	44.66 a	314.15 a	51.89 a	69.73 a	0.1595 b

D_1, D_2, Se, Se+D_1, and Se+D_2 indicates 10% PEG, 20% PEG, Na_2SeO_4, 10% PEG+Na_2SeO_4, and 20% PEG+Na_2SeO_4 treatment, respectively. Values with different letters are significantly different at $P<0.05$ applying LSD test.

3.2. Salinity

Soil salinity is one of the most severe abiotic stresses that negatively affects crop production worldwide (Hasanuzzaman et al., 2012). A plenty of research results have shown the ability of Se to protect plants from salt-stress induced damages when applied at low concentration. The interaction of Se with soil salinity has been studied earlier by Terry et al. (2000). It is hardly surprising that sulfate salinity drastically inhibits plant uptake of selenate (Zayed et al., 1998). Not all plant species are affected to the same extent of sulfate salinity. In Se-accumulators, selenate is taken up preferentially over sulfate. Chloride salinity had much less effect on selenate uptake than sulfate salinity (Wu and Huang, 1991). Generally, there is a small decrease in shoot accumulation of Se with increasing salt levels (Bañuelos et al., 1996). Kong et al. (2005) reported that at low concentrations (1–5 μM), Se tended to stimulate the growth, the activities of SOD and POD, as well as the accumulation of water-soluble sugar in leaves of sorrel (*R. patientia* × *R. tianshanicus*) seedlings. However, at higher concentrations (10–30 μM), Se exerted diminished beneficial effects on growth and enzyme activities. Results revealed that SOD and POD activity of salt-stressed seedlings increased when exposed to concentrations ranging 1–5 μM Se. At concentrations between 10 and 30 μM, there were adverse effects on both enzymes compared with that at 5 μM Se. In cucumber leaves, Se treatments at 5 and 10 μM significantly improved the growth rate and increased the photosynthetic pigments and proline contents when subjected to salt stress (Hawrylak-Nowak, 2009). Additionally, Se enhanced the salt tolerance of seedlings by protecting the cell membrane against lipid peroxidation (Djanaguiraman et al., 2005). Khedr et al. (2003) reported that proline induces the expression of salt-stress-responsive proteins and improves the salt-tolerance in the desert plant *Pancratium maritimum*. Increase of proline content in Se-treated soybean plants has also been reported by Djanaguiraman et al. (2005). However, the mechanisms and the reasons for proline accumulation in Se-supplied plants have not been fully investigated.

In our recent study, we investigated the regulatory role of exogenous Se in the antioxidant defense systems in rapeseed seedlings exposed to salt stress (Hasanuzzaman et al., 2011). Twelve-day-old seedlings, grown in Petri dishes, were supplemented with Se (25 μM Na_2SeO_4) and salt (100 and 200 mM NaCl) separately and in combination, and further grown for 48 h. The AsA content of the seedlings decreased significantly with increased salt stress, while the amount of GSH and GSSG increased (Table 3). In addition, the APX and GST activity increased significantly with increased salt concentration (both at 100 and 200 mM NaCl), while GPX activity increased only at moderate salt stress (100 mM NaCl). Glutathione reductase activity remained unchanged at 100 mM NaCl, while it was decreased under severe (200 mM NaCl) salt stress (Table 4). The CAT, MDHAR and DHAR activities decreased upon the imposition of salt stress, whereas a sharp decrease of these activities was observed under severe salt stress (200 mM NaCl). Concomitant increases in the levels of H_2O_2 and MDA were also measured. However, further investigation revealed that Se treatment had a synergistic effect: in salt-stressed seedlings, it increased the AsA and GSH contents; and the activities of CAT, APX, MDHAR, DHAR, GR, GST and GPX. As a result, addition of Se in salt-stressed seedlings led to a reduction in the levels of H_2O_2 and MDA as compared to salt stress alone (Table 3, 4). Phenotypic appearance was also better in Se-treated seedlings (Figure 2). We therefore, concluded that the exogenous application of Se rendered the plants more tolerant to salt stress-induced oxidative damage by enhancing their antioxidant defense and MG detoxification systems.

Table 3. Malondialdehyde (MDA), H_2O_2 content, reduced ascorbate (AsA), reduced glutathione (GSH) and oxidized glutathione (GSSG) in rapeseed seedlings induced by selenium under salt stress conditions (Hasanuzzaman et al., 2011)

Treatment	MDA content (nmol g^{-1} fresh weight)	H_2O_2 content (μmol g^{-1} fresh weight)	AsA content (nmol g^{-1} fresh weight)	GSH content (nmol g^{-1} fresh weight)	GSSG content (nmol g^{-1} fresh weight)
Control	25.58 d	3.70 d	5210.56 a	251.45 e	7.17 d
Na_1	43.34 b	6.50 ab	4064.22 b	432.56 c	13.48 b
Na_2	58.68 a	7.02 a	3141.51 c	358.68 d	16.95 a
Se	24.22 d	3.95 d	4890.04 a	261.28 e	8.14 cd
Se+Na_1	34.71 c	5.25 b	4973.40 a	568.79 a	9.43 c
Se+Na_2	43.29 b	6.06 c	3980.97 b	479.86 b	12.06 b

Na_1, Na_2, Se, Se+Na_1, and Se+Na_2 indicates 100 mM NaCl, 200 mM NaCl, 25 μM Na_2SeO_4, 100 mM NaCl+Na_2SeO_4, 200 mM NaCl+Na_2SeO_4 treatment, respectively. Values with different letters are significantly different at $P<0.05$ applying LSD test.

Figure 2. Phenological appearance of rapeseed seedlings induced by selenium under salt stress conditions. Na_1, Na_2, Se, Se+Na_1 and Se+Na_2 indicates 100 mM NaCl, 200 mM NaCl, Na_2SeO_4, 100 mM NaCl+Na_2SeO_4, 200 mM NaCl+Na_2SeO_4 treatment, respectively.

3.3. Metal toxicity

Although some metal elements have importance as micronutrients, at higher concentrations they are toxic to plants and other organisms (Hasanuzzaman et al., 2012). In recent decades, several studies have been conducted on the beneficial role of Se on metal stress tolerance in plants.

Table 4. Activities of antioxidant enzymes in rapeseed seedlings induced by selenium under salt stress conditions (Hasanuzzaman et al., 2011)

Treatment	CAT activity (μmol min^{-1} mg^{-1} protein)	APX activity (μmol min^{-1} mg^{-1} protein)	MDHAR activity (nmol min^{-1} mg^{-1} protein)	DHAR activity (nmol min^{-1} mg^{-1} protein)	GR activity (nmol min^{-1} mg^{-1} protein)	GST activity (nmol min^{-1} mg^{-1} protein)	GPX activity (nmol min^{-1} mg^{-1} protein)
Control	28.52 a	0.387 d	47.08 b	153.36 b	26.41 b	23.85 d	0.1628 d
Na_1	17.49 cd	0.509 bc	42.29 b	126.68 c	29.86 b	37.28 bc	0.1915 bc
Na_2	15.78 d	0.535 b	33.65 c	99.65 d	21.53 c	33.38 c	0.1502 d
Se	25.10 b	0.453 c	46.59 b	139.38 bc	30.04 b	25.35 d	0.1667 cd
Se+Na_1	26.79 ab	0.656 a	57.00 a	171.56 a	39.77 a	44.67 a	0.2317 a
Se+Na_2	20.75 c	0.628 a	47.69 b	142.72 bc	30.56 b	39.09 b	0.2026 b

Na_1, Na_2, Se, Se+Na_1, and Se+Na_2 indicates 100 mM NaCl, 200 mM NaCl, 25 μM Na_2SeO_4, 100 mM NaCl+Na_2SeO_4, 200 mM NaCl+Na_2SeO_4 treatment, respectively. Values with different letters are significantly different at $P<0.05$ applying LSD test.

Vorobets (2006) reported that during heavy metal stress Se might prevent its toxic effect in plants and they showed that the size of GSH pool shows a marked alteration in response to a combined lead (Pb) and Se treatment. It has been suggested that the protective effects of Se are due to the formation of non toxic Se-metal complexes (Vorobets and Mykiyevich, 2000). Pedrero et al. (2008) observed that the proportion of α-tocopherol was similar in the control plants and in those supplied with Se separately or in combination with cadmium (Cd). However, the percentage of α-tocopherol concentration increased to the level found in control and Se-enriched plants when Se was added simultaneously with Cd. It has been reported that an increase of α-tocopherol favors the stress tolerance of plants as it favors the scavenging of singlet oxygen ($^1O_2$) in chloroplasts (Munné-Bosch and Alegre, 2002; Munné-Bosch, 2005). Therefore, the increase of α-tocopherol in plants exposed to Se and Cd simultaneously, in comparison to those grown only in Cd, shows evidence that Se assists the plants in the adaptation. Pedrero et al. (2008) showed that, when plants were exposed simultaneously to Se and Cd, the MDA level noticeably decreased to the level found in the control. In the plants supplied only with Se, the level of MDA was the lowest. These findings can be attributed to the antioxidative effect of Se reported in previous studies (Hartikainen et al., 2000; Djanaguiram et al., 2005).

Under heavy metal stress, there are some possible mechanisms by which Se confers tolerance to stress. Filek et al. (2008) observed a protective role of Se ions on the Cd stress in the changes in growth and fresh weight of *B. napus* seedlings. Their results suggest that the Se effect is linked mainly to a reduction of oxygen radicals that are produced in the presence of Cd. Non-enzymatic deactivation of oxygen radicals by Se can also be enhanced at the low (2 μM) Se concentration. In that study, Se-induced production of antioxidants was high enough to ameliorate the toxicity of Cd at concentration levels of 400–600 μM. In contrast, compared with Cd concentration, Se concentration seemed to be very low to remove Cd from proteins, especially at higher concentration levels. Recently, Sun et al. (2010) observed the detoxification action of Se on garlic (*Allium sativum*) growth under Cd stress. They also considered the mechanisms for the protective role of Se ions on the Cd stress which were: (i) removal of Cd from metabolically active cellular sites, (ii) induction of Se to scavenge the Cd induced ROS, and (iii) the regulation of Se to phytochelatin activity. The three actions mitigated the effects of Cd on *A. sativum*. In our laboratory we observed the pre-treatment of *B. napus* seedling with exogenous Se showed better performance in response to enzymatic and non-enzymatic antioxidant when subjected to short-term Cd stress (Hasanuzzaman et al., unpublished data). The phenotypes of the Se supplemented seedlings also superiors to the seedlings subjected to Cd-stress without Se (Figure 3). We also postulated that pre-treatment of seedlings with Se increase the initial H_2O_2 level which acts as a signal and thus modulate the ROS scavenging enzymes activities when subjected to stress.

Srivastava et al. (2009) observed that Se acted as an antioxidant, inhibiting lipid peroxidation (reduced by 26–42%) via increased levels of thiols and GSH (increased by 24%). The results suggest that Se is either an antioxidant or it activates plant protective mechanisms, thereby alleviating oxidative stress and improving arsenic (As) uptake in *Pteris vittata*. Selenium supplementation significantly inhibited lipid peroxidation in the fronds. Compared to the controls, addition of 5 and 10 μM Se reduced the TBARS concentrations in the fronds by 26–42% and 27–35%. There were no significant differences in TBARS between two application rates of Se (5 and 10 μM) or two exposure times (5 and 10 d).

Figure 3. Phenological appearance of *Brassica napus* leaves induced by Se under Cd stress. a. control; b. 0.5 mM $CdCl_2$, 48 h; c. Se pretreatment (100 μM Na_2SeO_2, 24 h) followed by treatment with 0.5 mM $CdCl_2$, 48 h.

Cartes et al. (2010) first suggested that at low concentration Se alleviated the Al-induced oxidative stress by means of two mechanisms: (i) the improvement of the spontaneous dismutation of $O_2{\cdot}^-$ to H_2O_2, as previously hypothesised by Hartikainen et al. (2000), and (ii) through the activation of POD enzyme, an H_2O_2 scavenger, in response to the increased disproportion of O_2^- which possibly occurred by effect of the applied Se. Interestingly, Se application up to 2 μM improved root growth and steadily decreased TBARS accumulation in plants treated with 0 and 0.2 mM Al. However, above 2 μM, Se induced stress in plants grown with or without Al. Ribeiro et al. (2011) observed that the growth of seedlings of Townsville sytlo (*Stylosanthes humilis* H.B.K.) is inhibited by Al^{3+}, while their elongation being recovered with Na_2SeO_4 at 1.0 μM. Methyl viologen and H_2O_2, ROS-generating compounds, also inhibited seedling elongation and again growth was relieved by selenate. Qi et al. (2004) and Peng et al. (2002) reported that Se enhanced oxidizing ability and resistance and reduced membrane lipid peroxidation of rice roots under iron (Fe) stress.

3.4. Extreme Temperature

Different global circulation models predict that greenhouse gases will gradually increase the world's average ambient temperature and lead to global warming (Meehl et al., 2007). High temperature or heat stress results from temperatures high enough to damage plant tissues, substantially influencing the growth and metabolism of plants. On the other hand, in most regions around the world, plants are exposed to low temperature at least part each year which is a serious threat to the sustainability of crop yield (Hasanuzzaman et al., 2012). There are few reports regarding the protective role of Se under extreme temperature. Recently,

Djanaguiraman et al. (2010) investigated the effects of Se foliar spray (75 mg l^{-1}) on leaf photosynthesis, membrane stability and antioxidant enzymes activity and grain yield and yield components of grain sorghum (*Sorghum bicolor*) plants grown under high temperature (HT) stress (40/30°C). They observed that HT stress decreased chlorophyll content, chlorophyll a fluorescence, photosynthetic rate and antioxidant enzyme activities and increased oxidant production and membrane damage. Decreased antioxidant defense under HT stress resulted in lower grain yield compared with control temperature (32/22°C). However, application of Se decreased membrane damage by enhancing antioxidant defense resulting in higher grain yield. The increase in antioxidant enzyme activities and decrease in ROS content by selenium was greater in HT than in control. Overall, Se application was significantly increased photosynthetic rate, stomatal conductance and transpiration rate by 13.2%, 12.4% and 8.11%, respectively, compared with the unsprayed control. In addition, foliar spray of Se significantly reduced O_2^- content, H_2O_2 content, MDA level and membrane injury by 11.5%, 35.4%, 28.4% and 17.6%, respectively, compared with unsprayed plants. Moreover, Se application increased CAT activity in both control and HT stress; however, the maximum increase was observed in HT stress. Across the days of observation, Se application was increased CAT and POX enzyme activity by 25.9% and 23.6%, respectively, under HT stress and 9.2% and 3.3%, respectively, under control temperature. As a result, Se spray was significantly increased filled seed weight and seed size by 26.3% and 10.7%, respectively, over the untreated controls (Djanaguiraman et al., 2010). It has already been established that Se increases resistance and antioxidant capacity of plants, whereas there has been little effort to understand the role of Se in plants under low temperature stress. In a recent study on wheat (*Triticum aestivum*) plant, Chu et al. (2010) reported the protective role of Se under low temperature. They observed that Se treatments at 1.0 mg kg^{-1} significantly reduced MDA content and the rate of O_2^- production in seedlings grown under cold stress, which indicated that suitable Se treatment reduced MDA and oxidative stress in seedlings subjected to stressful condition. Additionally, compared with the control, Se treatments significantly increased anthocyanins, flavonoids, and phenolic compound contents of seedlings subjected to low temperature stress (Chu et al., 2010) which was mainly due to the ability to scavenge ROS. The effects of different Se treatments on POD and CAT activities in seedlings exposed to cold stress were also reported (Chu et al., 2010). They showed a significant increase in activities of POD and CAT in Se treated wheat seedlings under low temperature. In another study, Hawrylak-Nowak et al. (2010) reported that under short-term chilling, the contents of chlorophylls and carotenoids contents in cucumber (*Cucumis sativus*) showed no significant change due to Se treatment. However, they observed that, compared with the control, the Se-treated plants showed an increase of proline content in leaves, once after chilling and again after 7 days of re-warming. Upon stress, the MDA content in the root of plants decreased directly when it was treated with 2.5–10 μM Se. This was in comparable with the plants grown without Se, whereas MDA level increased in roots and leaves of plants exposed to 20 μM Se. Seven days later, the MDA level in the root of plants grown in the presence of Se was still lower than those of plants not treated with Se and generally expressed no significant change in leaves. Nevertheless, Se at concentrations of 2.5–10 μM modified the physiological response of *C. sativus* short-term chilling stress, causing an increase in proline content in leaves and diminishing lipid peroxidation in roots, the resistance of plants to low temperature was not clearly enhanced (Hawrylak-Nowak et al., 2010).

3.5. UV-Radiation

In the past few decades there has been a depletion of the stratospheric ozone (O_3) layer due to emissions of halogen-containing compounds of anthropogenic origin which has resulted in a concomitant increase in solar ultraviolet-B (UV-B) radiation. Enhanced UV-B radiation alters transpiration and photosynthesis, respiration potential, growth, development and morphology of plants. Several reports indicated the protective role of Se against UV-B radiation.

Valkama et al. (2003) investigated the possible ameliorative effects of Se addition (0.1 and 1 mg kg^{-1}) soil on the detrimental effects of enhanced UV-B radiation on strawberry (*Fragaria* × *ananassa*) and barley (*H. vulgare*) plants. They observed several effects of UV and Se as well as their interaction mostly for strawberry, but not for barley. The results provided evidence that the high Se concentration in soil had no ameliorative effect but increased the sensitivity of strawberry to enhanced UV-B radiation in the field. Under ambient radiation, Se did not alter leaf growth of strawberry, whereas under UV-B radiation, the high Se addition significantly decreased leaf growth. Strawberry runner biomass was affected by the interaction of Se and UV. Under ambient radiation Se did not change dry weight of runners, but in combination with UV-A or UV-B radiation the high Se dosage decreased dry weight of runners by about 30%. Although the high Se concentration positively influenced on quantum efficiency of PSII in strawberry leaves, it reduced runner biomass, leaf number and ratio of starch to chloroplast area. This suggests that the harmful effects of the high Se dosage on photosynthetic processes occurred as a result of changes in activity or/and biosynthesis of enzymes, rather than alteration of PSII. At the low concentration, Se effects were slight and variable. Although barley leaves accumulated higher Se concentrations than strawberry, there were no apparent changes in their growth, biomass or chlorophyll fluorescence due to Se effect either alone or in combination with UV-B.

Yao et al. (2010a) reported the effects of selenium (Se) on growth and some physiological traits of roots in wheat (*T. aestivum*) seedlings exposed to enhanced UV-B radiation are reported. Responses of roots were different depending on the Se concentration. Compared with the control, root weight of wheat seedlings treated with 1.0 and 2.0 mg Se kg^{-1} soil increased by 39.47% and 16.28%, respectively. The lower amount Se (0.5 mg kg^{-1}) and the higher amount of Se treatments (3.0 mg kg^{-1}) did not significantly affect on root weight. Se treatments significantly increased root activity, flavonoids and proline content, and activities of POD and SOD in wheat roots exposed to enhanced UV-B. In addition, the treatments with 0.5, 1.0, and 2.0 mg Se kg^{-1} significantly reduced MDA content and the rate of O_2^- production of roots, whereas the higher amount Se treatment only induced a decrease in the rate of O_2^- production. The results of this study demonstrated that optimal Se supply promoted roots growth of wheat seedlings, and that optimal Se supply could reduce oxidative stress in wheat roots under enhanced UV-B radiation. Same group (Yao et al., 2010b) further reported the effects of Se supply on antioxidant traits of *T. aestivum* seedlings (above ground portion) exposed to enhanced UV-B stress. Antioxidant responses of seedlings were different depending on the Se concentration. Compared with the control, the lower amount used (0.5 mg Se kg^{-1} soil) had no significant effect on biomass accumulation. The treatments with 1.0, 2.0, and 3.0 mg Se kg^{-1} promoted biomass accumulation of wheat seedlings, and the increased amount in biomass was the most at 1.0 mg Se kg^{-1} treatment. Se treatments with 1.0, 2.0, and

3.0 mg kg^{-1} also significantly increased activities of POD and SOD and reduced the rate of O_2^- production and MDA content of wheat seedlings. In addition, anthocyanins and phenolic compounds content in wheat seedlings were evidently increased by the treatments with 1.0 and 2.0 mg Se kg^{-1}. The lower Se treatment had no significant effect on MDA content, although it increased activities of antioxidant enzymes (POD, SOD, and CAT activities) and reduced the rate of O_2^- production. These results suggest that optimal Se supply is favorable for the growth of wheat seedlings and that optimal Se supply can reduce oxidative stress of seedlings under enhanced UV-B radiation.

Conclusion

The facts of Se is intriguing, enigmatic and challenging (even capricious) for researchers. Although a number of report indicated the protective role of Se under abiotic stress condition, to date the research works conducted on the physiological role of Se under stressful condition is scarce. Controversy exists over the question of whether Se is an essential plant micronutrient. On a cautionary note, the appropriate concentration of exogenous Se is still a matter of research. Complete elucidation of the role of Se as well as detailed protective mechanisms would be helpful for developing stress tolerance in plants. Despite the widespread occurrence of Se deficiency globally, Se toxicity (selenosis) is a problem in some areas. Some soils and mineral deposits are naturally Se rich, and exploitation of these seleniferous soils or fossil fuels can lead to toxic accumulation of Se in the environment. Se contamination of sediments, soils and drainage water particularly occurs in arid seleniferous areas with intensive crop irrigation. Effective enrichment of agricultural crops with Se via soil using Se-enriched fertilizers can be challenging due to varying soil Se concentrations, soil types, soil redox potentials, soil pH, microbiological activity etc. Furthermore, the high cost of Se fertilizer in combination with a modest incorporation rate should be considered. However, the appropriate application of exogenous can be an effective protectant for plants to combat with abiotic stress.

References

Aslam, M., Harbit, K.B., and Huffaker, R.C. (1990). Comparative effects of selenite and selenate on nitrate assimilation in barley seeding. *Plant Cell and Environment*, 13, 773–782.

Bañuelos, G.S., Zayed, A., Terry, N., Wu, L. and Akohoue, S. (1996). Accumulation of selenium by different plant species grown under increasing sodium and calcium chloride salinity. *Plant and Soil*, 183, 49–59.

Bekheta, M.A., Abbas, S., El-Kobisy, O.S. and Mahgoub, M.H. (2008). Influence of selenium and paclobutrazole on growth, metabolic activities and anatomical characters of *Gerbera jasmonii* L. *Australian Journal of Basic and Applied Science*, 2, 1284–1297.

Brown, T.A. and Shrift, A. (1982). Selenium-toxicity and tolerance in higher plants. *Biological Reviews of the Cambridge Philosophical Society*, 57, 59–84.

Cartes, P., Gianfera, L. and Mora, M.L. (2005). Uptake of selenium and its antioxidative activity in ryegrass when applied a selenate and selenite forms. *Plant and Soil*, 276, 359–367.

Cartes, P., Jara, A.A., Pinilla, L., Rosas, A. and Mora, M.L. (2010). Selenium improves the antioxidant ability against aluminium-induced oxidative stress in ryegrass roots. *Annals of Applied Biology*, 156, 297–307.

Carvalho, K.M., Gallardo-Williams, M.T., Benson, R.F. and Martin, D.F. (2003). Effects of selenium supplementation on four agricultural crops. *Journal of Agricultural and Food Chemistry*, 51, 704–709.

Chen, C.C. and Sung, J.M. (2001). Priming bitter gourd seeds with selenium solution enhances germinability and antioxidative responses under sub-optimal temperature. *Physiologia Plantarum*, 111, 9–16.

Chu, J., Yao, X. and Zhang, Z. (2010). Responses of wheat seedlings to exogenous selenium supply under cold stress. *Biological Trace Element Research*, 136, 355–363.

Dhillon, K.S. and Dhillon, S.K. (2003). Distribution and management of seleniferous soils. *Advances in Agronomy*, 79, 119–184.

Djanaguiraman, M., Devi, D.D., Shanker, A.K., Sheeba, A. and Bangarusamy, U. (2005). Selenium -an antioxidative protectant in soybean during senescence. *Plant and Soil*, 272, 77–86.

Djanaguiraman, M., Prasad, P.V.V. and Seppanen, M. (2010). Selenium protects sorghum leaves from oxidative damage under high temperature stress by enhancing antioxidant defense system. *Plant Physiology and Biochemistry*, 48, 999–1007.

Ekelund, N.G.A. and Danilov, R.A. (2001). The influence of selenium on photosynthesis and "light-enhanced dark respiration" (LEDR) in the flagellate *Euglena gracilis* after exposure to ultraviolet radiation. *Aquatic Sciences*, 63, 457–465.

Ellis, D.R. and Salt, D.E. (2003). Plants, selenium and human health. *Current Opinion in Plant Biology*, 6:1–7.

Filek, M., Keskinen, R., Hartikainen, H., Szarejko, I., Janiak, A., Miszalski, Z. and Golda, A. (2008). The protective role of selenium in rape seedlings subjected to cadmium stress. *Journal of Plant Physiology*, 165, 833–844.

Filek, M., Kościelniak, J., Łabanowska, M., Bednarska, E. and Bidzińska, E. (2010) Selenium-induced protection of photosynthesis activity in rape (*Brassica napus*) seedlings subjected to cadmium stress. Fluorescence and EPR measurements. *Photosynthesis Research*, 105, 27–37.

Germ, M. and Osvald, J. (2005). Selenium treatment affected respiratory potential in *Eruca sativa*. *Acta Agriculturae Slovenica*, 85, 329–335.

Germ, M., Kreft, I. and Osvald, J. (2005). Influence of UV-B exclusion and selenium treatment on photochemical efficiency of photosystem II, yield and respiratory potential in pumpkins (*Cucurbita pepo* L.). *Plant Physiology and Biochemistry*, 43, 445–448.

Germ, M., Stibilj, V., Osvald, J. and Kreft, I. (2007). Effect of selenium foliar application on chicory (*Cichorium intybus* L.). *Journal of Agricultural and Food Chemistry*, 55, 795–798.

Gregorio, S.D. (2008). Selenium: A versatile trace element in life and environment. In: Prasad, M.N.V. (ed.) *Trace elements as contaminants and nutrients: Consequences in ecosystems and human health*. Wiley and Sons, Inc., Hoboken, New Jersey.

Hamilton, S.J. (2004). Review of selenium toxicity in the aquatic food chain. *Science of the Total Environment*, 326, 1–31.

Hartikainen, H. and Xue, T. (1999). The promotive effect of selenium on plant growth as triggered by ultraviolet irradiation. *Journal of Environmental Quality*, 28, 1372–1375.

Hartikainen, H., Ekholm, P., Piironen, V., Xue, T., Koivu, T. and Yli-Halla, M. (1997). Quality of the ryegrass and lettuce yields as affected by selenium fertilization. *Agricultural and Food Science in Finland*, 6, 381−387.

Hartikainen, H., Pietola, L. and Simojoki, A. (2001). Quantification of fine roots responses to selenium toxicity. *Agricultural and Food Science in Finland*, 11, 53−58.

Hartikainen, H., Xue, T. and Piironen, V. (2000). Selenium as an antioxidant and pro-oxidant in ryegrass. *Plant and Soil*, 225, 193–200.

Hasanuzzaman, M. and Fujita, M. (2011). Selenium pretreatment upregulates the antioxidant defense and methylglyoxal detoxification system and confers enhanced tolerance to drought stress in rapeseed seedlings. *Biological Trace Element Research*, DOI: 10.1007/s12011-011-8998-9.

Hasanuzzaman, M., Hossain M.A. and Fujita, M. (2010). Selenium in higher plants: Physiological role, antioxidant metabolism and abiotic stress tolerance. *Journal of Plant Sciences*, 5, 354–375.

Hasanuzzaman, M., Hossain, M.A. and Fujita, M. (2011). Selenium-induced up-regulation of the antioxidant defense and methylglyoxal detoxification system reduces salinity-induced damage in rapeseed seedlings. *Biological Trace Element Research*, DOI: 10.1007/s12011-011-8958-4.

Hasanuzzaman, M., Hossain, M.A., da Silva, J.A.T. and Fujita, M. (2012). Plant response and tolerance to abiotic oxidative stress: antioxidant defense is a key factor. In: Venkateshwarulu, B., Shanker, A.K., Shanker, C., Mandapaka, M. (eds.) *Crop stress and its management: Perspectives and strategies*. DOI: 10.1007/978-94-007-2220-0_8, Springer, Berlin.

Hawrylak, B., Matraszek, R. and Szymańska, M. (2007). Reaction of lettuce (*Lactuca sativa* L.) to selenium in nutrient solution contaminated with nickel. *Vegetable Crops Research Bulletin*, 67, 63–70.

Hawrylak-Nowak, B. (2009). Beneficial effects of exogenous selenium in cucumber seedlings subjected to salt stress. *Biological Trace Element Research*, 132, 259–269.

Hawrylak-Nowak, B., Matraszek, R. and Szymańska, M. (2010). Selenium modifies the effect of short-term chilling stress on cucumber plants. *Biological Trace Element Research*, 138:307–315.

Hu, Q.H., Xu, J. and Pang, G.X. (2003). Effect of selenium on the yield and quality of green tea leaves harvested in early spring. *Journal of Agricultural and Food Chemistry*, 51, 3379–3381.

Khedr, A.H.A., Abbas, M.A., Wahid, A.A.A., Quick, W.P. and Abogadallah, G.M. (2003). Proline induces the expression of salt-stress-responsive proteins and may improve the adaptation of *Pancratium maritimum* L. to salt-stress. *Journal of Experimental Botany*, 54, 2553–2562.

Kong, L., Wang, M. and Bi, D. (2005). Selenium modulates the activities of antioxidant enzymes, osmotic homeostasis and promotes the growth of sorrel seedlings under salt stress. *Plant Growth Regulation*, 45, 155−163.

Konze, J.R., Schilling, N. and Kende, H. (1978). Enhancement of ethylene formation by selenoamino acids. *Plant Physiology*, 62, 397-401.

Kuznetsov, V.V., Kholodova, V.P., Kuznetsov, V.V. and Yagodin, B.A. (2003). Selenium regulates the water status of plants exposed to drought. *Doklady Biological Sciences*, 390, 266–268.

Lyons, G.H., Stangoulis, J.C.R. and Graham, R.D. (2005). Tolerance of wheat (*Triticum aestivum* L.) to high soil and solution selenium levels. *Plant and Soil*, 270, 179−188.

MacLeod, J.A., Gupta, U.C., Milburn, P. and Sanderson, J.B. (1998). Selenium concentration in plant material, drainage and surface water as influenced by Se applied to barley foliage in a barley red clover potato rotation. *Canadian Journal of Soil Science*, 78, 685–688.

Martin-Tanguy, J. (2001). Metabolism and function of polyamines in plants: recent development (new approaches). *Plant Growth Regulation*, 34, 135−148.

Meehl, G.A., Stocker, T.F., Collins, W.D., Friedlingstein, P., Gaye, A.T., Gregory, J.M., Kitoh, A., Knutti, R., Murphy, J.M., Noda, A., Raper, S.C.B., Watterson, I.G., Weaver, A.J., and Zhao, Z.C. (2007) Global Climate Projections. In: Solomon, S., Qin, D., Manning, M., Chen, Z., Marquis, M., Averyt, K.B., Tignor, M., Miller, H.L. (eds.) *Climate Change 2007*: The Physical Science Basis. Contribution of Working Group I to the Fourth Assessment Report of the Intergovernmental Panel on Climate Change, pp. 749–845, Cambridge University Press, Cambridge, New York.

Munné-Bosch, S. and Alegre, L. (2002). The functions of tocopherols and tocotrienols in plants. *Critical Reviews in Plant Sciences*, 21, 31–57.

Munné-Bosch, S. (2005). The role of α-tocopherol in plant stress tolerance. *Journal of Plant Physiology*, 162, 743–748.

Munshi, C.B. and Mondy, N.I. (1992). Glycoalkaloid and nitrate content of potatoes as affected by method of selenium application. *Biological Trace Element Research*, 33, 121–127.

Padmaja, K., Prasad, D.K.K. and Prasad, A.R.K. (1989). Effect of selenium on chlorophyll biosynthesis in mung bean seedling. *Phytochemistry*, 28, 3321−3324.

Parker, D.R. and Page, A.L. (1994). Vegetation management strategies for remediation of selenium-contaminated soils. In: Frankenberger, W.T. Jr. and Benson, S. (eds.) *Selenium in the Environment*, pp. 327–341, Marcel Dekker, New York.

Pedrero, Z., Madrid, Y., Hartikainen, H. and Cámara, C. (2008). Protective effect of selenium in broccoli (*Brassica oleracea*) plants subjected to cadmium exposure. *Journal of Agricultural and Food Chemistry*, 56, 266–271.

Peng, X.L., Liu, Y.Y. and Luo, S.G. (2002). Effects of selenium on lipid peroxidation and oxidizing ability of rice roots under ferrous stress. *Journal of Northeast Agricultural University*, 19: 9–15.

Pennanen, A., Xue, T. and Hartikainen, H. (2002). Protective role of selenium in plant subjected to severe UV irradiation stress. *Journal of Applied Botany*, 76, 66–76.

Qi, X., Liu, Y.Y. and Song, T.X. (2004). Effects of selenium on root oxidizing ability and yield of rice under ferrous stress. *Journal of Northeast Agricultural University*, 11, 19–22.

Rani, N., Dhillon, K.S. and Dhillon, S.K. (2005). Critical levels of selenium in different crops grown in an alkaline silty loam soil treated with selenite-Se. *Plant and Soil*, 277, 367–374.

Ribeiro, D.M., Mapeli, A.M., Antunes, W.C. and Barros, R.S. (2011). A dual role of selenium in the growth control of seedlings of *Stylosanthes humilis*. *Agricultural Sciences*, 2, 78–85.

Seppänen, M., Turakainen, M. and Hartikainen, H. (2003). Selenium effects on oxidative stress in potato. *Plant Science*, 165, 311–319.

Shanker, A.K. (2006). Countering UV-B stress in plants: does selenium have a role? *Plant and Soil*, 282, 21–26.

Simojoki, A., Xue, T., Likkari, K., Pennanen A. and Hartikainen, H. (2003). Allocation of added selenium in lettuce and its impact on root. *Agricultural and Food Science in Finland*, 12, 155–164.

Singh, M., Singh, H. and Bhandari, D.K. (1980). Interaction of selenium and sulphur on the growth and chemical composition of raya. *Soil Science*, 129, 238–244.

Smrkolj, P., Germ, M., Kreft, I. and Stibilj, V. (2006). Respiratory potential and Se compounds in pea (*Pisum sativum* L.) plants grown from Se-enriched seeds. *Journal of Experimental Botany*, 57, 3595–3600.

Sors, T.G., Ellis, D.R. and Salt, D.E. (2005). Selenium uptake, translocation, assimilation and metabolic fate in plants. *Photosynthesis Research*, 86, 373–389.

Srivastava, M., Maa, L.Q., Rathinasabapathib, B. and Srivastava, P. (2009). Effects of selenium on arsenic uptake in arsenic hyperaccumulator *Pteris vittata* L. *Bioresource Technology*, 100, 1115–1121.

Sun, H.W., Ha, J., Liang, S.X. and Kang, W.J. (2010). Protective role of selenium on garlic growth under cadmium stress. *Communication in Soil Science and Plant Analysis*, 41, 1195–1204.

Tadina, N., Germ, M., Kreft, I., Breznik, B. and Gaberščik, A. (2007). Effects of water deficit and selenium on common buckwheat (*Fagopyrum esculentum* Moench.) plants. *Photosynthetica*, 45, 472–476.

Terry, N., Zayed, A.M., de Souza, M.P. and Tarun, A.S. (2000). Selenium in higher plants. *Annual Review of Plant Physiology and Plant Molecular Biology*, 51, 401–432.

Turakainen, M. (2007). Selenium and its effects on growth, yield and tuber quality in potato. University of Helsinki, Department of Applied Biology, Publication No. 30. Helsinki, pp-50.

Turakainen, M., Hartikainen, H. and Seppänen, M.M. (2004). Effects of selenium treatments on potato (*Solanum tuberosum* L.) growth and concentrations of soluble sugars and starch. *Journal of Agricultural and Food Chemistry*, 52, 5378–5382.

Valadabadi, S.A., Shiranirad, A.H. and Farahani, H.A. (2010) Ecophysiological influences of zeolite and selenium on water deficit stress tolerance in different rapeseed cultivars. *Journal of Ecology and Natural Environments*, 2, 154–159.

Valkama, E., Kivimäenpää, M., Hartikainen, H. and Wulff, A. (2003). The combined effects of enhanced UV-B radiation and selenium on growth, chlorophyll fluorescence and ultrastructure in strawberry (*Fragaria* × *ananassa*) and barley (*Hordeum vulgare*) treated in the field. *Agricultural and Forest Meteorology*, 120, 267–278.

Vorobets, N. and Mykiyevich, I. (2000). Single and combined effects of lead and selenium on sunflower seedlings. *Scientific Workshop on Horticulture and Vegetable Growing*, 19, 390.

Vorobets, N. (2006). Glutatione peroxidase activity in sunflower shoots exposed to lead and selenium. *Annals of University of Mariae Curie - SK Łodowska Lublin*, XIX: 151–154.

Wang, C.-Q., Xu, H.-J. and Liu, T. (2011). Effect of Selenium on Ascorbate–Glutathione metabolism during PEG-induced water deficit in *Trifolium repens* L. *Journal of Plant Growth Regulators*, DOI 10.1007/s00344-011-9206-z.

White, P.J., Bowen, H.C., Parmaguru, P., Fritz, M., Spracklen, W.P., Spidy, R.E., Meachamn, M.C., Mead, A., Harriman, M., Trueman, L.J., Smith, B.M., Thomas, B. and Broadley, M.R. (2004). Interactions between selenium and sulphur nutrition in *Astragalus thaliana. Journal of Experimental Botany*, 55: 1927−1937.

Wu, L. and Huang, Z.Z. (1991). Chloride and sulfate salinity effects on selenium accumulation by tall fescue. *Crop Science*, 31:114–18.

Xiaoqin, Y., Jianzhou, C. and Guangyin, W. (2009). Effects of drought stress and selenium supply on growth and physiological characteristics of wheat seedlings. *Acta Physiologiae Plantarum*, 31, 1031–1036.

Xue, T. and Hartikainen, H. (2000). Association of antioxidative enzymes with synergistic effect of selenium and UV irradiation in enhancing plant growth. *Agricultural and Food Science in Finland*, 9:177−186.

Xue, T., Hartikainen, H. and Piironen, V. (2001). Antioxidative and growth-promoting effect of selenium in senescing lettuce. *Plant and Soil*, 27:55−61.

Yao, X., Chu, J. and Ba, C. (2010a). Responses of wheat roots to exogenous selenium supply under enhanced ultraviolet-B. *Biological Trace Element Research*, 137:244–252.

Yao, X., Chu, J. and Wang, G. (2009). Effects of selenium on wheat seedlings under drought stress. *Biological Trace Element Research*, 130, 283–290.

Yao, X.Q., Chu, J.Z. and Ba, C.J. (2010b). Antioxidant responses of wheat seedlings to exogenous selenium supply under enhanced ultraviolet-B. *Biological Trace Element Research*, 136, 96–105.

Zayed, A., Lytle, C.M. and Terry, N. (1998). Accumulation and volatilization of different chemical species of selenium by plants. *Planta*, 206, 284–292.

Zayed, A., Pilon-Smits, E., de Souza, M., Lin, Z.Q. and Terry, N. (2000). Remediation of selenium polluted soils and waters by phytovolatilization. In: Terry, N.N. and Bañuelos, G, (eds.). *Phytoremediation of Contaminated Soil and Water*, pp. 61–83, Lewis Publishers, Boca Raton, FL.

In: Selenium: Sources, Functions and Health Effects
Editors: Chinatsu Aomori and Megumi Hokkaido
ISBN: 978-1-61942-061-8

Chapter 5

Selenium Supplementation of Diet Affects Antioxidant Activity and Protects against Antibiotic Resistance and Avian Influenza in Poultry

Hirotada Tsujii
Shinshu University Faculty of Agriculture, Japan

Abstract

The worldwide increase in the use of antibiotics as an integral part of the poultry and livestock production industry to treat and prevent infectious bacterial diseases and as growth promoters at subtherapeutic levels in feed has led to the problem of the development of bacterial antibiotic resistance during recent years. These uses promote the selection of antibiotic resistance in bacterial populations. The resistant bacteria from agricultural environments may be transmitted to humans, in whom they cause disease that cannot be treated by conventional antibiotics. Poultry that are given antibiotics often carry antibiotic-resistant strains of *Salmonella* or antibiotic-resistant transposons, which eventually reach humans through poultry meat, eggs and other food. On the other hand, avian influenza is an infection caused by influenza viruses that occur naturally in birds. The virus that causes this infection in birds can change (mutate) to infect humans. Selenoenzyme glutathione blocks the replication of *Salmonella* and influenza virus. Glutathione plays an important role in regulating viral replication and infectivity. Given the current interest in the use of safe alternatives to antibiotics, natural antioxidants including selenium and vitamin E are important for animal health. Regarding nutritional supplementation, selenium and vitamin E can decrease the effects of stress and infection on feed intake and body weight gain in chicken. Nutrition has some effects on both innate and cellular immunities if the feed is severely restricted.

Keywords: antibiotic resistance, antioxidant, avian influenza, poultry, selenium

1. INTRODUCTION

The meat and eggs of poultry provide the protein necessary for the body to build and maintain all its structural components. Protein is also required to manufacture the body's hormones, enzymes, and antibodies. This food group also supplies many essential vitamins and minerals required by the body. The global poultry industry is expanding as the human population increases. Per capita consumption is also increasing.

As consumption of poultry products increases, the microbial safety of these types of food becomes more important for consumers, producers, and public health officials.

Chicken meat and egg producers have fed growth-promoting antibiotics to food animals for years.

It is increasingly possible that poultry meat purchased at supermarkets has been contaminated by multiple-antibiotic-resistant superbugs of *Salmonella* and *E. coli.*

Antibiotics are used in human and veterinary medicine to eliminate microorganisms causing infections, such as bacteria. In food-producing animals, the antibiotics used to treat various infectious diseases may be the same or similar to those used for humans.

Highly pathogenic avian influenza A (H5N1) viruses are entrenched among poultry in parts of Asia, Africa, and perhaps the Middle East. The occurrence of human influenza A (H5N1) in Southeast Asia has paralleled large outbreaks of avian influenza A (H5N1) (Pandemic Influenza 2011).

2. ANTIBIOTIC-RESISTANCE

Salmonella of Chicken Meat and Egg

Of the various kinds of *Salmonella*, Enteritidis, the most common type, has now mutated into a strain that occurs more frequently in both eggs and chicken meat. However, in the past few years, the bacterium has been observed even inside eggs that have been disinfected and intact. Experts suggest that a new strain of *Salmonella* Enteritidis can develop inside the ovaries of hen, spreading the bacteria to the egg even prior to the shell being formed (The Poultry Site Latest News 2006).

Enterococci are not the most important pathogens for animals; in humans, however, they have been implicated in infective endocarditis and urinary tract infections for nearly a century (*Murray* 1990). *Salmonella* is one of the most important pathogens responsible for human food poisoning in the developed world (Cerro et al. 2002) and chicken products are widely acknowledged to be a significant reservoir for *Salmonella*. They have frequently been identified as a source of *Salmonella* contamination and consequently thought to be major sources of the pathogen in humans (Baeumler et al. 2000). In spite of the improvement in hygiene, food processing, education of food handlers, and information to the consumers, foodborne diseases still dominate as the most important public health problem in most countries (Domínguez et al. 2002).

Antibiotics in Chicken Feed

Penicillin was discovered by Scottish scientist and Nobel laureate Alexander Fleming in 1928. This serendipitous finding began the modern era of antibiotic discovery. In 1942, the first U.S. patient with streptococcal infection was miraculously cured with a small dose of penicillin. In veterinary practice, antibiotics are used in livestock production, disease prevention, and as growth-promoting feed additives (Swartz 2002, Sun 1984). The average growth improvement was estimated to be between 4 and 8%, and feed utilization was improved by 2 to 5% (Ewing and Cole 1994). In the poultry industry, common drug additives to chicken feed are the antibiotics amprolium, penicillin, and chlortetracycline.

Sixty years later, penicillin-resistant Streptococcus is widespread. Recently, scientists have discovered strains of bacteria in meat that are resistant to antibiotics used in animal feed. The seriousness of this finding has been underscored by a case report of a man becoming paralyzed from an infection acquired by eating contaminated chicken. The bacteria that caused the illness are known to have developed resistance to antibiotics. The use of antibiotics in animals disrupts the normal flora of the intestine, resulting in the emergence of antibiotic-resistant salmonellae and their prolonged fecal shedding into the environment (Threlfall 2002, Helms et al. 2003).

In the United States, 240 million doses of antibiotics are prescribed every year, almost one per person. Altogether, medical use accounts for 60 percent of antibiotic use. The other 40 percent is used in livestock feed to promote rapid growth. By 1980, 75 percent of all cattle in the United States received antibiotics, 90 percent of swine and veal calves, 50 percent of sheep, and nearly 100 percent of chickens and poultry. The drugs were used not only to prevent infection but also to fatten up the animals and ensure maximum growth, and thus profits (Macrobiotic Articles 2009).

Many public health advocates say that the use of antibiotics in poultry causes disease germs to become resistant not only to those drugs but also to the closely related drugs used to treat human diseases. The theory is that stronger, more drug-resistant strains of bacteria grow when competing organisms are killed off. Strong resistance to a drug may cause it and others in its chemical class to become ineffective for treating some diseases. The use of antibiotics in animals disrupts the normal flora of the intestine, resulting in the emergence of antibiotic-resistant *Salmonella* and its prolonged fecal shedding into the environment (Threlfall 2002, Helms et al. 2003). The fatality rate in people infected with antibiotic-resistant *Salmonella* is 21 times greater than that for those infected with non-antibiotic-resistant *Salmonella* strains (Altekruse et al. 1999).

Agricultural Antibiotic Use

Antimicrobial agents are also still used for growth promotion in animal husbandry in many countries (Aarestrup 1999, Schwarz et al. 2001). The appearance of antibiotic resistance (AR) has been directly linked to the use and overuse of antibiotics (McGowan 1983, Tenover and McGowan 1996, Shlaes et al. 1997, Mellon and Fondriest 2001). AR bacteria have been found in farm animals where antibiotics are heavily used (Aarestrup 1995, 1998, 1999), in associated food products (Chadwick et al. 1996, Bates 1994), in environments contaminated by animal waste (Chee-Sanford et al. 2001, Linton 1988), and in farm workers (van den Bogaard 1997, Simonsen et al. 1998, Levy et al. 1976).

In general, when an antibiotic is applied in poultry farming, the drug eliminates the sensitive bacterial strains, leaving behind or selecting those variants with unusual traits that can resist it. These resistant bacteria then multiply, increasing their numbers a million-fold a day, becoming the predominant microorganism in the population. Such bacteria transmit their genetically defined resistance characteristics to subsequent progeny of the strains and to other bacterial species via mutation or plasmids (Gould 2008).

Smith et al. (2002) described that agricultural antibiotic use (AAU) may cause AR bacterial infections in humans by two different processes. First, AAU increases the frequency of AR in zoonotic pathogens such as Campylobacter or Salmonella. These pathogens are typically acquired through exposure to contaminated animal food products. Therefore, the incidence of AR in zoonotic infections of humans is directly related to the prevalence of AR bacteria in food animals. Second, AR bacteria from food animals may facilitate the development of AR in human commensal bacteria, which ordinarily colonize humans without causing infection. Commensal bacteria typically have long persistence times, frequently undergo human-to-human transmission, and have high bacterial loads that are associated with good health but not disease.

Antimicrobial-Resistant *Salmonella*

Foodborne diseases caused by non-typhoid salmonella represent an important public health problem worldwide. Intestinal salmonellosis typically resolves in five to seven days and does not require treatment with antibiotics. However, bacteremia occurs in 3 to 10 percent of reported, culture-confirmed cases and is particularly common among patients at the extremes of age and those who are immuno compromised (White et al. 2001). The emergence of antimicrobial-resistant *Salmonella* is associated with the use of antibiotics in animals raised for food; resistant bacteria can be transmitted to humans through food, particularly that of animal origin.

The ability of bacteria to acquire AR genes and subsequently spread them to many different bacterial species is well known (Hall et al. 1997). A multistate outbreak of *Salmonella* occurred in the United States in 2007, which was associated with frozen turkey pot pies. This outbreak involved 401 cases in 41 states, with 32% resulting in hospitalization (CDC, 2008). In Europe, an international outbreak of *Salmonella* occurred in 2008, which was associated with meat products including chicken in various forms. This outbreak involved at least 119 cases, which occurred in the United Kingdom, Ireland, and Finland (O'Flanagan et al., 2008).

Multidrug Resistance

White et al. (2001) identified, in the isolates of *S. enteric,* serotypes Typhimurium DT104 and DT104b, which were identical to those characterized in strains of Typhimurium DT104 found in many other outbreaks. The ability of bacteria to acquire AR genes and subsequently spread them to many different bacterial species is well known (Hall et al. 1997). The emergence of multidrug-resistant serotypes, especially *Salmonella* Typhimurium definitive phage type (DT)104, has become a potential problem in animal husbandry and human medicine. S. Typhimurium DT104 initially emerged among cattle in England and Wales, but was later isolated from poultry, sheep, swine, and humans in the European continent and the USA (*Besser et al. 2000, Imberecht et al. 1998, Baggesen et al. 2000, Poppe et al. 1998, Leegaard et al. 2000*). In the 1990s, the frequency of isolation of *S.* Typhimurium DT104 in cattle worldwide increased, and has been detected in Japan (*Sameshima et al. 2000*), where it has also spread. Yoshimura et al. (2000) reported that *S. dublin* isolated in Japan during 1992–1995 showed a high frequency of resistance to ampicillin, dihydrostreptomycin, nalidixic acid, and oxytetracycline.

Serotype Typhimurium

Serotype Typhimurium DT104 has been identified in numerous types of food, including beef, pork, poultry, and unpasteurized dairy products, and its presence is thought to be due to the widespread clonal dissemination of multidrug-resistant isolates (Glynn *et al.* 1998, Baggesen *et al.* 2000, Davis *et al. 1999*). S. Typhimurium and S. Infantis are serotypes frequently isolated from food-producing animals and food poisoning cases in Japan (Esaki et al. 2003). S. Typhimurium DT104 and 104B were isolated from feces and diagnostic submissions mainly from cattle; some were also isolated from diagnostic submissions from swine. With the AR genes integrated in the chromosome, most DT104 isolates show MDR to five drugs, commonly referred to as resistance (R)-type ACSSuT (Threlfall et al. 1994). Esaki et al. (2004) described that Infantis was frequently isolated from patients suffering from foodborne illness, and presents a significant public health concern related to poultry possessing strains with some resistance determinants.

Integrons, one type of mobile DNA element, have been associated with the transfer of resistance and often contain one or more linked antimicrobial-resistance gene (Hall and Stokes 1993). Integrons are therefore particularly important, since a strong selection pressure exerted by antibiotics can potentially result in the mobilization and dissemination of linked multidrug-resistance phenotypes.

3. Selenium Deficiency and Avian Influenza

Se-Deficiency in the Soil

The selenium (Se) content in food (plants, and indirectly animals, since these feed on the plants) depends on the level of the nutrient in the soil. Soils in some parts of the world, such as China, Russia, and New Zealand, have very low levels of Se. In the United States, soil is Se-deficient in parts of the Pacific Northwest, from the Great Lakes region to the New England states, and along the Eastern Seaboard into Florida. Soil is considered Se-deficient when there is less than 0.5 mg of Se per kg of soil. There are varius areas around the world where the soil is deficient in Se; in those areas, it was discovered that the lack of Se was related to autoimmune diseases (Laz and Feher 2009, Toulis et al. 2010, McKenzie et al. 2006), a form of heart disease called Keshan disease (Neve 1996), and a joint disease called Kashin-Beck disease (Combs 2001).

Recently, a renaissance in Se research has found low Se levels to be a significant risk for several degenerative diseases.

Avian Influenza

Influenza is one of the oldest and most common infections and now new fears are spreading about the dangers of a new pandemic of avian influenza (AI). AI was first recorded in Italy in 1878, (Perroncito 1878). The disease, originally known as fowl plague, frequently caused massive outbreaks in poultry, including two outbreaks in the United States (1924 and 1929) (Mohler1926). In 1955, Schäfer discovered that the virus causing fowl plague is an influenza virus. The highly pathogenic H5N1 is descended from a strain that first appeared in Scotland in 1959 (Pereira et al. 1965). The H5N1 virus subtype, a highly pathogenic AI virus, first infected humans in 1997 during a poultry outbreak in Hong Kong SAR, China. Since its widespread re-emergence in 2003 and 2004, this avian virus has spread from Asia to Europe and Africa and has become entrenched in poultry in various countries, resulting in millions of poultry infections, several hundred human cases, and many human deaths (WHO: Avian Influenza, 2011).

Avian Influenza and Se Deficiency

Many novel virues have been found in Se-deficient populations. Yuan et al. (2006) reported on the relationship between the avian influenza's internal factors and Se deficiency. Avian influenza broke out in Italy, a Se-deficient country in Europe, followed by Spain, Russia, and Holland, which are also deficient in Se. The earliest, largest, and five most wide spread instances avian influenza outbreak all occurred in Se-deficient areas, and the areas where avian influenza occurred in 2005 are also short of Se (Yuan et al. 2006). Jaspers et al. (2007) described that Se deficiency has a significant impact on the morphology and influenza-induced host defense responses in human airway epithelial cells. The spread of avian influenza over the world is related to Se deficiency: over 40 countries and regions are deficient in Se, including parts of Europe, USA, Canada and Asia, China is seriously deficient in Se (Yuan et al. 2006).

Mutations

In recent history, China has experienced a human epidemic of Keshan disease, which manifests as cough, difficult breathing, pulmonary edema, and eventually death by heart failure. There was some controversy over whether the disease is caused by Coxsackie virus, which is normally quite benign, or by a deficiency of Se. Recent studies have also demonstrated a possible etiological role of enterovirus infection in a particular form of heart muscle disease, selenium deficiency-related endemic cardiomyopathy (Keshan disease) (Li et al. 1995, 2000). Coxsackie virus B3 RNA can survive and replicate in heart muscle in Keshan disease, which may play an important role in the occurrence of Keshan disease. The possible mechanism of occurrence of Keshan disease is associated with a point mutation in the Coxsackie virus B3 genome (Ren et al. 2004).

Beck et al. (2001) reported that sequencing of viral isolates recovered from selenium-deficient mice revealed mutations in the viral genome of both Coxsackie virus and influenza viruses, which are associated with their increased pathogenesis. Furthermore, the antioxidant, selenoenzyme glutathione peroxidase-1, was found to be critically important, as glutathione peroxidase knockout mice developed myocarditis, similar to the Se-deficient mice, when infected with a benign strain that causes myocarditis. When they recovered the virus from the mice, they found an increased number of viral mutations, resulting in a more virulent phenotype. Therefore, host deficiency in Se leads to increased viral mutations in the influenza virus genome, resulting in a more virulent phenotype (Beck. et al. 2004). Akaike et al. (2000) reported increased rates of mutation of an RNA (Sendai) virus that had been exposed to reactive nitrogen species (RNS), such as nitric oxide (NO) and peroxynitrite ($ONOO^-$). Both NO and O^-_2 have been shown to increase the pathogenesis of influenza virus infection in laboratory experiments (Akaike et al. 1996, 2003, Maeda and Akaike 1991).

4. Substitute for Antibiotics

Reduce the Resistant Salmonella

Most infections with antimicrobial-resistant salmonella are acquired by eating contaminated food of animal origin (Angulo et al. 2000, Fey et al. 2000). The routine practice of giving antimicrobial agents to domestic livestock as a means of preventing and treating diseases, as well as promoting growth, is an important factor in the emergence of antibiotic-resistant bacteria that are subsequently transferred to humans through the food chain (Tollefson et al. 1997, Witte 1998).

Efforts are needed to reduce the prevalence of resistant salmonella in food, including the adoption of guidelines for the prudent use of antimicrobial agents in animals used for food, the passage of new food-safety regulations, and a reduction in the number of pathogens present on farms and in slaughterhouses.

Probiotics

The use of the term 'probiotic' to describe food supplements specifically designed to improve health, however, dates from 1974 when Parker used it to describe growth-promoting animal feed supplements. He defined the term as 'organisms and substances which contribute to intestinal microbial balance' (Fuller 1990).

The claims made for microbial feed supplementation of farm animals mainly include improvement of growth rate and feed utilization, disease resistance, and reduction of gut shedding of enteropathogenic bacteria. The use of probiotics for control of food-borne pathogens, such as *Salmonella*, and as an alternative to antimicrobial growth promoters in poultry production has recently drawn interest.

Probiotics are defined as live microorganisms, which, when administered in adequate amounts, confer a health benefit on the host through improvements to the intestinal microbial balance (FAO/WHO, 2009).

Lactic Acid Bacteria as Probiotics

Lactic acid bacteria and bifidobacteria are the most common types of microbes used as probiotics. Ljungh and Wadström (2006) described that lactic acid bacterial strains commonly produce antimicrobial substance(s) with activity against the homologous strain, but lactic acid bacterial strains also often produce microbicidal substances with effects against gastric and intestinal pathogens and other microbes, or compete for cell surface and mucin binding sites. This could be the mechanism behind reports that some probiotic strains inhibit or decrease translocation of bacteria from the gut to the liver. A protective effect against cancer development can be ascribed to binding of mutagens by intestinal bacteria, reduction of the enzymes beta-glucuronidase and beta-glucosidase, and deconjugation of bile acids, or merely to enhancement of the immune system of the host.

Oral dosing of poultry with native gut microorganisms to prevent infection with salmonellae was first reported by Nurmi and Rantala (1973). Although *Lactobacillus* established in chickens as monoassociates will not protect against colonization by *Salmonella*, its omission from the protective microflora results in decreased resistance (Impey et al. 1982). Prophylactic inoculation of germ-free chicks with *Lactobacillus acidophilus* was shown to reduce shedding of pathogenic *Escherichia coli* from 100 to 47% (Watkins et al. 1982). Moreover, an increased weight gain in chicks was detected after oral doses of *Lact. acidophilus* (Tortuero 1973).

Tellez et al. (2001) described that a combination of *Lactobacillus acidophilus*, *Streptococcus faecium*, and *Salmonella Enteritidis, Salmonella Typhimurium*, and *Salmonella* Heidelberg-specific antibodies have a beneficial effect in reducing the colonization of *Salmonella Enteritidis* in market-aged broilers.

A variety of microbial species have been used as probiotics, including species of *Bacillus*, *Bifidobacterium*, *Enterococcus*, *E. coli*, *Lactobacillus*, *Lactococcus*, *Streptococcus*, a variety of yeast species, and undefined mixed cultures. *Lactobacillus* and *Bifidobacterium* species have been used most extensively in humans, whereas species of *Bacillus*, *Enterococcus*, and *Saccharomyces* yeast have been the most common organisms used in livestock (Simon et al. 2001).

In chickens, probiotic lactic acid bacteria have been shown to enhance intestinal mucosal immunity (Yurong et al. 2005), increase the serum antibody response (Haghighi et al. 2005, 2006), and alter expression of a number of genes associated with the immune response (Brisbin et al. 2008, 2010). Probiotics have a role in alleviating post antibiotic treatment syndromes. More trials are needed to establish the efficacy of the probiotics that are currently on the market.

Prebiotics

The most commonly used prebiotics are carbohydrate substrates with the ability to promote the components of the normal intestinal microflora, which may evince a health benefit to the host. The dominant prebiotics are fructooligosaccharide products (FOS, oligofructose, inulin). However, trans-galactooligosaccharides, glucooligosaccharides, glycooligosaccharides, lactulose, lactitol, maltooligosaccharides, xylo-oligosaccharides, stachyose, raffinose, and sucrose thermal oligosaccharides have also been investigated (Monsan and Paul 1995, Orban et al. 1997, Patterson et al. 1997, Piva 1998, Collins and Gibson 1999). Oligosaccharides have been shown to increase beneficial bacteria like bifidobacteria or lactobacilli in the intestinal tract of broilers (Patterson et al. 1962). Mannan oligosaccharides (MOS) are derived from mannans on yeast cell surfaces. From the literature, MOS improve antibody response in broilers and layers (Cotter et al. 2000) and modulate the immune response in chickens (Cotter et al. 2002, Savage et al. 1996; Shashidhara and Devegowda 2003). Alternatively, they are thought to act by binding and removing pathogens from the intestinal tract and stimulation of the immune system (Spring et al. 2000).

Furthermore, synbiotics involves the combined administration of specific prebiotics with probiotics to provide definite health benefits by synergistic action (Harish and Varghese 2006).

5. Selenium and Vitamin E as Antioxidants

Antioxidants

Dietary composition has an impact on the immune function of chicken. Nutritional immunology is becoming popular. Nutrition has some effects on both innate and cellular immunities if the feed is severely restricted (Hangalapura et al. 2005). Antibody responses, blood lymphocyte proliferative responses, and production of antioxidants in chicken at different levels of feed restrictions that result in lower spleen weight cause a marked reduction in natural antibodies that bind lipoteichoic acid. Oxidation reactions can produce free radicals. Free radicals are atoms, molecules, or any compounds containing one or more unpaired electrons. Free radicals are highly unstable, very reactive, and are capable of damaging molecules such as DNA, proteins, lipids, and carbohydrates. Damage to DNA is associated with mutations, translation errors, and disruption of protein synthesis. Damage to proteins causes modifications in ion transport and receptor functions, as well as altered enzymatic activities. Polyunsaturated fatty acid oxidation alters membrane composition, structure, and properties (fluidity, permeability, etc.) as well as the activity of membrane-bound enzymes. The damage to biological molecules ultimately compromises growth, development, immunocompetence, and reproduction. Cells are under constant attack by free radicals, many of which are formed as a natural consequence of normal metabolic activity and as part of the immune system's strategy for destroying invading microorganisms (Surai and Dvorska 2007). In nature, there are many compounds possessing antioxidant properties, such as Se, vitamin E (VE), vitamin C, and zinc.

Vitamin E

Konjufca et al. (2004) suggested that the addition of VE may have a substantial positive impact with respect to disease resistance and improved broiler health. In the case of humoral immune response, the effect of added VE depended on the nature of the antigen and quantity. Typically, the addition of VE in amounts between 25 and 50 IU to the basal diet containing the National Research Council suggested requirement of 10 IU was the most immunomodulatory. This emphasizes that higher levels of VE were often less effective (Leshchinsky and Klasing, 2001).

Selenium

Se is an essential trace element for human and animal health. Se is known to have important roles in reproductive functions and development, immunocompetence, and aging. As a constituent of selenoproteins, Se has structural and enzymatic roles, in the latter context being best known as an antioxidant and catalyst for the production of active thyroid hormone. Se is a component of the cell enzyme glutathione peroxidase (GPx) (Mills, 1957).

Numerous lines of evidence suggest that Se plays an important role in many aspects of health, including chemopreventive effects (Combs et al. 2001, Li et al. 2010), muscle metabolism (Chariot and Bignani 2003, Terry et al. 2009), oxidant defense (Rayman 2000), neurobiology (Schweizer et al. 2004), aging (Martin-Romero et al. 2001), and reproduction (Kaur and Bansal 2005). It has been found that Se is essential for the normal function of the immune system (Rayman 2000, McKenzie et al. 1998, Hoffmann and Berry 2008). Se can improve the immune status and anti-inflammatory action (Leng et al. 2003, Shilo et al. 2008). Se compounds have also been reported to regulate the function of neutrophils, NK cells, B lymphocytes, and T cells (Kiremidjian-Schumacher et al. 1992) and affect the incorporation of Se into organs that are important for the immune system, such as spleen and lymph nodes (Hawkes et al. 2001). Se-deficient diets were found to inhibit bursal and thymic growth (Marsh et al. 1986) and reduce T-cell numbers (Kiremidjian- Schumacher et al. 1994). However, excess dietary Se caused growth retardation of lymphoid organs (Green and Albers 1997, Peng et al. 2009), resulted in B-cell toxicity (Vega et al. 2007), and inhibited Con A-induced T-cell mitogenesis *in vitro* (Ueno et al. 2008).

The biological function of Se is primarily implemented through selenoproteins (Stadtman 2000), a group of proteins that contain the amino acid selenocysteine as an integral part of their polypeptide chain. There are at least 25 identified mammalian selenoproteins and the chicken selenoproteome consists of at least 23 selenoproteins (Lobanov et al. 2008). The GPx family was the first selenoprotein family identified with known function (Rotruck et al. 1973). Other selenoproteins include thioredoxin reductases, iodothyronine deiodinases, SeW, SeP, SeM, selenoprotein N, selenoprotein H and selenoprotein T (Lu and Holmgren 2009). Although the function for many identified selenoproteins is still not confirmed, such as SeW, most selenoproteins are found to contain specific structure motifs that suggest their biological function to be involved in antioxidative or redox-related reactions (Ferguson et al. 2006, Aachmann et al. 2007, Dikiy et al. 2007). SeW has been suggested to be involved in antioxidation and cell cycle progression (Jeong et al. 2002, Chung et al. 2009, Hawkes et al. 2009, Wang et al. 2010). SeW mRNA expression is high in skeletal muscle, followed by that in brain. In addition, the SeW gene is ubiquitously expressed in heart, skeletal muscle, brain, testis, spleen, kidney, lung, liver, stomach, and pancreas in chickens fed a commercial diet supplemented with sodium selenite (Ou et al. 2010).

Glutathione Peroxidase

Se is incorporated as selenocysteine at the active site of a wide range of selenoproteins. Under physiological conditions, the Se in selenocysteine is an extremely efficient biological catalyst. Among the selenoproteins with identified biological functions are the antioxidant enzymes GPx and thioredoxin reductase (Trx) (Li and Beck 2007).

In general, the roles of selenoenzymes, GPx, in the intracellular and extracellular antioxidant defense systems have been well characterized. GPx isoenzymes use GSH as a donor of reducing equivalents to detoxify H_2O_2, peroxi-derivatives of various organic compounds (e.g., fatty acids, cholesterol, and phospholipids) (Arthur 2000), and peroxynitrite (Briviba et al. 1998). Thus, GPx is essential for maintenance of the cellular redox status and for the regulation of redox-sensitive signaling and transcription mechanisms. While higher levels of Se have generally been shown to provide protection against viruses, the relationship between host Se status and immunity to bacterial pathogens is not so one-sided (Hoffmann and Berry 2008).

The Effect of Se Chicken Meat

Organic Se (Se enriched yeast, Se enriched algae) is deposited into the animal tissue more efficiently than inorganic Se (selenite, selenate, selenide) (Choct et al., 2004). Edens (2001) and Surai (2006) reported that the improved growth rate and improved feed conversion (Naylor et al. 2000) of broilers fed an organic Se- supplemented diet could be related to the increased concentrations of the active form of thyroid hormone in the serum of chickens supplemented with organic Se as well as to the immunomodulating properties of Se. Mahmoud and Edens (2005) evaluated the effect of organic Se on performance and different physiological parameters of broilers in either pathogenic *Escherichia coli*-challenged or heat-stressed conditions. In chickens, adequate intake of organic forms of Se may be linked to reduced drip loss in the final meat and to improved feather production (Peric et al. 2009).

Organic Se and VE improved growth in broilers (Özkan et al. 2007) and layer (Hossain et al. 2010). Dietary Se significantly increased the VE contents of egg yolk (Surai 2000), and VE concentrations in plasma of chickens (Thompson and Scott, 1970). The lower lipid peroxidation in egg yolks with dietary Se supplementation could be related to a corresponding higher α-tocopherol concentration (Qi and Sim 1998; Galobart et al. 2001).Se supplementation, especially in an organic form, of laying hen and broiler diets is effective in increasing the α-tocopherol and Se contents of eggs and broiler meat.It has been known for a long time that morbidity and mortality in chicken from several infections (including Salmonella and influenza) depend heavily on nutrition status, with protein-energy malnutrition and several micronutrient deficiencies all being important.Considerable evidence indicates that high doses of Se, and VE, along with the other vitamins and minerals dramatically improves one's immune system and hence, protection against these epidemics.

References

Aarestrup FM. Occurrence of glycopeptide resistance among Enterococcus faecium isolates from conventional and ecological poultry farms. Microb. Drug. Resist., 1995; 1: 255-257.

Aarestrup FM, Carstensen B. Effect of tylosin used as a growth promoter on the occurrence of macrolide resistant enterococci and staphylococci in pigs. *Microb. Drug. Resist.,* 1998; 4: 307–312.

Aarestrup FM. Association between the consumption of antimicrobial agents in animal husbandry and the occurrence of resistant bacteria among food animals. *Int. J. Antimicrob. Agents.,* 1999; 12: 279–285.

Altekruse SF, Cohen ML, Swerdlow DL. Centers for Disease Control and Prevention. 1999; www.cdc.gov/ncidod/eid/vol3no3/cohen.htm.

Aachmann FL, Fomenko DE, Soragni A, Gladyshev VN, Dikiy A. Solution structure of selenoprotein W and NMR analysis of its interaction with 14-3-3 proteins. *J. Biol. Chem.,* 2007; 282: 37036–37044.

Akaike T, Noguchi Y, Ijiri S, Setoguchi K, Suga M, Zheng YM, Dietzschold B, Maeda H. Pathogenesis of influenza virus-induced pneumonia: involvement of both nitric oxide and oxygen radicals. *Proc. Natl. Acad. Sci. USA.,* 1996; 93: 2448–2453.

Akaike T, Fujii S, Kato A, Yoshitake J, Miyamoto Y, Sawa T, Okamoto S, Suga M, Asakawa M, Nagai Y, Maeda H. Viral mutation accelerated by nitric oxide production during infection in vivo. *FASEB J.* 2000; 14: 1447–1454.

Akaike T, Okamoto S, Sawa T, Yoshitake J, Tamura F, Ichimori K, Miyazaki K, Sasamoto K, Maeda H. 8-Nitroguanosine formation in viral pneumonia and its implication for pathogenesis. *Proc. Natl. Acad. Sci. USA*., 2003; 100: 685–690.

Arthur JR. The glutathione peroxidases. *Cell Mol. Life Sci.,* .2000; 57: 1825–1835.

Angulo FJ, Johnson KR, Tauxe RV, Cohen ML. Origins and consequences of antimicrobial-resistant nontyphoidal Salmonella: implications for the use of fluoroquinolones in food animals. *Microb. Drug Resist.,* 2000; 6: 77-83.

Baeumler AJ, Hargis BM, Tsolis RM. Tracing the origins of Salmonella outbreaks. *Science.* 2000; 287: 50-52.

Baggesen DL, Sandvang D, Aarestrup FM. Characterization of Salmonella *enterica serovar typhimurium DT104 isolated from Denmark and comparison with isolates from Europe and the United States. J. Clin. Microbiol.,* 2000*;* 38*:* 1581*–1586.*

Bates J, Jordens JZ, Griffiths DT. Farm animals as a putative reservoir for vancomycin-resistant enterococcal infection in man. J. Antimicrob. Chemother., 1994; 34: 507-514.

Beck MA, Nelson HK, Shi Q, Van Dael P, Schiffrin EJ, Blum S, Barclay D, Levander OA. Selenium deficiency increases the pathology of an influenza virus infection, *FASEB J.,* 2001; 15:1481–1483.

Beck MA, Handy J, Levander OA. Host nutritional status: the neglected virulence factor. *Trends Microbiol.,* 2004; 12: 417-423.

Besser TE, Goldoft M, Pritchett LC, Khakhria R, Hancock DD, Rice DH, Gay JM, Johnson W, Gay CC. *Multiresistant* Salmonella *Typhimurium DT104 infections of humans and domestic animals in the Pacific Northwest of the United States. Epidemiol. Infect.,* 2000; 124*:* 193*–200.*

Brisbin JT, Zhou H, Gong J, Sabour P, Akbari MR, Haghighi HR, Yu H, Clarke A, Sarson AJ, Sharif S. Gene expression profiling of chicken lymphoid cells after treatment with Lactobacillus acidophilus cellular components. *Dev. Comp. Immunol.*, 2008; 32: 563-574.

Brisbin JT, Gong J, Parvizi P, Sharif S. Effects of Lactobacilli on Cytokine Expression by Chicken Spleen and Cecal Tonsil Cells. *Clin. Vaccine. Immunol.*, 2010; 17: 1337–1343.

Briviba K, Kissner R, Koppenol WH, Sies H. Kinetic study of the reaction of glutathione peroxidase with peroxynitrite. *Chem Res Toxicol.* 1998; 11: 1398–1401.

CDC. Multistate outbreak of Salmonella infections associated with frozen pot pies—United States, 2007. *Morb. Morta. Wkly. Rep.*, 2008; 57: 1277–1280.

Cerro A, Soto SM, Mendoz MC. Virulence and antimicrobial-resistance gene profiles determined by PCR-based procedures for Salmonella isolated fromsamples of animalorigin. *Food Microbiol.*, 2002; 20: 421-438.

Chadwick PR, Woodford N, Kaczmarski EB, Gray S, Barrell RA, Oppenheim BA. Glycopeptide-resistant enterococci isolated from uncooked meat. *J. Antimicrob. Chemother.*, 1996; 38: 908–909.

Chariot P, Bignani O. Skeletal muscle disorders associated with selenium deficiency in humans. *Muscle Nerve.*, 2003; 27: 662-668.

Chee-Sanford JC, Aminov RI, Krapac IJ, Garrigues-Jeanjean N, Mackie RI. Occurrence and diversity of tetracycline resistance genes in lagoons and groundwater underlying two swine production facilities. *Appl. Environ. Microbiol.*, 2001; 67: 1494-1502.

Choct M, Naylor AJ, Reinke N. Selenium supplementation affects broiler growth performance, meat yield and feather coverage. *Br. Poult. Sci.*, 2004; 45: 677-683.

Chung YW, Jeong D, Noh OJ, Park YH, Kang SI, Lee MG, Lee TH, Yim MB, Kim IY.

Antioxidative role of selenoprotein W in oxidant-induced mouse embryonic neuronal cell death. *Mol. Cells.* 2009; 27: 609–613.

Combs GF Jr. Selenium in global food systems. Br. J. Nutr., 2001; 85: 517-547.

Combs GF Jr, Clark LC, Turnbull BW. An analysis of cancer prevention by selenium. *Biofactors,* 2001; 14: 153-159.

Collins M D, Gibson GR. Probiotics, prebiotics, and synbiotics: approaches for modulating the microbial ecology of the gut. *Am. J. Clin. Nutr.*, 1999; 69: S1042–1057.

Cotter PF, Malzone A, Paluch B, Lilburn MS. Sefton AE. Modulation of humoral immunity in commercial laying hens by a dietary prebiotic. *Poult. Sci.*, 2000; 79: 38-38.

Cotter PF, Sefton AE, Lilburn MS. Manipulating the Immune System of Layers and Breeders: Novel Applications of Mannan Oligosaccharides. In: Nutritional Biotechnology in the Feed and Food Industries, Lyons, T.P. and K.A. Jacques (Eds.). Nottingham University Press, Nottingham, England, 2002. p 21-27.

Davis MA, Hancock DD, Besser TE, Rice DH, Gay JM, Gay C, Gearhart L, DiGiacomo R. Changes in antimicrobial resistance among Salmonella enterica Serovar typhimurium isolates from humans and cattle in the Northwestern United States, 1982-1997. *Emerg. Infect. Dis.*, 1999; 5: 802-806.

Dikiy A, Novoselov SV, Fomenko DE, Sengupta A, Carlson BA, Cerny RL, Ginalski K, Grishin NV, Hatfield DL, Gladyshev VN. SelT, SelW, SelH, and Rdx12: genomics and molecular insights into the functions of selenoproteins of a novel thioredoxin-like family. *Biochemistry,* 2007; 46: 6871–6882.

Domínguez C, Gómez I, Zumalacárregui J. Prevalence of Salmonella and Campylobacter in retail chicken meat in Spain. *Int. J. Food Microbiol.*, 2002. 30; 72: 165-168.

Edens FW. Involvement of Sel-Plex in physiological stability and performance of broiler chickens. Biotechnology in Feed Industry. Proc. 17th, Alltech Ann. Symp. Nottingham, *Nottingham University Press*. 2001; 349-376.

Esaki H, Morioka A, Ishihara K, Kojima A, Shiroki S, Tamura Y, Takahashi T. Antimicrobial susceptibility of Salmonella isolated from cattle, swine and poultry (2001-2002): report from the Japanese Veterinary Antimicrobial Resistance Monitoring Program. *J. Antimicrob. Chemother.*, 2004; 53: 266-270.

Ewing WN, Cole DJA. The living gut. An introduction to microorganisms in nutrition. 1994; www.cabdirect.org/abstracts/ 19951403230.htmlFAO/ WHO. Health and Nutritional Properties of Probiotics in Food including Powder Milk with Live Lactic Acid Bacteria. 2009; http://www.who. int/entity/ foodsafety/ publications/fs_management/en/probiotics.

Ferguson AD, Fomenko DE, Labunskyy VM, Arac D, Chelliah Y, Amezcua CA, Rizo J, ladyshev VN, Deisenhofer J. NMR structures of the selenoproteins Sep15 and SelM reveal redox activity of a new thioredoxin-like family. *J. Biol. Chem.*, 2006; 281: 3536–3543.

Fey PD, Safranek TJ, Rupp ME, Dunne EF, Ribot E, Iwen PC, Bradford PA, Angulo FJ, Hinrichs SH. Ceftriaxone-resistant Salmonella infection acquired by a child from cattle. *N. Engl. J. Med.* 2000; 342: 1242-1249.

Fuller R. Probiotics in human medicine *Gut*, 1991; 32: 439-442.

Galobart J, Barroeta AC, Baucells MD, Codony R, Ternes W. Effect of Dietary Supplementation with Rosemary Extract and α-Tocopheryl Acetate on Lipid Oxidation in Eggs Enriched with ω3-Fatty Acids. *Poult. Sci.*, 2001; 80:460–467.

Glynn MK, Bopp C, Dewitt W, Dabney P, Mokhtar M, Angulo FJ. Emergence of multidrug-resistant Salmonella enterica serotype typhimurium DT104 infections in the United States. *N. Engl. J. Med.*, 1998; 338: 1333-1338.

Gould IM. The epidemiology of antibiotic resistance. *Int. J. Antimicrob. Agents.*, 2008; 32: S2-9.

Green DE, Albers PH. Diagnostic criteria for selenium toxicosis in aquatic birds: histologic lesions. *J. Wildl. Dis.*, 1997; 33: 385-404.

Haghighi HR, Gong J H, Gyles CL, Hayes MA, Sanei B, Parvizi P, Gisavi H, Chambers JR, Sharif S. Modulation of antibody-mediated immune response by probiotics in chickens. *Clin. Diagn. Lab. Immunol.*, 2005; **12:** 1387-1392.

Haghighi HR, Gong JH, Gyles CL, Hayes MA, Zhou HJ, Sanei B, Chambers JR, Sharif S. Probiotics stimulate production of natural antibodies in chickens. *Clin. Vaccine. Immunol.*, 2006; **13 :** 975-980.

Hall RM. Mobile gene cassettes and integrons: moving antibiotic resistance genes in gram-negative bacteria. Ciba Found Symp. 1997; 207: 192-202.

Hall RM, Stokes HW. Integrons: novel DNA elements which capture genes by site-specific recombination. *Genetica*. 1993; 90: 115-132.

Helms M, Vastrup P, Gerner-Smidt P, Mølbak K. Short and long term mortality associated with foodborne bacterial gastrointestinal infections: registry based study. BMJ. 2003 ; 15: 326: 357.

Hangalapura BN, Nieuwland MG, De Vries Reilingh G, Buyse J, Van Den Brand H, Kemp B, Parmentier HK. Severe feed restriction enhances innate immunity but suppresses cellular immunity in chicken lines divergently selected for antibody responses. *Poult. Sci.,* 2005; 84: 1520-1529.

Harish KI, Varghese T. Probiotics in humansevidence based review. *Calicut. Med.,* 2006; J.4: e3.

Hawkes WC, Wang TTY, Alkan Z, Richter BD, Dawson K. Selenoprotein W modulates control of cell cycle entry. *Biol. Tace. Elem. Res.,* 2009; 131: 229–244.

Hawkes WC, Kelley DS, Taylor PC. The effects of dietary selenium on the immune system in healthy men. *Boil. Trace. Elem. Res.,* 2001; 81: 189-213.

Hoffmann PR, Berry MJ. The influence of selenium on immune responses. *Mol. Nutr. Food Res.,* 2008; 52: 1273-1280.

Hossain MdS, Afrose S, Takeda I, Tsujii H. Effect of Selenium-enriched Japanese Radish Sprouts and *Rhodobacter capsulatus* on the Cholesterol and Immune Response of Laying Hens. Asian-Aust. *J. Anim. Sci.,* 2010; 23: 630–639.

Imberechts H, De Filette M, Wray C, Jones Y, Godard C, Pohl P. Salmonella *typhimurium phage type DT104 in Belgian livestock. Vet. Rec.* 1998*;* 143*:* 424*–425.*

Impey CS, Mead GC, George SM. Competitive exclusion of salmonella from the chick caecum using a defined mixture of bacterial isolates form the caecal microflora of an adult bird. *J. Hygiene.,* 1982; 89: 479-490.

Jaspers I, Zhang W, Brighton LE, Carson JL, Styblo M, Beck MA. Selenium deficiency alters epithelial cell morphology and responses to influenza. *Free Radic Biol. Med.,* 2007; 42: 1826-1837.

Jeong D, Kim TS, Chung YW, Lee BJ, Kim IY. Selenoprotein W is a glutathione- dependent antioxidant in vivo. *FEBS Lett.* 2002; 517: 225–228.

Kaur P, Bansal MP. Effect of selenium-induced oxidative stress on the cell kinetics in testis and reproductive ability of male mice. *Nutrition.* 2005; 21: 351–357.

Kiremidjian-Schumacher L, Roy M, Wishe HI, Cohen MW, Stotzky G.Regulation of cellular immune responses by selenium. *Biol. Trace. Elem. Res.,* 1992; 33: 23–35.

Kiremidjian-Schumacher L, Roy M, Wishe HI, Cohen MW, Stotzky G. Supplementation with selenium and human immune cell functions, II: effect on cytotoxic lymphocytes and natural killer cells. *Biol. Trace. Elem. Res.,* 1994; 41: 115-127.

Konjufca VK, Bottje WG, Bersi TK, Erf GF. Influence of dietary vitamin E on phagocytic functions of macrophages in broilers. *Poult. Sci.,* 2004; 83: 1530-1534.

Laz C, Feher J. The effect of selenium therapy on autoimmune thyroiditis. *CEMED* 2009; 3: 269-277.

Leegaard TM, Caugnat DA, Frøholm LO, Høiby EA, Lassen J. *Emerging antibiotic resistance* in **Salmonella** *typhimurium in Norway. Epidemiol Infect.* 2000*;* 125*:* 473*–480.*

Leng L, Bobček R, Kuricova S, Boldizarova K, Grešakova L, Ševcikova Z, Revajova V, Levkutova M, Levkut M. Comparative metabolic and immune responses of chickens fed diets containing inorganic selenium and Sel-PlexTM organic selenium. In: Lyons T.P., Jacques K.A. (eds.): Proc. Altech's 19th International Symposium. *Nutritional Biotechnology in the Feed and Food Industries.* 2003; p131-145.

Leshchinsky TV, Klasing KC. Relationship between the level of dietary vitamin E and the immune response of broiler chickens. *Poult. Sci.,* 2001; 80: 1590-1599.

Levy SB, FitzGerald GB, Macone AB. Changes in intestinal flora of farm personnel after introduction of a tetracycline-supplemented feed on a farm. *N. Engl. J. Med.,* 1976; 295: 583–588.

Li JL, Gao R, Li S, Wang JT, Tang ZX, Xu SW. Testicular toxicity induced by dietary cadmium in cocks and ameliorative effect by selenium. *Biometals*. 2010; 23: 695-705.

Li W, Beck MA. Selenium deficiency induced an altered immune response and increased survival following influenza A/Puerto Rico/8/34 infection. *Exp. Biol. Med.*, 2007; 232: 412-419.

Li Y, Yang Y, Chen H. Detection of enteroviral RNA in paraffinembedded myocardial tissue from patients with Keshan by nested PCR. *Zhonghua. Yixue. Zazhi.*, 1995; 75: 344-345.

Li Y, Peng T, Yang Y, Niu C, Archard LC, Zhang H. High prevalence of enteroviral genomic sequences in myocardium from cases of endemic cardiomyopathy (Keshan disease) in China. *Heart*. 2000; 83: 696-701.

Linton AH. Plasmids in the environment.Schriftenr Ver Wasser Boden Lufthyg. 1988; 78:197-224.

Ljungh A, Wadström T. Lactic acid bacteria as probiotics. *Curr. Issues. Intest. Microbiol.*, 2006; 7: 73-89.

Lobanov AV, Hatfield DL, Gladyshev VN. Reduced reliance on the trace element selenium during evolution of mammals. *Genome. Biol.*, 2008; 9: R62.

Lu J, Holmgren A. Selenoproteins. *J. Biol. Chem.*, 2009; 284: 723–727.

Maeda H, Akaike T. Oxygen free radicals as pathogenic molecules in viral diseases. *Proc. Soc. Exp. Biol. Med.*, 1991; 198: 721-727.

Mahmoud KZ, Edens FW. Influence of organic selenium on hsp 70 response of heat-stressed and enteropathogenic *Escherichia coli*-challenged broiler chickens (*Gallus gallus*). *Comp. Biochem. Physiol. Part C*. 2005; 141: 69-75.

Marsh JA, Combs GF Jr, Whitacre ME, Dietert RR. Effect of selenium and vitamin E dietary deficiencies on chick lymphoid organ development. P *Soc. Exp. Biol. Med.*, 1986; 182: 425-436.

Macrobiotic Articles. 2009; www.macrobiotic-way.com/antibiotics.

Martin-Romero FJ, Kryukov GV, Lobanov AV, Carlson BA, Lee BJ, Gladyshev VN, Hatfield DL. Selenium metabolism in Drosophila: selenoproteins, selenoprotein mRNA expression, fertility, and mortality. *J. Biol. Chem.*, 2001; 276: 29798-29804.

McGowan JE Jr. Antimicrobial resistance in hospital organisms and its relation to antibiotic use. *Rev. Infect. Dis.*, 1983; 5: 1033–1048.

McKenzie RC, Rafferty TS, Beckett GJ. Selenium: an essential element for immune function. *Trends Immunol Today*. 1998; 19: 342–345.

Mellon M, Fondriest S. Hogging It. Estimates of antimicrobial abuse in livestock. *Union of Concerned ScientistsMagazine*. 2001; 23: No.1.

Mills GC. Hemoglobin catabolism. I. Glutathione peroxidase, an erythrocyte enzyme which protects hemoglobin from oxidative breakdown. *J. Biol. Chem.*, 1957; 229 :189-197.

Monsan P, Paul F. Oligosaccharide feed additives. Biotechnology in Animal Feeds and Animal Feeding. R. J. Wallace and A. Chesson, ed. *VCH*, New York. 1995. p 233–245.

McKenzie RC, Beckett GJ, Arthur JR. Effects of selenium on immunity and aging. In: Hatfield DL, Berry MJ, Gladyshev VN, eds. Selenium: Its Molecular Biology and Role in Human Health. 2nd ed. New York: *Springer*. 2006; p311-323 .

Mohler JR. Fowl pest in the United States. *J. Am. Vet. Med. Assoc.*, 1926; 21: 549–559.

Murray BE. The life and times of the **Enterococcus**. *Clinical Microbiology Reviews., 3. 1990; 46–65.*

Neve J. Selenium as a risk factor for cardiovascular diseases. *J. Cardiovasc. Risk.* 1996;3: 42-47.

Nurmi E, Rantala M. New aspects of Salmonella infection in broiler production. *Nature,* 1973; 241: 210–211.

O'Flanagan D, Cormican M, McKeown P, Nicolay N, Cowden J, Mason B, Morgan D, Lane C, Irvine N, Browning L. A multi-country outbreak of Salmonella Agona, February - August 2008. *Eurosurveillance.* 2008; 13: 1–2.

Orban JI, Patterson JA, Sutton AL, Richards GN. Effect of sucrose thermal oligosaccharide caramel, dietary vitamin-mineral level, and brooding temperature on growth and intestinal bacterial populations in broiler chickens. *Poult. Sci.*, 1997; 76: 482–490.

Ou BR, Jiang MJ, Lin CH, Liang YC, Lee KJ, Yeh JY. Characterization and expression of chicken selenoprotein W. *Biometals.* 2011; 24: 323–333.

Özkan S, Basmacıoğlu-Malayoğlu H, Yalçın S, Karadaş F, Koçtürk S, Çabuk M, Oktay G, Özdemir S, Özdemir E, Ergül M. Dietary vitamin E (α-tocopherol acetate) and selenium supplementation from different sources: performance, ascites-related parameters and antioxidant status in broilers reared at low and optimum temperatures. Br Poult Sci. 2007; 48: 580-593.Pandemic Influenza; 2011 www.cidrap.umn.edu/idsa/influenza /panflu/biofacts/panflu. html -.

Patterson R, Younger JS, Weigle WO, DixonFJ. The metabolism of serum proteins in the hen and chick and secretion of serum proteins by the ovary of the hen. *J. Gen Physiol.,* 1962; 45: 501-513.

Patterson JA, Orban JI, Sutton AL, Richards GN. Selective enrichment of bifidobacteria in the intestinal tract of broilers by thermally produced kestoses and effect on broiler performance. *Poult. Sci.,* 1997; 76: 497–500.

Peng X, Cui Y, Cui W, Deng J, Cui H. The decrease of relative weight, lesions, and apoptosis of bursa of Fabricius induced by excess dietary selenium in chickens. *Boil. Trace Elem. Res.,* 2009; 131: 33-42.

Pereira HG, Tumova B, Law VG. Avian influenza A viruses. Bull World *Health Org.,* 1965; 32: 855–860.

Perić L, Milošević N, Žikić D, Kanački Z, Džinić N, Nollet L, Spring P. Effect of selenium sources on performance and meat characteristics of broiler chickens.*J. Appl. Poult. Res.,* 2009; 18: 403-409.

Perroncito E. Epizoozia tifoide nei gallinacei. *Annali. Accad. Agri. Torino.* 1878; 21: 87–126.

Poppe C, Smart N, Khakhria R, Johnson W, Spika J, Prescott J. Salmonella typhimurium DT104: a virulent and drug-resistant pathogen. Can. Vet. J., 1998*; 39, 559–65.*

Qi, GH, Sim JS. Natural tocopherol enrichment and its effect in n-3 fatty acid modified chicken eggs. *J. Agric. Food Chem.,* 1998; 6: 1920–1926.

Rayman MP. The importance of selenium to human health. *Lancet.* 2000; 356: 233-241.

Ren LQ, Li XJ, Li GS, Zhao ZT, Sun B, Sun F. Coxsackievirus B3 infection and its mutation in Keshan disease. *World J. Gastroenterol.,* 2004; 10: 3299-3302.

Rotruck JT, Pope AL, Ganther HE, Swanson AB, Hafeman D, Hoekstra WG. Selenium: biochemical role as a component of glutathione peroxidase. *Scien.ce.* 1973; 179: 588–590.

Sameshima T, Akiba M, Izumiya H, Terajima J, Tamura K, Watanabe H, Nakazawa M. Salmonella typhimurium DT104 from livestock in Japan. Jpn. J. Infect. Dis., 2000*;* 53*:* 15-*16.*

Savage TF, Cotter PF, Zakrzewska EI. The effect of feeding a mannan oligosaccharide on Immunoglobulins, plasma IgG and bile IgA of Wrolstad MW male turkeys. *Poult. Sci.,* 1996; 75: 143-143.

Schwarz S, Kehrenberg C, Walsh TR. Use of antimicrobial agents in veterinary medicine and food animal production. Int. J. Antimicrob. Agents., 2001; 17: 431-437.

Schweizer U, Schomburg L, Savaskan NE. The neurobiology of selenium: lessons from transgenic mice. *J. Nutr.,* 2004; 134: 707-710.

Schäfer W. Vergleichender sero-immunologische Untersuchungen über die Viren der Influenza und klassischen Geflügelpest, *Z. Naturf.,* 1955; 10b: 81–91.

Shashidhara RG, Devegowda G. Effect of dietary mannan oligosaccharide on broiler breeder production traits and immunity. *J Poult. Sci.,* 2003; 82: 1319-1325.

Shilo S, Pardo M, Aharoni-Simon M, Glibter S, Tirosh O. Selenium supplementation increases liver MnSOD expression: Molecular mechanism for hepato-protection. *J. Inorg. Biochem.,* 2008; 102: 110-118.

Shlaes DM, Gerding DN, John JF Jr, Craig WA, Bornstein DL, Duncan RA, Eckman MR, Farrer WE, Greene WH, Lorian V, Levy S, McGowan JE Jr, Paul SM, Ruskin J, Tenover FC, Watanakunakorn C. Society for Healthcare Epidemiology of America and Infectious Diseases Society of America Joint Committee on the Prevention of Antimicrobial Resistance: guidelines for the prevention of antimicrobial resistance in hospitals. *Clin. Infect. Dis.,* 1997; 25: 584-599.

Simon, O, Jadamus A, Vahjen. W. Probiotic feed dditives—effectiveness and expected modes of action. *J. Anim. Feed Sci.,* 2001; 10: 51–67.

Simonsen GS, Haaheim H, Dahl K H, Cruse H, Lovseth A, Olsvik O, Sundsfjord A. Transmission of VanA-type vancomycin-resistant enterococci and vanA resistance elements between chicken and humans at avoparcin-exposed farms. *Microb. Drug Res.,* 1998; 4: 313–318.

Smith DL, Harris AD, Johnson JA, Silbergeld EK. Animal antibiotic use has an early but important impact on the emergence of antibiotic resistance in human commensal bacteria. 2002; *www.pnas.org/content/99/9/6434.full.*

Sun M. In search of Salmonella's smoking gun. Science. 1984; 226: 30-32.

Swartz MN. Human diseases caused by foodborne pathogens of animal origin. Clin. Infect Dis., 2002; 34: S111-122.

Spring P, Wenk C, Dawson KA, Newman KE. The effect of dietary mannanoligosaccharides on cecal parameters and the concentrations of enteric bacteria in the ceca of *Salmonella*-challenged broiler chicks. *Poult. Sci.,* 2000; 79: 205–211.

Stadtman TC. Selenium biochemistry. Mammalian selenoenzymes. *Ann. N. YAcad. Sci., 2000*; 899: 399-402.

Surai PF, Dvorska J. Strategies to enhance antioxidant protection and implications for the wellbeing of companion animals.2007; *en.engormix.com › › Technical articles.*

Surai PF. Effect of selenium and vitamin E content of the maternal diet on the antioxidant system of the yolk and the developing chick. Br. Poult. Sci., 2000; 41: 235- 243.

Surai PF. Selenium in Nutrition and Health, Nottingham University Press, Nottingham, United Kingdom.*Thompson and Scott*, 2006. p1970.

Tenover FC, McGowan JEJ. Reasons for the emergence of antibiotic resistance. *Am. J. Med. Sci.,* 1996; 311: 9–16.

Tellez G, Petrone VM, Escorcia M, Morishita TY, Cobb CW, Villaseñor L, Promsopone B. Evaluation of avian-specific probiotic and Salmonella enteritidis-, Salmonella typhimurium-, and Salmonella heidelberg-specific antibodies on cecal colonization and organ invasion of Salmonella enteritidis in broilers. *J. Food Prot.*, 2001; 64: 287-291.

Terry EN, Michal JJ, Hostetler CE, Kincaid RL. Levels of mRNA for three selenoproteins in skeletal muscle of fetal and newborn pigs. *Livestock Science,* 2009; 124: 21-25.The poultry site Latest News. Salmonella bacteria affects chicken meat; infections on the rise. 2006; www.thepoultrysite.com.

Threlfall EJ, Frost JA, Ward LR. Rowe B. Epidemic in cattle and humans of Salmonella typhimurium DT 104 with chromosomally integrated multiple drug resistance. *Vet. Rec.,* 1994; 134: 577.

Threlfall EJ. Antimicrobial drug resistance in Salmonella: problems and perspectives in food- and water-borne infections. *FEMS Microbiol. Rev.,* 2002; 26: 141-148.

Tollefson L, Altekruse SF, Potter ME. Therapeutic antibiotics in animal feeds and antibiotic resistance. *Rev. Sci. Tech.,* 1997; 16: 709-715.

Tortuero, F. Influence of implantation of Lactobacillus acidophilus in chicks on the growth, feed conversion, malabsorption of fats syndrome and intestinal flora. *Poult. Sci.,* 1973; 52: 197–203.

Toulis KA, Anastasilakis AD, Tzellos TG, Goulis DG, Kouvelas D. Selenium supplementation in the treatment of Hashimoto's thyroiditis: A systematic review and a meta-analysis. *Thyroid.,* 2010; 20:1163-1173 .

Ueno H, Hasegawa G, Ido R, Okuno T, Nakamuro K. Effect of selenium status and upplementary selenoi-chemical sources on mouse T-cell mitogenesis. *J. Trace. Elem. Biol.,* 2008; 22: 9-16.

Van den Bogaard AE, Jensen LB, Stobberingh EE. Vancomycin-resistant enterococci in turkeys and farmers. *N. Engl. J. Med.,* 1997; 337: 1558-1559.

Vega L, Rodríguez-Sosa M, García-Montalvo EA, Del Razo LM, Elizondo G. Non- optimal levels of dietary selenomethionine alter splenocyte response and modify oxidative stress markers in female mice. *Food Chem. Toxicol.,* 2007; 45: 1147-1153.

Xiao-Long W, Chuan-Ping Y, Kai X, Ou-Jv Q. Selenoprotein W depletion in vitro might indicate that its main function is not as an antioxidative enzyme. *Biochemistry.,* 2007; 75: 201-207.

Watkins BA, Miller BF, Neil DH. In vivo inhibitory effects of Lactobacillus acidophilus against pathogenic Escherichia coli in gnotobiotic chicks. *Poult. Sci.,* 1982; 61: 1298–1308.

White DG, Zhao S, Sudler R, Ayers S, Friedman S, Chen S, McDermott PF, McDermott S, Wagner DD, Meng J. The isolation of antibiotic-resistant salmonella from retail ground meats. *N. Engl. J. Med.,* 2001; 18: 345:1147-1154.

WHO: Avian Influenza. 2011; *www.who.int/mediacentre /factsheets/avian_influenza/en/.*

Witte W. Medical consequences of antibiotic use in agriculture. *Science*. 1998; 279: 996-997.

Yuan Q, Lai Q, Shaozhong Q. Avian influenza's "internal factor of selenium deficiendy" and its epidemic prevention "strategy of treating cause" 2006 www.369.com.cn/ En/ qinliugan.htm.

Yurong Y, Ruiping S, Shimin Z, Yibao J. Effect of probiotics on intestinal mucosal immunity and ultrastructure of cecal tonsils of chickens. *Arch. Anim. Nutr.,* 2005; 59: 37-246.

Yoshimura H, Endoh Y S, Ishimaru M, Takagi M. Antimicrobial resistance patterns of Salmonella typhimurium and Salmonella dublin isolated from cattle in Japan. *Vet. Rec.*, 2000; 147: 48–50.

In: Selenium: Sources, Functions and Health Effects
Editors: Chinatsu Aomori and Megumi Hokkaido
ISBN: 978-1-61942-061-8

Chapter 6

SELENIUM IN CRITICAL ILLNESS

V. Mishra and A. Shenkin
Department of Clinical Chemistry, Royal Liverpool University Hospital, Liverpool, UK

ABSTRACT

The role of Selenium (Se) in critical illness has been an issue of great interest and anxiety for researchers. Sepsis and the systemic inflammatory response syndrome form the pathogenesis of many cases of critical illness, and these are characterised by increased free radical induced oxidative stress. If this inflammatory process is not controlled it can lead to multiple organ dysfunction and contribute to high morbidity and mortality in patients in the intensive care unit (ICU). Therefore it is important to limit the free radical induced oxidative stress process in critical illness. It has been hypothesised that antioxidants, by preventing or removing free radicals, may reduce the oxidative stress in critical illness. Se is an important part of the antioxidant system, and is known to mediate its antioxidant action through the enzyme glutathione peroxidase. Besides this, it also reduces inflammation by inhibiting NF-κB (nuclear factor-kappa B), which is involved in the transcription of several pro inflammatory cytokines such as IL-6 (interleukin-6). Se also improves the immune response of the body by enhancing humoral and cellular immunity. Thus Se may play an important role in improving the clinical outcome in critical illness. Several studies have assessed the role of high dose parenteral Se (as sodium selenite) in critical illness. Although studies have shown an improvement in antioxidant status, there is no clear beneficial effect on overall clinical outcome, in terms of mortality rate in ICU, infection rate, length of stay in ICU or renal replacement therapy. However some studies have shown that high dose Se may improve the mortality rate in patients with particularly severe sepsis or septic shock. Se levels in blood are influenced by the acute phase response, which reduces plasma Se levels through redistribution of selenoproteins. Therefore low plasma Se levels in critical illness may be due to inflammation or severity of illness, rather than Se deficiency. Hence Se levels in blood in critical illness should be interpreted with caution. Further studies are needed to identify Se requirements in critical illness.

INTRODUCTION

Selenium (Se) has been the focus of intense clinical research in the past decade as it is of fundamental importance to human health. The role of Se and its supplementation in critical illness has been an issue of wide debate. Critical illness is characterized by increased oxidative stress contributing to high morbidity and mortality.

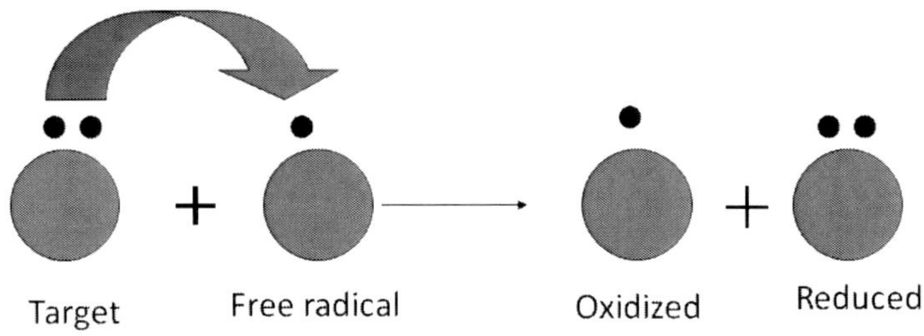

Figure 1. Production of free radicals (Reactive oxygen species).

Oxidative stress is an imbalance between increased free radical production and reduced antioxidant status. Free radicals(FR) or reactive oxygen species are molecules or species in biological systems that contain one or more unpaired electrons. They achieve stability by gaining an electron (reduction) or donating an electron (oxidation) to surrounding molecules to produce an electron pair and in the process the attacked molecule then becomes a FR (Fig. 1). FR such as superoxide ($O_2^{-\bullet}$) and hydroxyl radicals ($OH^{\bullet}$) may be produced by normal intracellular processes (Fig. 2) or by inflammatory cells in several conditions including infection, injury, trauma, stress or surgery where they are likely to be produced in an exaggerated manner [1]. FR inflict cell injury either directly by interacting with its structure, function and genetic activity or by damaging lipids, proteins, carbohydrates and nucleic acids [2]. Lipid peroxidation is one of the major consequences of FR-induced oxidative damage (Fig. 3). Lipid hydroperoxides, formed as a consequence also inhibit the circulating anti-proteases leading to extracellular molecular damage. This FR-induced tissue damage in the body is limited by various antioxidants (Fig. 4),which when present even at relatively low concentrations, significantly limit the formation of, or scavenge, FR generated in the human body [3].

Haber-Weiss reaction

- $O_2^{\bullet} + O_2^{\bullet} + 2H^+ \rightarrow H_2O_2 + O_2$
- $O_2^{\bullet} + H_2O_2 \rightarrow OH^{\bullet} + OH^- + O_2$

Fenton reaction

- $O_2^{\bullet} + Fe^{3+} \rightarrow Fe^{2+} + O_2$
- $Fe^{2+} + H_2O_2 \rightarrow OH^{\bullet} + OH^- + Fe^{3+}$

Figure 2. Sources of free radical production.

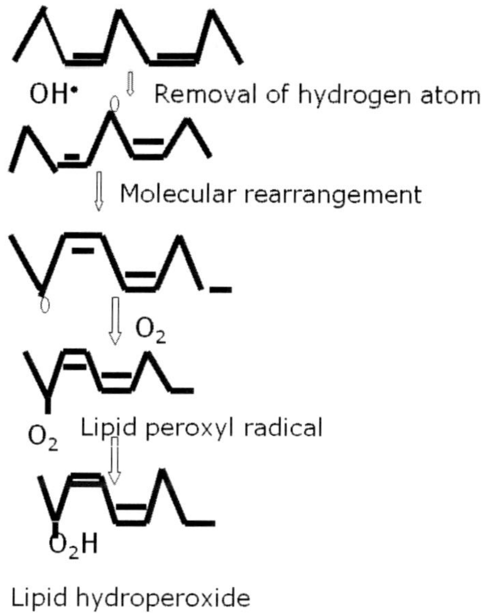

Figure 3. Free radical induced lipid peroxidation.

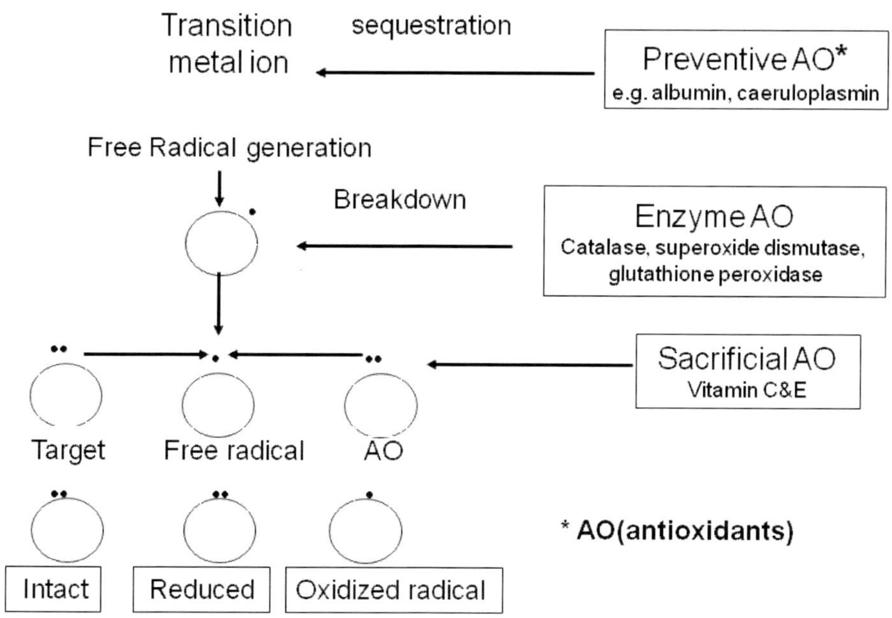

Figure 4. Role of Antioxidants.

Pathogenesis of Critical Illness

Sepsis or SIRS (the systemic inflammatory response syndrome) is the systemic inflammatory response to infection (commonly gram-negative bacteria) or to non-infectious stimuli such as trauma, surgery, pancreatitis or ischemia [4]. When the systemic inflammatory response is due to a confirmed infectious process it is termed sepsis [5]. If this inflammatory response is not controlled, it can lead to the development of inadequate organ perfusion or multiple organ dysfunctions [6]. Further, it may be accompanied by hypotension unresponsive to fluid therapy, referred to as septic shock, which develops in approximately 40% of septic patients and adversely affects the prognosis (Fig. 5) [6]. Sepsis or SIRS, along with its consequences, is an increasingly common cause of morbidity and mortality, particularly in critically ill patients, accounting for 10% to 50% of the deaths in intensive care unit (ICU) [7]. The mortality rate for those patients who develop septic shock and multiple organ failure ranges from 20% to 60% [7].

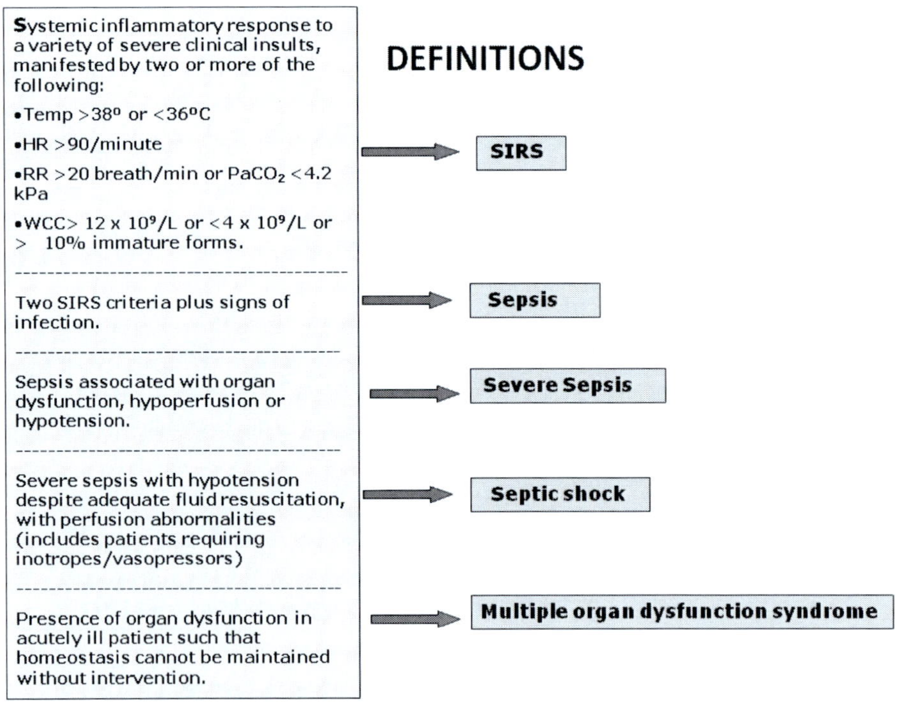

Figure 5. Definitions of systemic inflammatory response.

Oxidative stress has an important role in the development and manifestations of SIRS and its consequent multiple organ failure [8]. When a host tissue is attacked by a pathological insult of either an infectious or non-infectious nature, an inflammatory reaction occurs, resulting in subsequent clearance of the stimulus by the phagocytic cells. Upon recognition of a phagocytic or soluble stimulus, both neutrophils and macrophages experience a "respiratory burst", which is characterized by an increase in oxygen consumption and activation of the hexose monophosphate shunt (HMP) releasing $O_2^{-\bullet}$ and further formation of hydrogen peroxide (H_2O_2) and $OH^\bullet$ (Fig.6) [9]. The generation of $OH^\bullet$ represents an important mechanism in the potentiation of cell and tissue injury by phagocyte-derived $O_2^{-\bullet}$ [10]. This is a normal physiological response to defend the body against pathogenic stimuli where there is a localization of inflammatory cells such as neutrophils and macrophages to the site of infection to initiate the killing of microorganisms. This FR generation is a critical component of microbial killing by activated neutrophils.

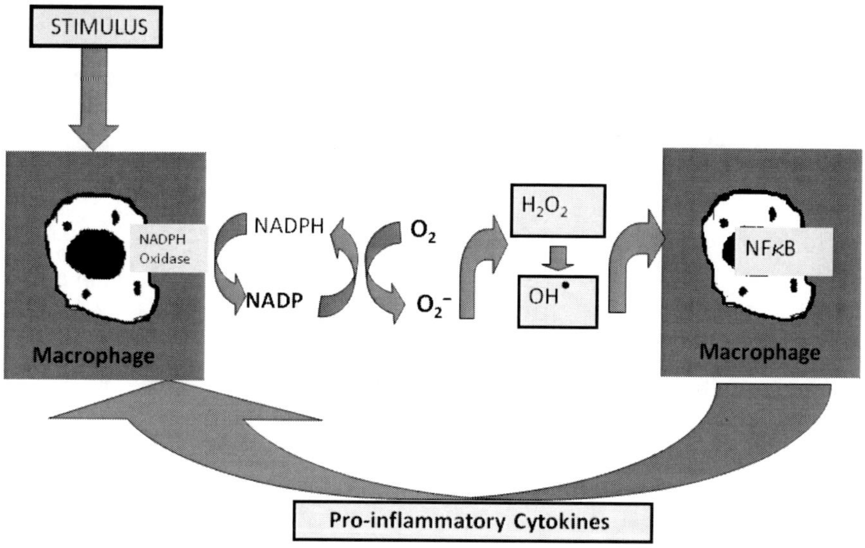

Figure 6. Oxidative stress in critical illness.

The septic cascade of critical illness is initiated by the release of endotoxin (by gram-negative bacteria) into the circulation (Fig. 5). Endotoxin (lipopolysaccharide), thus released from the bacterial cell wall, induces the production of soluble mediators from macrophages and mononuclear cells and these mediators contribute to various sequelae of the septic cascade [8]. Once in the circulation, endotoxin stimulates the release of a variety of powerful mediators and cytokines such as tumor necrosis factor (TNF) α, interleukins (IL)-1, IL-6, IL-8 and platelet activating factor (PAF) from mononuclear phagocytes and other cells, including the endothelial cells themselves [11,12]. These cytokines not only activate macrophages releasing FR but also result in the activation of NF-κB, a transcription factor, which promotes the synthesis and release of a variety of cytokines [13], thus setting up a vicious circle of increased FR production and continuous release of cytokines (Fig. 6). Thus, in critical illness this process of inflammation is aggravated, resulting in increased production of FR, which can cause extensive tissue damage, leading to multiple organ failure [8]. If infection is not brought under control, these inflammatory mediators may leak into the circulation and activate coagulation and the complement cascade, resulting in vascular abnormality [14]. Once this happen the signs of organ failure set in. As additional inflammatory mediators are released into the circulation, blood vessels are damaged, more sites including liver and kidney are affected, blood pressure drops, and the patient goes into septic shock [15]. Thus the characteristic pathophysiological changes seen in sepsis are caused not by the invading pathogen itself but by an overwhelming inflammatory mediator response to infective or non-infective stimuli.

The action of FR is normally limited by the antioxidant (AO) defense system of the body, but in critically ill patients the AO capacity is likely to be compromised. There are several reasons for poor AO status in critical illness, including increased utilization in the face of increased FR production, poor nutritional status, redistribution due to acute phase response, and inadequate provision of optimal nutrition [8]. The patients in the intensive care unit (ICU) are usually malnourished due to chronic illness, advanced age and the long duration of stay in ICU that contribute to further reduction in the AO status.

ROLE OF SE IN CRITICAL ILLNESS

Se is known to mediate its antioxidant action through selenoproteins (Fig.7) where it is incorporated as selenocysteine. The glutathione peroxidase enzyme (GSH-Px), which represents a major class of functionally important selenoproteins, was the first to be characterized [16] The antioxidant effects of Se were suggested to be mediated through the GSH-Px enzymes that remove potentially damaging lipid hydroperoxides and H_2O_2 to less toxic products and play a unique role in protecting cells against free radical induced oxidative stress (Fig.8) [17]. At least six of these peroxidases have now been identified as operating in different cell and tissue compartments: GSH-Px 1 in cell cytosol, GSH-Px 2 in GI tract, GSH-Px 3 in plasma, GSH-Px 4 in cell membrane [18]. The role of plasma GSH-Px as an antioxidant enzyme is not fully understood. Although its preferred substrate, glutathione (reduced, GSH), usually has a low plasma concentration [16], increased cellular oxidative stress can export oxidized glutathione (GSSG) to the plasma compartment, where a combination of glutathione reductase and GSH-Px can restore redox balance and either return reduced glutathione to the cell, or use it as a substrate for plasma GSH-Px. GSH-Px roughly corresponds to only 39% of plasma Se [19]. Thus, Se can act as an antioxidant when incorporated as selenoenzymes in the extracellular space, the cell cytosol, in association with cell membranes and specifically in the gastrointestinal tract, all with potential to influence immune processes. Additionally the selenoenzyme thioredoxin reductase may also act as an antioxidant [16] by decreasing the nucleus redox potential. It also regenerates other antioxidants such as ascorbic acid and α-tocopherol. Other selenoproteins that may have antioxidant functions are selenoproteins P and W [16]. Besides its role as an antioxidant, Se also improves the immune response of the body by enhancing humoral and cellular immunity [20]. Moreover, Se as GSH-Px inhibits NF-κB (nuclear factor-kappa B), which is involved in the transcription of several inflammatory mediators [21] such as IL-6. Thus Se not only reduces oxidative stress by removing FR but also reduces inflammation and improves the defense mechanism of the body.

Glutathione peroxidases Cytosolic GPx Plasma GPx PH-GSH-Px	Antioxidant
Deiodinases Type I 5'-deiodinase Type II 5'-deiodinase	T4 T3
Thioredoxin reductase	Detoxification
Selenoprotein P	Unknown, cell defense ?

Figure 7. Selenoproteins.

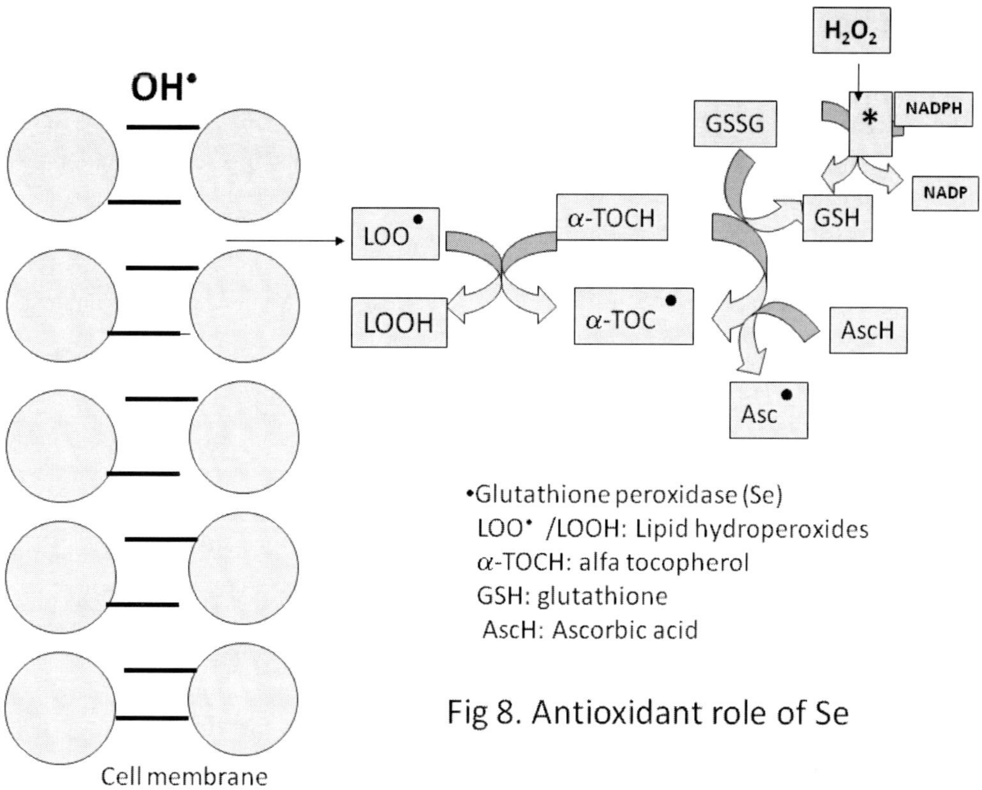

Figure 8. Antioxidant rose of Se.

Nearly all patients with sepsis or SIRS admitted to ICU have low plasma Se levels and GSH-Px activity [22,23,24]. This is either due to increased trans-capillary escape of selenoproteins to interstitial fluid, utilisation of selenoproteins in the face of increased FR production, loss of Se through drains, fistula and skin as seen in surgical or burn patients or poor intake of Se in diet due to chronic illness. Moreover, plasma Se levels correlate inversely with the severity of SIRS and mortality rate [23].

Therefore improving the Se status may help in achieving better clinical outcomes by reducing the oxidative stress in critical illness. Hence Se supplementation may be beneficial in critically ill patients.

Effect of Acute Phase Response on Plasma Se

The acute phase response (APR) or SIRS is defined as non-specific systemic response to inflammation or infection. The acute-phase response is a coordinated sequence of reactions involving biochemical and physiological changes whose aim is to limit damage and aid repair. APR is initiated by any kind of injury i.e. infective or non-infective, which further initiates the release of various cytokines including interleukin 1, 6 and TNFα. These cytokines by activating macrophages remove the causative factor and thereby contain the local damage, eliminate necrotic materials and repair the damage process. Therefore APR primarily defends the body against inflicted injury. However if this process gets aggravated as seen in critical illness, it leads to multiple organ dysfunction.

APR is characterized by the increased production of certain proteins such as C-reactive protein (CRP), fibrinogen, anti-proteases (positive acute phase reactants). At the same time there is increased capillary permeability which results in the transfer of certain proteins such as albumin, prealbumin and transferrin into the extravascular space with a decrease in their concentration in blood (negative acute phase reactants)[25]. During APR there are marked alterations in the distribution of trace metals such as Fe, Zn, Cu, Se. It is hypothesized that the alteration in these trace metals is a beneficial response to defend body against the invading injury.

Se is a negative acute-phase reactant. It has been seen that Se levels in blood are low in patients admitted to ICU by at least 40-60% than in healthy individuals [22,23]. Similar findings were observed in burn patients [24] suggesting that the distribution of body Se may change during critical illness. Similarly studies in critical illness have reported a striking reciprocal relationship between plasma Se and CRP concentrations [22]. Besides critical illness Se levels in blood also fall after minor inguinal hernia repair [26]. An important observation was that the low plasma Se in critically ill patients returns to normal as the inflammation subsides and clinical condition improves, without Se supplements (23)

The reduction in Se concentrations in blood appears to be accounted for by decrease in both the selenoprotein-P fractions and in the amount of Se associated with plasma GSHPx.

Much of this reduction in circulating Se concentrations is therefore probably due to redistribution, similar to the fall in plasma albumin in acute injury. This results in selective uptake of Se by body tissues for metabolic use such as production of selenoenzymes as part of FR defense. The redistribution or trans-capillary redistribution of selenoproteins into extravascular space (tissues) therefore supports its role as an antioxidant, defending the body by reducing the free radical, induced oxidative damage.

Since trauma or SIRS causes an obligatory decrease in Se levels in blood, and resolution of this leads to an increase in Se, its levels in blood should be interpreted cautiously. Hence, as with other trace elements, caution must be used in interpreting a single plasma Se result in a patient with ongoing acute illness. Measuring plasma CRP levels at the same time, and monitoring sequential Se concentrations relative to changes in CRP can achieve greater accuracy of interpretation.

Thus measurement of Se levels in plasma is an unpredictable marker of Se status in seriously ill patients. [22]

Se Supplementation in Critical Illness

Several studies have assessed the role of Se supplementation in critically ill patients. These studies have either supplemented Se alone or in combination with other antioxidants. The primary end point of these studies was clinical outcome in the form of 28 day ICU mortality. Besides this, studies also assessed the outcome on length of stay in ICU, rate of infection, requirement for renal replacement therapy and duration of mechanical ventilation.

The clinical assessments were carried out at the time of admission to ICU, usually by APACHEII and during the stay in ICU by SOFA score.

Acute Physiology and Chronic Health Evaluation II (APACHE II): Determined within 24hrs of admission to the ICU to provide a general measure of the severity of the disease which includes the assessment of respiratory, cardiac, haematology , renal and neurological systems. This score assesses the severity of illness at the time of admission. This score is based on physiological variables which include body temperature, mean arterial pressure, heart rate, respiratory rate, arterial pH, arterial oxygen, plasma sodium, plasma potassium, plasma creatinine, hematocrit, WBC count, Glasgow coma score. Each variable has a score from 0 to 4, which are added together. Besides this, it also takes into consideration the age of patient and chronic health evaluation (CHE). The higher is the score, the more severe is the illness and a score above 15 is considered to have poor prognosis [27].

Sepsis related organ failure assessment (SOFA) score (Fig. 9): Determines the clinical outcome during the patient's stay in ICU. This score is derived from physiological and laboratory variables taking into account the various systems of the body such as respiratory, cardiovascular, renal, neurological, coagulations and hepatic functions. SOFA score is used for the evaluation of organ dysfunction or failure for patients in ICU and has been shown to correlate well with mortality in the ICU. A high SOFA score is associated with a worse prognosis and vice versa [28].

SOFA score >>>	1	2	3	4
Respiratory PaO2/FiO2 kPa [mmHg]	53 [< 400]	40 [< 300]	27 [<200] with support	< 13 [< 100] with support
Coagulation platelets x 10^3/mm^3	<150	<100	<50	<20
Liver bilirubin µmol/L	20–32	33–101	102–204	>204
CVS hypotension, or drugs (µg/kg/min) for > 1 hr	MAP <70	DA $\leq$ 5 or DB any amount	DA > 5 or Adr < 0.1 or NA < 0.1	DA > 15 or Adr > 0.1 or NA > 0.1
CNS: GCS	13–14	10–12	6–9	<6
Renal creatinine µmol/L or UOP mL/day	110–170	171–299	300–440 <500 mL/day	>440 <200mL/day

Figure 9. Sepsis related organ failure assessment (SOFA)score.

Studies, which Supplemented Se Only (Table 1)

The studies supplemented high dose of Se (sodium selenite parenterally) and compared its effect against the standard dose of Se 0.44µmol (35 µg).

Table 1. Studies which supplemented only Se

STUDY	POPULATION	INTERVENTION	MORTALITY Study	Control
Kuklinski (1991)	Acute pancreatitis (N=17)	Se 6.77μmole (500 μg)	0/8	8/9
Zimmermann (1997)	Post-surgical/trauma (N=40)	Se 12.66 μmole (1000 μg)	3/20	8/20
Angstwurm (1999)	Patients with SIRS (N=42)	Se 6.77 μmole (535μg)X 3 days, 3.61 μmole(285μg) X 3 days,1.96 μmole (155μg)X3 days and 0.44 μmole (35μg) thereafter.	7/21	11/21
Mishra(2007)	Patients with SIRS (N=40)	Se 6 μmole (474μg)X 3 days, 4 μmole (316μg) X 3 days,2 μmole (158μg)X3 days and 0.4μmole (31.6μg) thereafter.	8/18	11/22
Angstwurm (SIC study, 2007)	Multicentre mixed ICU (N=249)	Se 12.66 μmole (1000 μg) as bolus on day 1 followed by 12.66 μmole Se from day 2 to day 14	46/116	61/122
Forceville (2007)	Septic Shock patients (N=60)	Se 4000μg on day 1 followed by 1000 μg for 9 days	14/31	13/29

Early Clinical Trials

There have been a small number of early trials [29,30], which evaluated the effect of high dose Se supplementation.

The first study by Kuklinski 1991[29] administered Se 6.77μmole (500 μg) daily to patients (n=17) with acute pancreatitis. Mortality was markedly lower in the study group as compared to the control group. Zimmermann 1997[30] carried out a prospective placebo controlled study in post surgical or trauma patients (n=40) with APACHEII score >14 admitted to ICU. The study patients were supplemented a high dose of Se 12.66 μmole (1000 μg) whereas the control group was supplemented a standard dose of Se daily over a period of 14 days. The end point of the study was mortality rate which was significantly lower in the study (15%) as compared to the control (40%) group.

Although these two studies [29, 30] showed a significant reduction in mortality, suggesting a beneficial effect of Se supplementation, they were carried out in a small number of patients and did not explain the effect of Se on other parameters of clinical outcome, such as infectious complications, length of stay on ventilator or duration of length of stay in ITU, renal replacement therapy, or its effect on antioxidant status. Since these studies were poorly designed, the reliability of their results remains uncertain.

Angstwurm 1999

This was a landmark study, which suggested a particular benefit with high dose Se supplementation to ICU patients. Angstwurm et al [31] carried out a prospective double blind, randomized controlled trial to determine the affect of Se replacement on morbidity and mortality in patients with SIRS. This study was carried in patients (n=42) with SIRS and

APACHEII score >15 on admission to ICU. The study group received Se over a period of 9 days; 6.77 μmole (535μg) for 3 days, 3.61 μmole (285μg) for 3 days and 1.96 μmole (155μg) for 3 days and thereafter 0.44 μmole (35μg) daily. The control group received 0.44 μmole (35 μg) of Se throughout the treatment period.

This study showed a significant increase in antioxidant status, as assessed by plasma Se and GSHPx activity in the study group. In addition, it showed a significant reduction in the requirement of renal replacement therapy in the study group as compared to the control group. Although there was no significant reduction in mortality in the study group (33.5%) as compared to the control (52%) group, a sub analysis of those patients with the most severe sepsis revealed a significant reduction in mortality. Based upon these findings they concluded that Se replacement in patients with SIRS seems to improve the clinical outcome and reduces the incidence of acute renal failure requiring haemodialysis.

Mishra 2007

Mishra et al [32] replicated the Angstwurm study to assess the effect of high dose Se supplementation on Se status in blood, oxidative stress and possible effects on requirement for renal replacement therapy in patients admitted to the ICU. This prospective randomized controlled trial was carried in septic patients (n=40) admitted to ICU with APACHEII score >15. The study patients were randomized to high dose Se (n=18) that received 6μmole (474 μg), 4μmole (316 μg), 2 μmole (158 μg) daily, each for 3 consecutive days followed by 31.6 μg (0.4 μmole) daily,whereas the control group (n=22) received the standard dose of Se 31.6 μg (0.4 μmole) daily throughout the study period. The study period was 14 days during which blood was collected to assess the antioxidant status and clinical outcome. The SOFA score performed in all the study patients assessed the clinical outcome.

The plasma Se and plasma GSH-Px activity increased significantly in patients who received a high dose of Se as compared to the control group. The results confirmed that trace element supplementation could achieve normalization of plasma Se concentration and enhance the antioxidant activity. Over the 14-day study, as CRP levels fell from the initial high concentration, there was a corresponding increase in the plasma Se levels in both groups, suggesting that redistribution may be the major cause of low plasma Se in these patients [23] and that the acute phase response may be responsible for low plasma Se levels in the study patients at the time of admission.

The primary clinical end point of the study was requirement for renal replacement therapy. In contrast to Angstwurm et al [31] who found a reduction in acute renal failure and need for renal replacement therapy, this study did not reduce the requirement for renal replacement therapy. A further possible end point of the study was 28-day ICU or hospital mortality, which did not show a significant difference between high dose Se group (44%) and control group (50%).

The study did not find any significant difference between the two groups with respect to duration of stay in ICU and infection rate. The likely reason for Se supplementation in not improving the clinical outcome was that the study was not adequately powered for this end point due to small sample size.

An important secondary clinical outcome was the SOFA score, which was high at the time of admission to ICU and reduced in both groups throughout the study period. However, this reduction only reached statistical significance in the high dose Se group although there was a strong trend to significance in the control group. This therefore suggests that high dose

Se may lead to a more rapid resolution of organ dysfunction. It was particularly interesting to note a significant negative correlation between plasma Se and SOFA at the time of admission and subsequent days in both groups. This indicates that a comparable amount of redistribution of Se took place in each group in spite of the fact that one group received a high dose of Se as compared to the other which received a standard dose of Se.

This study therefore suggested that severity of illness has a similar effect on redistribution of Se whatever the level of Se intake. Thus measurement of plasma Se levels in plasma is an unpredictable marker of Se status because levels fall with acute phase response provoked by injury or infection and increase with improvement in inflammation. Hence Se levels in blood are influenced by inflammation and severity of illness as well as by the dose of Se administered.

This study was therefore unable to confirm the findings of Angstwurm et al [31] that high dose Se supplementation improves the clinical outcome in critically ill patients, in spite of the fact that both studies had similar number of patients and used an identical regimen of Se supplementation.

Angstwurm (Selenium in Intensive Care or SIC Study) 2007

Angstwurm et al [33] in sequel to their previous study carried a large double blind, placebo-controlled, multicenter Se trial in ICU patients (n=249) with SIRS or sepsis. The study administered a very high dose of Se 12.66 µmole (1000 µg), as bolus (within 30 min) on day 1 followed by 12.66 µmole Se daily, as a continuous infusion from day 2 to day 14. The control group which received the standard dose of Se throughout the study period.

The study found a reduction of 10.3% in the 28-day mortality in the study patients as compared to the controls, but this reduction was not statistically significant. This study did, however, find that high-dose Se therapy reduced mortality significantly in patients in the highest quartile of APACHEII score (55.6% in high dose versus 81.5% in standard dose, $p=0.04$) and in those with septic shock with disseminated intravascular coagulation (40.5% in high dose versus 66.7% in standard dose, $p=0.01$). Somewhat surprisingly, despite the apparent benefit on outcome, this study did not confirm the previous [30] benefit on renal function, nor on infection, pneumonia or liver function. The study therefore suggested that high dose selenium supplementation reduces mortality rate in patients with particularly severe sepsis or septic shock.

Forceville 2007

This study [34] supplemented a very high dose of Se (4000 µg) on day 1 followed by 1000 µg daily for 9 days in septic shock patients (n=31), whereas the control group received a standard dose of Se (n=29). There was no significant reduction in mortality in the study group (mortality 50% in both study and control group).

Forceville 2009

This animal study [35] assessed the effect of high dose Se on septic shock. The sheep (n=21) with experimental peritonitis were randomized into (i) bolus injection (2 mg Se, followed by 0.06 µg/ kg/hour, n = 7); (ii) continuous infusion (4 µg/kg/hour Se, n = 7), or (iii) control (n = 7). Compared with the other groups, sheep given a bolus of Se had delayed hypotension with better maintained cardiac index, delayed hyperlactatemia, fewer sepsis-induced microvascular alterations, and a prolonged survival time. The study suggested that the administration of a large bolus of Se (rather than a continuous administration) results in beneficial effects, possibly by a transient oxidative effect.

Studies, Which Supplemented Se in Combination with Other Antioxidants (Table 2)

These studies supplemented Se in combination with other antioxidants such as trace metals Zn and Cu or vitamins such as vitamin E, C and βcarotene.

Table 2. Studies which supplemented Se with other antioxidants

STUDY	POPULATION	INTERVENTION	MORTALITY		Infection/Patient	
			Study	Control	Study	Control
Berger (1998)	Burns >50% burn surface area (N=20)	Cu 40.4 μmole, Se 159 μg, Zn 406 μmole or standard supplementation			1.9 ± 0.9	3.1 ± 1.1
Berger (2001)	Major Trauma (N=32)	Se (500 μg/day) alone or Se (500 μg), Zn(13mg), α-tocopherol (150mg) or placebo	Se alone 2/9 Se+Zn+ α-tocopherol 0/11	1/11	NA	
Berger (2007)	Burns > 20 % burn surface area (N=21)	Cu (59μmole) + Se (375 μg+ Zn (574μmole) for 5-15 days.	1/11	1/21	2.1 ± 1.0	3.6 ± 1.0
Berger (2008)	Mixed ICU (N = 200)	Se 270μg , Zn 30 mg, vitamin C 1.1 g, and vitamin B1 100 mg with a double loading dose on day 1,2	8/102	5/98	36/102	34/98
SIGNET (2010)	Surgical ICU (N=502)	Glutamine 20.2 g daily and/or Se 500 μg daily for 7 days	NA		Glutamine (odds ratio,OR, 1.07, 95% CI 0.75–1.53) Se (OR 0.81, 95% CI 0.57–1.15)	

Berger 1998

Berger et al [36] undertook a study in severely burned patients. It was carried out in patients (n=20), burned on 48±17% of their body surface, APACHEII score 11±4, and who were studied for 30 days after injury. Patients received either the usually recommended standard intravenous trace element supplies (Cu 20 μmole, Se 32 μg, Zn 100 μmole) or additional large supplements (Cu 40.4 μmole, Se 159 μg, Zn 406 μmole) plus standard trace elements.

The number of infections per patient was significantly lower in the group which received additional large trace element supplementation (1.9 ± 0.9 per patient) than in the group, which received standard supplementation (3.1 ± 1.1 per patient).

Berger 2001a and 2001b

The objective of this study [37,38] was to determine the effect of early micronutrient supplementation on the antioxidant status in trauma patients. 32 patients admitted with major trauma were randomized to receive, either Se (500 μg) alone or Se (500 μg), Zn (13mg), α-tocopherol (150mg) or placebo for 5 days after injury. There was no significant difference in duration of mechanical ventilation, infection, length of stay in ICU in either group.

Berger 2002 And 2007

These studies [39,40] involved burn patients (n=21) with more than 20 % burn surface area, supplemented with intravenous Cu (59 μmole) plus Se (375 μg) plus Zn (574 μmole) versus placebo from admission for 5-15 days. The infection rate reduced significantly in group, which received trace element supplementation (2.1 ± 1.0 per patient) versus placebo (3.6 ± per patient).

Berger et al [39,40] showed a reduced infection rate in burn patients treated with a combination of trace element supplementation (Zn, Cu and Se), especially with a reduction in nosocomial pneumonia. The assumption is that these supplements are not only preventing the deficiency caused by losses through the skin, but they are also maintaining immune function, especially in the lungs which get first pass of the intravenous supplements used. The burned skin are found to take up the supplements provided, perhaps accounting for the the improved rate of skin healing in these patients and the reduced need for skin grafting (40)

Berger 2008

The aim of this study [41] was to investigate the impact of early antioxidant micronutrients on clinical outcome in ICU patients with conditions characterized by oxidative stress. This study was carried in (n=102,study) and (n=98,control) patients admitted to ICU with organ failure after complicated cardiac surgery, major trauma, or subarachnoid hemorrhage. The study group received intravenous supplements for 5 days (Se 270 μg , Zn 30 mg, vitamin C 1.1 g, and vitamin B1 100 mg) with a double-loading dose on days 1 and 2 or placebo. There was no significant difference in organ function endpoints, incidence of acute kidney failure, infectious complications, length of stay. However plasma concentration of Se, Zn and GSHPx increased significantly in the study group.

SIGNET (Scottish Multicentre Trial of Glutamine and SE Supplemented Parenteral Nutrition for Critically Ill Patients) 2010

There is considerable interest in glutamine and Se in critical care as both offer the potential to enhance host defenses, through different but complimentary mechanisms and may reduce subsequent infections and mortality. The SIGNET trial (randomized controlled factorial trial)[42] is the largest, critical care study of both supplements. The study enrolling participants (n=502) from 10 hospitals in Scotland was completed in December 2008. The majority of patients were surgical and the mean age was 64 years. More than half of the patients had sepsis at recruitment, and just over a quarter were under-nourished.

The SIGNET trial was a randomized, factorial design, controlled trial of parenteral glutamine 20.2 g daily and/or Se supplementation 500 μg daily in adult patients in ICU. Supplementation was for up to seven days, unless death occurred or parenteral nutrition was discontinued. The primary outcomes were whether each participant had an infection within the first 14 days, and all cause mortality. Secondary outcomes were ICU and acute hospital length of stay; days of antibiotic use; sepsis-related organ failure assessment score and any serious adverse events.

In an intention to treat analysis there was no statistically significant reduction in participants with infections from glutamine (odds ratio, OR, 1.07, 95% CI 0.75–1.53) or Se (OR 0.81, 95% CI 0.57–1.15). In the pre specified subgroup analysis of those people who received at least 5 days of supplementation or died before this time, the corresponding figures for infection were glutamine (OR 1.08,95% CI 0.66–1.76) and Se (OR 0.61,95% CI 0.37–0.99). There was no significant effect on mortality or length of stay. The study reported a significant reduction (10-15%) in infection rates, such as pneumonia and C.difficile, with Se supplemented parenteral nutrition in critically ill patients as long as the parenteral nutrition was administered for at least 5 days. In the Se arm, no clear benefit on mortality was observed but a significant reduction in new infection rates was observed as described above.

Meta-Analysis of Studies, Which Supplemented Se Either Alone or in Combination

A meta-analysis published by Heyland et al [43] assessed the affect of antioxidants in critically ill patients. They reviewed the data of eleven studies that replaced either trace elements or vitamins or a combination of two in critically ill patients. The primary outcome of this meta-analysis was mortality and infection.

The Se studies were compared with non-Se studies, which did not supplement Se but vitamins such as C, E or trace metals such as Zn. The Se studies were associated with a trend towards lower mortality (RR 0.59, p=0.09) while non-Se was found to have no effect on mortality (RR 0.73, p=0.3). Only two Se studies and three non-Se studies reported the infectious complications, and neither of the studies showed any significant effect on infectious complications. They therefore suggested that trace elements, particularly high doses of parenteral Se (more than 500μg per day), either alone or in combination with other antioxidants, are safe or may be associated with a reduction in mortality in critically ill patients.

A Cochrane review [44] pooled the mortality data of studies, which supplemented Se in critically ill patients. It was found that the quality of these trials was poor, the studies had small numbers of subjects and data related to clinical outcome was limited. Also, no clear evidence emerged for the benefits of Se supplementation with respect to days on ventilator, length of stay in ITU or hospital, and infection rate. This review further stated that there was statistical heterogeneity in the data, and reanalysis of the data by random fixed model did not show a significant effect of Se supplementation on clinical outcome.

Hence the review suggested that although a beneficial effect is possible, the studies are small, poorly designed or carried out on a specific patient population and show statistical heterogeneity. Thus there was insufficient evidence at that time to recommend supplementation of Se in critically ill patients and further trials were required

Canadian Practice Guidelines (2009)

The Canadian committee established guidelines for parenteral Se supplementation in critically ill patients by analysing the mortality data from all studies, which either supplemented Se alone or in combination to patients in ICU (Fig.10). It was noted that the evidence from these trials suggested that the effect of Se supplementation with respect to reduction is small and that the confidence interval overlap's 1. This remained unchanged even after the exclusion of the small study by Kuklnsky et al [29]. A similar result was shown from Se alone studies (Fig.11). There also was concern regarding the heterogeneity in the trial designs, the negative safety reports and inconsistency in dosing ranges in the critically ill patients.

It was therefore felt that there was not enough evidence to support the use of parenteral Se supplementation, alone or in combination with other antioxidants in critically ill patients.

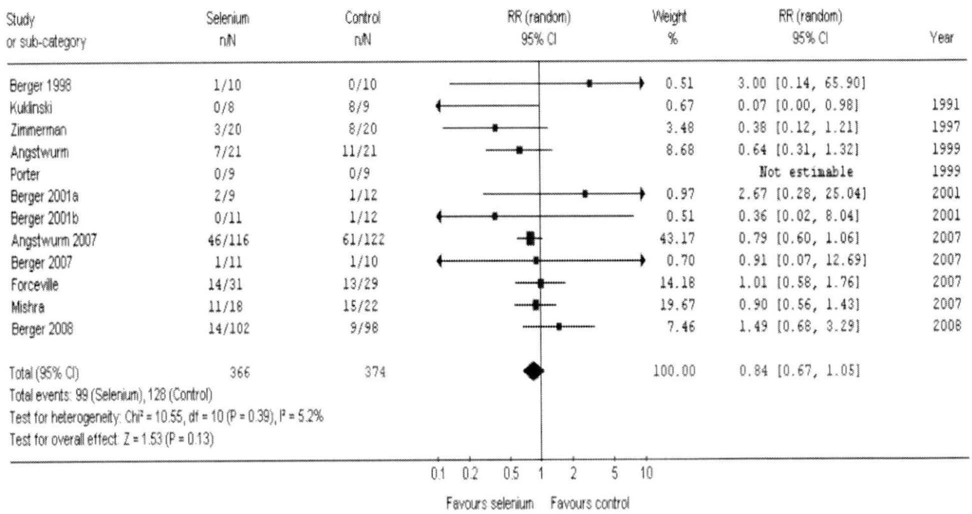

Figure 10. Mortality analysis of studies with supplemented Se with other antioxants.

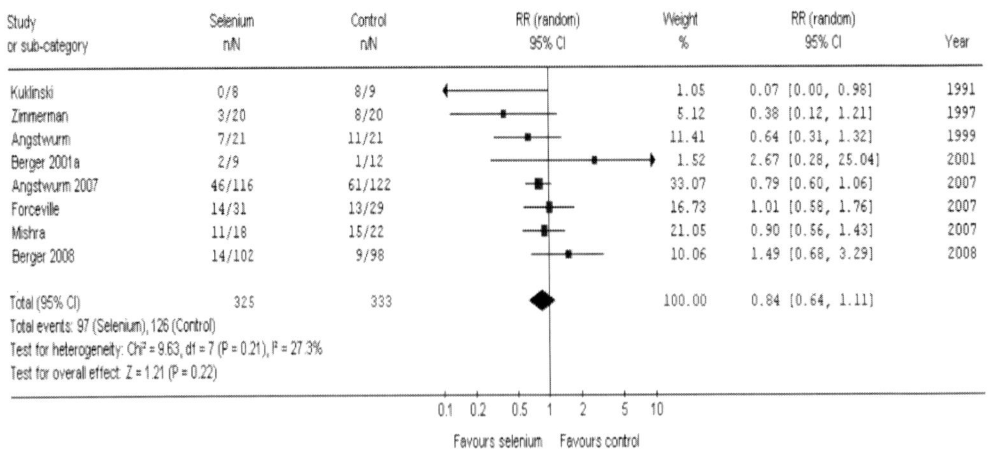

Figure 11. Mortality analysis of studies which supplemented only Se.

CONCLUSION

- There is some supportive data for high dose parenteral Se supplementation in very severe sepsis or septic shock but this needs to be investigated further through well-designed randomised controlled trials.
- There is evidence that high dose Se supplementation (together with high dose zinc and copper) reduces the rate of infection in patients with burns.
- The acute phase response and severity of illness results in a redistribution of plasma Se leading to low plasma levels
- Se levels in blood should be interpreted with caution, and preferably on more than one occasion together with C-reactive protein.
- Despite some promising results, there are no definitive answers regarding the effects of Se supplementation on mortality or morbidity of critically ill patients.

Further prospective clinical trials are required of different doses of Se, and in patients with a a range of severity of illness.

REFERENCES

[1] Halliwell B, Gutteridge JMC. *Free radicals in biology and medicine*. Second edition.Oxford, Clarendon Press, 1989.
[2] Slater TF. *Free radical mechanisms in tissue injury*. London, Pion, 1972.
[3] Gutteridge JMC, Halliwell B. Free radicals and antioxidants in the year 2000. A historical look to the future. *Ann N Y Acad Sci*. 2000; 899: 136-47.
[4] American College of Chest Physicians/ Society of Critical Care Medicine Consensus Conference: Definitions for sepsis and organ failure and guidelines for the use of innovative therapies in sepsis. Members of the American college of chest physicians/ Society of Critical Care Medicine, Consensus Conference Committee. *Critical Care Medicine* 1992; 6: 864-874.
[5] Dellinger RP, Bone RC. To SIRS with love. *Crit Care Med*. 1998 Jan; 26:178-9.
[6] Bone RC, Grodzin CJ, Balk RA. Sepsis: a new hypothesis for pathogenesis of the disease process.*Chest*.1997;112:235-243.
[7] Lowry SF. Sepsis and its complications: The clinical problem. *Crit Care Med*1994; 22: S1-S2.
[8] Bulger EM, Maier RV. Antioxidants in critical illness. *Archives of Surgery* 2001; 136: 1201-7.
[9] Babior BM. The respiratory burst oxidase. *Curr Opin Hematol*.1995;2:55-60.
[10] Sbarra AJ, Selvaraj RJ, Cetrulo CL, Thomas G, Louis F, Kennison R. Phagocytosis and onset of human labor. *Am J Obstet Gynecol*. 1983 Jul 15; 146:622-629.
[11] Johnston RB Jr. Monocytes and macrophages. *N Engl J Med*. 1988; 318: 747-752.
[12] Carrigan SD, Scott G, Tabarizian M. Toward resolving the challenges of sepsis diagnosis. *Clinical Chemistry* 2004; 50: 1301-1314.
[13] Balk R A, Bone R C.The septic syndrome: Definition and clinical implications. *Crit Care Clin* 1989; 5:1-8.

[14] Texereau J, Pene F, Chiche JD, Rousseau C, Mira JP. Importance of hemostatic gene polymorphisms for susceptibility to and outcome of severe sepsis. *Crit Care Med.* 2004; 32: S313-S319.

[15] Taylor DE, Piantadosi CA. Oxidative metabolism in sepsis and sepsis syndrome. *J Crit Care* 1995; 10: 122-135.

[16] Brown KM, Arthur JR. Selenium, selenoproteins and human health: a review. *Public Health Nutr* 2001;4(2B):593–9.

[17] Arthur JR. The glutathione peroxidases. *Cell Mol Life Sci* 2000;57: 1825–35.

[18] Arthur JR, Beckett GJ. Selenium and endocrine systems. *J Endocrinol* 2005; 184: 455–65.

[19] Harrison I, Littlejohn D, Fell GS. Distribution of selenium in human blood and plasma and serum. *Analyst* 1996; 121(2): 189–94.

[20] Arthur JR, Mckenzie RC, Beckett GC. Selenium in the immune system. *Am Soc Nutr Sci* 2003:457S–9S.

[21] Mihm S, Droge W. Intracellular glutathione levels controls DNA binding activity of NFkB like proteins. *Immunology* 1990; 181: 245-7

[22] Hawker FH, Stewart PM, Snitch PJ. Effects of acute illness on selenium homeostasis. *Crit Care Med* 1990; 18: 442–6.

[23] Forceville X, Vitoux D, Gauzit R, Combes A, Lahilaire P, Chappuis P. Selenium, systemic immune response syndrome, sepsis and outcome in critically ill patients. *Crit Care Med* 1998;26 :1536–44.

[24] Berger MM, Cavadini C, Bart A, Blondel A, Bartholdi I, Vandervale A. Selenium losses in 10 burned patients. *Clin Nutr* 1992; 11: 75–82.

[25] Fleck A. Clinical and nutritional aspects of changes in acute-phase proteins during inflammation.*Proc Nutr Soc* 1989;Sep 48(3):347-54.

[26] Nichol C, Herdman J, Sattar N, O'Dwyer PJ, O'Reilly SJ, Littlejohn D, Fell G. Changes in the concentrations of plasma selenium and selenoproteins after minor elective surgery: Further evidence for a negative acute phase response*? Clin Chem* 1998; 44(8): 1764–6.

[27] Knaus A, Draper EA, Wagner DP, Zimmerman JE. APACHEII: A severity of disease classification system. *Crit Care Med* 1985; 13: 818-29.

[28] Vincent J-L, Moreno R, Takala J, Willatts S, Mendonca A, Bruining H. The SOFA (sepsis related organ failure assessment). *Intens Care Med* 1996; 22: 707–10. /

[29] Kuklinski B, Buchner M, Schweder R,Nagel R. Akute Pancreatitis—eine "free radical disease. Letalitatssenkung durch Natriumselenit (Na2SeO3)-Therapie. Z Gesamte. *Inn Med* 1991; 46:S145–S149.

[30] Zimmermann T, Albrecht S, Kuhne H, Vogelsang U, Grutzmann R, Kopprasch S. Selensubstitution bei sepsis-patienten. *Med Klin* 1997; 92 (Suppl III): 3–4.

[31] Angstwurm MW, Schottdorf J, Schopohl J, Gaertner R. Se replacement in patients with severe systemic inflammatory response syndrome improves clinical outcome. *Crit Care Med* 1999; 27: 1807–1813.

[32] Mishra V, Baines M, Parry S, Mclaughlin J, Carson J, Wenstone R, Shenkin A.Effect of Selenium supplementation on biochemical markers and outcome in critically ill patients" *Clinical Nutrition* 2007; 26: 41-50.

[33] Angstwurm MW, Engelmann L, Zimmermann T, Lehmann C, Spes CH, Abel P, Straub R, Meier-Hellmann A, Insel R, Radke J, Schuttler J, Gartner R. Selenium in intensive

care (SIC) study: Results of a prospective randomized, placebo-controlled, multiple-center study in patients with severe systemic inflammatory response syndrome, sepsis, and septic shock. *Crit Care Med.* 2007 ; 35(1): 1-9.

[34] Forceville X, Laviolle B, Annane D, Vitoux D, Bleichner G, Korach JM, Cantais E, Georges H, Soubirou JL, Combes A, Bellissant E. Effects of high doses of selenium, as sodium selenite, in septic shock: a placebo-controlled, randomized, double-blind, phase II study. *Crit Care*. 2007;11(4):R73.

[35] Wang Z, Forceville X, Van Antwerpen P, Piagnerelli M, Ahishakiye D, Macours P, De Backer D, Neve J, Vincent JL.A large-bolus injection, but not continuous infusion of sodium selenite improves outcome in peritonitis. *Shock* 2009 Aug; 32(2): 140-6.

[36] Berger MM, Spertini F, Shenkin A, Wardle C, Wiesner L, Schindler C, Chioléro RL.Trace element supplementation modulates pulmonary infection rates after major burns: a double –blind, placebo-controlled trial. *Am J Clin Nutr* 1998; 68:365-71

[37] Berger MM, Recmond MJ, Shenkin A, Rey F, Wardle C, Cayeux C, Schindler C, Chiolero. Influence of early trace element and vitamin E supplements on antioxidant status after major trauma: a controlled trial. *Nutrition Research* 2001; 21: 41-54.

[38] Berger MM, Recmond MJ, Shenkin A, Rey F, Wardle C, Cayeux C, Schindler C, Chiolero. Influence of selenium supplements on the post-traumatic alterations of the thyroid axis: a placebo- controlled trial. *Intensive Care Med* 2001; 27:91–100

[39] Berger MM, Eggimann P, Heyland DK, Chiolero RL ,Revelly JP,Day A, Raffoul W , Shenkin A. Reduction of nosocomial pneumonia after major burns by trace element supplementation: aggregation of two randomised trials. *Critical Care* 2006; 10(6): R153 (Epub)

[40] Berger MM, Baines M, Raffoul W, Benathan M, Chiolero R, Reeves, Revelly P, Cayeux C, Sénéchaud I, Shenkin A. Trace element supplementation after major burns modulates antioxidant status and clinical course by way of increased tissue trace element concentrations. *American Journal of Clinical Nutrition* 2007; 85(5), 1293-1300.

[41] Berger MM, Soguel L, Shenkin A, Revelly JP, Pinget C, Baines M, Chiolero RL. Influence of early antioxidant supplements on clinical evolution and organ function in critically ill cardiac surgery, major trauma, and subarachnoid hemorrhage patients. *Crit Care*. 2008;12(4):R101.

[42] Andrews PJ, Avenell A, Noble DW, Campbell MK, Croal BL, Simpson WG, Vale LD, Battison CG, Jenkinson DJ, Cook JA. Randomised trial of glutamine, selenium, or both, to supplement parenteral nutrition for critically ill patients. *BMJ* 2011;342:d1542

[43] Heyland DK, Dhaliwal R, Suchner U, Berger MM. Antioxidant nutrients: a systematic review of trace elements and vitamins in the critically ill patient. *Intensive Care Med.* 2005; 31: 327-337.

[44] Avenell A, Noble DW, Barr J, Engelhardt T. Se supplementation for critically ill adults. *Cochrane Review The Cochrane Library* 2005, Issue 1:1-18.

In: Selenium: Sources, Functions and Health Effects
Editors: Chinatsu Aomori and Megumi Hokkaido
ISBN: 978-1-61942-061-8

Chapter 7

SELENIUM AND TROPICAL DISEASES

Daniel Plano[1,2]***, María Font***[1,2]***,***
Juan Antonio Palop[1,2] ***and Carmen Sanmartín***[1,2]
[1]Instituto de Salud Tropical, Universidad de Navarra, Pamplona, Spain
[2]Departamento de Química Orgánica y Farmacéutica, University of Navarra, Pamplona, Spain

1. ABSTRACT

According to World Health Organization, neglected tropical diseases encompass all diseases that occur solely, or principally, in the tropics. In practice, the term is often taken to refer to infectious diseases that thrive in hot, humid conditions. Clinical and epidemiological studies suggest that selenium (Se) play an important role in tropical diseases, such as tuberculosis, leishmaniasis, filariasis and chagas, acting as preventive agent or in diagnosis and prognosis. Recent studies have evinced the importance of selenium in oxidative status and antioxidant defense capabilities during the course of infection and progression of the illness in human patients and experimental models. For this reason, one of the most relevant mechanism of action proposed involve selenoproteins, i.e. glutathione peroxidase (GPx), an enzyme that protects against oxidative stress and modulates the redox processes. In addition, it was observed that low Se levels were positively correlated with an increased susceptibility to infections. Besides, Se supplementation is proposed as an adjuvant therapy for treatment of these chronic diseases. However, there is a lack in the literature references related to synthesis and biological evaluation of novel derivatives containing selenium moiety against these diseases. During the last years our research group is interested in the design and synthesis of organoselenium compounds as new class of agents for treatment of neglected tropical diseases.

In the present year we have reported two general structures with leishmanicidal activity, corresponding both of them to symmetrical compounds. The firsts are alkyl imidoselenocarbamates (alkyl isoselenourea) which possessed a moderate effect *in vitro* and the second ones are selenocyanates and diselenides. It is remarkable that some of them showed stronger *in vitro* antileishmanial activity than edelfosine and miltefosine, used as reference drugs, and combined high potency and low cytotoxicity against Jurkat and THP-1 cells.

2. Introduction

2.1. Tropical Diseases: An Overview

According to World Health Organization (WHO), neglected tropical diseases (NTDs) encompass all diseases that occur solely, or principally, in the tropics. In practice, the term is often taken to refer to infectious diseases that thrive in hot, humid conditions. The WHO recommends five public-health strategies for the prevention and control of NTDs: preventive chemotherapy; intensified case-management; vector control; the provision of safe water, sanitation and hygiene; and veterinary public health (that is, applying veterinary sciences to ensure the health and well-being of humans).

2.1.1. Chagas

Chagas disease, caused by *Trypanosoma cruzi*, is presented in endemic geographic regions in Middle and South America, affecting 15 million infected people. Treatment options are limited by toxicity of available drugs, parasite resistance and poor drug activity during the chronic phase of the disease. The drugs of choice for treatment of chagas disease are benznidazole (Rochagan®; Roche) and nifurtimox (Lampit®; Bayer) [1], but resistances are increasing and therapeutic success is phase-dependent. Clinically, the pathogenesis of the disease can be divided into three phases, namely a short acute phase, a long-lasting latent phase, and a chronic phase emerging in 10–30% of patients. Up to 30% of patients will develop heart damage, and up to 10% will develop damage to their oesophagus, colon or autonomic nervous system, or all of these, in the late chronic phase of the disease. Patients ultimately die, usually from sudden death caused by arrhythmias; this often occurs in early adulthood. More than 10,000 deaths are estimated to occur annually from chagas disease. The incapacitating effects and mortality of chagas disease have been one of the biggest public-health problems in Latin America. According to a published study [2], the 10-year mortality rate may range from 9% to 85%, depending on the cardiac damage.

2.1.2. Filariasis

One hundred and twenty million people in 83 countries of the world are infected with lymphatic filarial parasites, and it is estimated that over 1.3 billion (20% of the world's population) are at risk of acquiring infection. Ninety percent of these infections are caused by *Wuchereria bancrofti* and most of the remainder by *Brugia malayi*. Brugian parasites are confined to areas of east and south Asia, especially India, Malaysia, Indonesia, the Philippines, and China [3]. Approximately 10–30% of the infected individuals develop clinical findings such as hydrocoele, adenolymphangitis and lymphoedema that can progress to elephantiasis.

2.1.3. Leishmaniasis

Leishmaniasis is a disease caused by protozoan parasites transmitted through the bites of infected sandflies. This disease has a wide range of clinical symptoms. The differing manifestations of the disease arise from infections with different species of *Leishmania*. Leishmaniasis is a predominantly rural disease, and is prevalent in 88 countries on 4 continents. The disease is estimated to cause 1.6 million new cases annually, of which an estimated 500,000 are visceral and 1.1 million cutaneous or mucocutaneous.

Visceral leishmaniasis, also known as kala-azar, attacks the internal organs and is the most severe form of the disease. Left untreated, it is usually fatal within 2 years. Furthermore, a percentage of cases may evolve to skin dissemination of parasites. Visceral leishmaniasis is characterized by irregular bouts of fever, substantial weight loss, swelling of the spleen and liver, and pancytopenia. After treatment, this form may evolve into a cutaneous form known as post-kala-azar dermal leishmaniasis, which requires lengthy treatment.

The cutaneous form is the most common. It usually causes ulcers on the face, arms and legs. Although the ulcers heal spontaneously, they cause serious disability and leave severe and permanently disfiguring scars. The cutaneous form may produce up to 200 lesions and lead to disability. The patient is left permanently scarred, and so may become socially stigmatized. Diffuse cutaneous leishmaniasis produces chronic skin lesions that never heal spontaneously. Recidivans cutaneous leishmaniasis is a relapsing form that appears after treatment.

The most disfiguring form is the mucocutaneous, which invades the mucous membranes of the upper respiratory tract, causing gross mutilation as it destroys the soft tissues of the nose, mouth and throat. Patients with this form of the disease may also suffer from discrimination and prejudice.

Coinfection with *Leishmania* and HIV is an emerging problem that requires urgent attention. Patients, who are coinfected, may relapse repeatedly despite receiving proper treatment, and frequently the outcome is fatal [4].

Conventional therapies are based on treatment with pentavalent antimonials (sodium stibogluconate and meglumine antimonate), pentamidine, or amphotericin B [5]. The high resistance rates developed against pentavalent antimonials, their high dose regimens, and the long treatment courses of parentheral administration required -as well as their high renal and cardiac toxicity- are all major drawbacks. Two recent papers [5, 6] summarize available data about antileishmanial drugs, with specific information about those that are currently included in clinical trials, their status and future perspectives in the area of combinational therapy, plant product research, and novel drug targets. The alkyl-phospholipid miltefosine has emerged as an oral drug that is highly effective in children but occasionally presents gastrointestinal complications and also shows teratogen effects [7]. Edelfosine, another alkyl-phospholipid, has also been tested and was found to display a higher *in vitro* activity than miltefosine [8].

2.1.4. Malaria

Approximately 3.3 billion people live in 108 malarious countries and are seasonally at risk of infection with *Plasmodium* spp. For 2008, the WHO estimated that there were 243 million cases of malaria worldwide, accounting for an estimated 863,000 deaths [9]. The vast majority of cases (85%) occurred in Africa, followed by 10% in Southeast Asia and 4% in Eastern Mediterranean regions.

Four *Plasmodium* spp. are recognised as human pathogens, namely *Plasmodium falciparum* (the causative agent of severe tropical malaria), *Plasmodium malariae*, *Plasmodium vivax* and *Plasmodium ovale*.

Currently, malaria control relies on a limited number of tools; in particular malaria treatment with artemisinin derivatives and so-called artemisinin-based combination therapy

(ACT) as well as vector control with insecticidal pyrethroids, but both could be lost to resistance at any time [10-12]. Sustainability of effective programmes through training of clinical employees and institution strengthening of malaria clinics as well as improved surveillance and drug development are necessary for malaria eradication [13].

2.1.5. Trypanosomiasis

Human African Trypanosomiasis (HAT), or sleeping sickness, is caused by infection with protozoan parasites belonging to the genus *Trypanosoma*. It is a vector-borne disease and is usually fatal if left untreated. Parasites are transmitted through the bites of tsetse flies (*Glossina* spp.) that have acquired their infection from humans or animals. Following the bite of the infected fly, the parasite multiplies in the lymph and blood, causing headaches, fever, weakness and pain in the joints. Over time, the parasite crosses the blood–brain barrier, migrates to the central nervous system (CNS) and causes severe neurological and psychiatric disorders leading to death. Following continued control efforts, the number of cases reported in 2009 by the WHO has dropped below 10,000 for the first time in 50 years, however the estimated number of actual cases is currently 30,000.

The human disease takes two forms, depending on the subspecies of *Trypanosome* involved. *Trypanosoma brucei gambiense* causes a chronic infection that may persist for months or even years without major signs or symptoms of the disease. *Trypanosoma brucei rhodesiense* causes an acute infection; signs and symptoms are observed a few weeks after the infective bite. The acute form develops rapidly, soon invading the CNS. Both forms begin with a haemolymphatic stage and become fatal when the *Trypanosomes* invade the CNS. HAT is one of the very few infectious diseases with a mortality rate of 100% if left untreated.

Four drugs are registered for the treatment of sleeping sickness and are provided free of charge to endemic countries through a WHO private partnership with Sanofi-Aventis (pentamidine, melarsoprol and eflornithine) and Bayer AG (suramin). During the first stage, pentamidine is used for *T. b. gambiense* infections; it is also effective against the haemolymphatic stage of *T. b. rhodesiense*. Second-stage treatment melarsoprol, which is used for both forms of HAT, is an arsenic compound and has severe side effects, including reactive encephalopathy that can be fatal in up to 10% of cases. Moreover, upcoming resistance has been reported, particularly from Central Africa. A combination treatment of nifurtimox and eflornithine (NECT) was introduced in 2009. After safety and efficacy approval of NECT in clinical trials, the drug was listed as an 'Essential Medicine' by the WHO. Furthermore, NECT is provided free for the treatment of HAT by the WHO. Fexinidazole, a 2-substituted 5-nitroimidazole, is the first new clinical drug candidate in 30 years with the potential to treat advanced-stage sleeping sickness. After promising pre-clinical pharmacological and safety studies, fexinidazole as oral treatment has entered first-in-human phase I studies in September 2009.

2.1.6. Tuberculosis

Tuberculosis (TB) is caused by infection with *Mycobacterium tuberculosis*, which is transmitted through inhalation of aerosolized droplets. Tuberculosis constitutes a serious global health problem with nearly 10 million new cases of tuberculosis and 1.7 million deaths every year [14]. Furthermore, it is estimated that two billion people live with latent *M. tuberculosis* infection and represent a potential source of future active tuberculosis [15]. The US Centers for Disease Control (CDC) defines multi-drug resistant (MDR) TB as TB infection that is resistant to multiple anti-TB medications, including at least two of the first line agents used to treat TB (isoniazid and rifampin) [16]. While the number of drug-resistant strains of *M. tuberculosis* in the US has been declining since 1997, the number of drug-resistant strains in other countries has risen from 25.5% in 1993, to 80% in 2007 [17].

The drugs used in the TB treatment can be classified into two main groups:

First Line anti-TB Agent

Rifampicin has long been considered the backbone of treatment for TB. Data suggests that resistance to rifampicin leads to more treatment failures and patient deaths than resistance to any other agent [18]. Rifampicin is a bactericidal antibiotic that inhibits the bacterial DNA-dependent RNA polymerase. It targets the β-subunit of the RNA polymerase, physically blocking elongation of the RNA chain.

Isoniazid is another first line agent for the treatment of TB infection. The results of numerous studies suggest that isoniazid primarily affects cell wall mycolic acid synthesis through targeting of InhA, an NADH-dependent enoyl acyl carrier protein reductase [19].

Pyrazinamide is similar to isoniazid in that it is a prodrug that requires activation before exerting its effects. Pyrazinamide is converted to pyrazinoic acid via pyrazinamidase, a bacterial enzyme encoded by the 561 nucleotide pncA gene [20, 21].

Ethambutol is a bacteriostatic agent, and similar to isoniazid, inhibits mycobacterium cell wall synthesis [22]. More specifically, ethambutol disrupts arabinogalactan synthesis by inhibiting arabinosyl transferase. The inhibition of arabinosyl transferase prevents the formation of the mycolyl-arabinogalactan–peptidoglycan complex, which functions to increase cell wall permeability [22].

Second Line anti-TB Agents

Ethionamide is a thioamide antibiotic that is widely used as a second line agent for MDR TB. Ethionamide is activated via S-oxidation following administration and can be bacteriostatic or bactericidal. The precise mechanism by which ethionamide exerts its effects is somewhat unclear, but a common theory is that it acts by inhibiting peptide synthesis. Once activated, ethionamide is metabolized into 4-pyridylmethanol [23].

Fluoroquinolones, particularly third generation fluoroquinolones such as gatifloxacin and moxifloxacin, have an important place in the treatment of MDR TB. All fluoroquinolones work in a similar fashion by inhibiting the function of mycobacterial DNA gyrase, which is encoded gyrAB.

Streptomycin is another common second line agent used in the treatment of MDR TB. Classified as an aminocyclitol glycoside, streptomycin binds to the 16S ribosomal (r)RNA and interferes with translational proofreading, thereby inhibiting protein synthesis [24].

2.2. Selenoproteins in the Metabolism of the Parasites

Proteins containing the 21st naturally occurring amino acid, selenocysteine (Sec), are called selenoproteins. Most selenoproteins belong to one of two main groups according to the location of the Sec residue. Selenoproteins of the first group contain Sec very close to the C-terminus of the protein. Mammalian selenoproteins K, S, O, I, R and thioredoxin reductases are examples of such proteins. Other proteins, such as mammalian selenoproteins H, M, T, V, W, Sep15, selenophosphate synthetase 2, thyroid hormone deiodinases and glutathione peroxidases, have Sec in the *N*-terminal segments of proteins.

2.2.1. Plasmodium

By searching computationally for evolutionarily conserved Sec insertion sequence (SECIS) elements, which are RNA structures involved in Sec insertion, Lobanov *et al.* identified four unique *P. falciparum* selenoprotein genes [25]. The *Plasmodium* SECIS elements are functional. So, four unique selenoproteins were found in *Plasmodium*, including two predicted to be located in the apicoplast (a cell organelle that contains several unique metabolic pathways and is absent in the host). Besides, full genome analysis of *Anopheles gambiae* mosquito –an important vector of *P. falciparum*- revealed that the enzyme glutathione reductase is absent and functionally substituted by the thioredoxin system [26].

Most protozoan parasites are fast replicant organisms. It needs an adequate supply of deoxyribonucleotides for rapid DNA synthesis and protection from reactive oxygen metabolites generated by the host's immune system. The thioredoxin system can fulfill both tasks and therefore seems to be an essential part of the parasite's metabolism [27].

2.2.2. Kinetoplastida and other Parasites

A Sec biosynthesis pathway is present in Kinetoplastida (*Leishmania* and *Trypanosoma*). These include the existence of SECIS-containing coding sequences in *L. major* and *L. infantum*, the incorporation of ^{75}Se into *Leishmania* proteins, the occurrence of selenocysteine-tRNA (tRNAsec uca) in both *Leishmania* and *Trypanosoma* and in addition the finding of all genes necessary for selenocysteine synthesis such as SELB, SELD, PSTK and SECp43 [28]. Furthermore, *L. major* and *T. Brucei* selenophosphate synthetases (SELD) are a class I selenide-dependent enzyme, involved in the formation of the Se-containing substrate for the Sec biosynthesis pathway [29].

The selenoproteomes of Kinetoplastida protozoa have three selenoproteins, including distant homologs of mammalian SelK and SelT, and a novel multidomain selenoprotein designated SelTryp. In all three proteins Sec is located within predicted redox motifs (CxxU in the case of SelT and SelTryp and CxxxU in SelK), which is suggestive of their possible role in maintaining the redox equilibrium in these parasites [30].

Unlike mammalian cells that rely upon two separate enzymes, glutathione reductase and thioredoxin reductase, to maintain the level of reduced thiols, *Schistosomes* utilize a hybrid of the two systems known as thioredoxin glutathione reductase [31]. It is a unique selenoenzyme that is essential for growth of the organism and possesses a C-terminal Sec residue that is required for its activity [32].

3. SELENIUM AND SELENODERIVATIVES IN TROPICAL DISEASES

3.1. Chagas

In 2002, a study of advanced chronic chagasic patients reported that low Se levels were positively correlated with cardiac insufficiency [33]. Several *in vivo* experiments have been designed to investigate the role of Se in chagas disease. These studies demonstrated an increased susceptibility to infection in mice fed Se-deficient chow [34] and increased severity of myopathy was observed in Se-depleted chronically infected mice [35]. Se supplementation alleviated heart muscle damage [36] and had previously been shown to reduce parasitemia and mortality in infected mice [37].

A study conducted by de Souza *et al.* showed that right ventricular dilatation is a marker for severity of the cardiac disease in infected mice [38]. Recently, it has been demonstrated that Se supplementation (2 ppm Se supplied as sodium selenate in the drinking water) beginning two weeks before infection of mice with *T. cruzi* prevented the extreme dilation of the right ventricle, whereas Se supplementation of uninfected mice had no effect on cardiac dimensions [39].

Another recent study showed the effect of Se on the gastrointestinal manifestations of mice chronically infected with *T. cruzi* [40]. In chronically infected mice, enlargement of the intestine associated with decreased intestinal motility was observed. Se supplementation (as 2 ppm Se supplied as sodium selenate in the drinking water or as 3 ppm Se in the chow as Se-methylselenocysteine) beginning two weeks before infection reduced intestinal dilation and preserved intestinal motility.

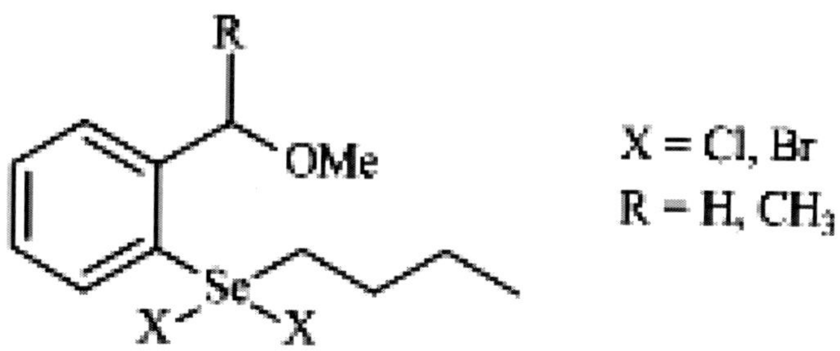

Figure 1. Organoselenanes compounds with cysteine proteases inhibitory effect.

Recent developments in the study of the basic biochemistry of *T. cruzi* have allowed the identification of novel targets for chemotherapy that include sterol metabolism, enzymes such as trypanothione reductase, cysteine proteinase, hypoxanthine-guanine phosphoribosyltransferase, glyceraldehyde-3-phosphate dehydrogenase, DNA topoisomerases, dihydrofolate reductase and farnesyl pyrophosphate synthase [41, 42].

Cruzipain, also known as cruzain or GP57/51, is a member of the papain C1 family of cystein proteinases. This *T. cruzi* enzyme consists of a catalytic moiety with high homology to cathepsins S and L [43]. Irreversible inhibitors of cruzipain, such as several peptidyl diazomethylketones, peptidyl fluoromethylketones and peptidyl vinyl sulphones interfered with the in vitro intracellular cycle of *T. cruzi*, killing the parasite [44]. Recently, Piovan *et al.* have reported the synthesis and evaluation as cysteine protease inhibirtors of novel organoselenium (IV) compounds (organoselenanes) [45]. These compounds (Figure 1) inhibited the cathepsin S faster than cathepsin V. So, these Se compounds could be considered as potential candidates for cysteine proteases inhibition in the treatment of chagas disease.

3.2. Filariasis

Currently, a documented relationship between selenium levels in the organism and the filariasis disease is lacking, as well as none compound containing Se has demonstrated anti-filariasis effect either. Nevertheless, the last year Singh *et al.* suggested that selenium glutathione peroxidase could be an alternative diagnostic marker for the detection of *Bancroftian filariasis* in an endemic area [46].

3.3. Leishmaniasis

The Se increased concentration in plasma has been recognized as a new defensive strategy against *Leishmania* infection [47, 48]. Recently, a study has implicated the human selenium-containing selenoprotein P in the protection against illness caused by *Trypanosoma* due to its protective action against oxidative damage [49]. The biochemical pathways that constitute the cellular targets for Se are still under investigation.

A study conducted by Mukhopadhyay *et al.* showed a moderate inhibitory effect for sodium selenite (Na_2SeO_3) against *L. donivani* promastigotes [50]. The LD_{50} values (concentration of drug which cause inhibition by 50%) for this compound were 0.125 mM for the strains UR6 and AG83.

3.3.1. A Case Study: Novel Selenoderivatives with in Vitro Leishmanicidal Activity against Leishmania Infantum

Various reports have showed that trypanosomatids, in comparison with mammals, have a different strategy to withstand oxidative stresses because they lack classical glutathione reductase, glutathione-dependent peroxidase, and thioredoxin reductase activities. It is well known [51] that the glutathione system that controls the redox status in higher eukaryotes is replaced in trypanosomatids by a trypanothione system where the key enzyme is trypanothione reductase which catalyzes the reduction of trypanothione disulfide to trypanothione [52]. Based on the chemical analogy of sulfur and Se, we decided to explore the relevance of this trace element to generate new molecules with leishmanicidal activity. Besides, several of the most effective antiprotozoal agents were originally developed as anticancer drugs [53, 54]. These facts prompted us to test some Se compounds of our chemical library as antileishmanial agents. The rationale behind the design of these compounds was to maintain molecular symmetry -a structural property that is frequently present in antitumoral [55] and antiprotozoal drugs, such as suramon, pentamidine and ethambutol- and the presence of at least one Se atom.

In the present year we have reported two general structures with leishmanicidal activity [56, 57]. The firsts present a central nucleus made up of an alkyl imidothiocarbamate (alkyl isothiourea) or alkyl imidoselenocarbamate (alkyl isoselenourea) connected by a carbonyl group on each side to two identical lateral aromatic or heteroaromatic rings (Figure 2), which possessed a moderate effect *in vitro* [56]. The second ones are selenocyanates and diselenides (Figure 2) and it is remarkable that some of them showed stronger *in vitro* antileishmanial activity than edelfosine and miltefosine, used as reference drugs. Furthermore, they combined high potency and low cytotoxicity against Jurkat and THP-1 cells [57].

Methyl *N,N´*-bis(pyridine-3-carbonyl)imidoselenocarbamate highlights as the most potent leishmanicidal agent among the evaluated alkyl isoselenourea derivatives. It is able to reduce *in vitro* infection rates by 62%, showing negligible toxicity over the differentiated THP-1 cells [56].

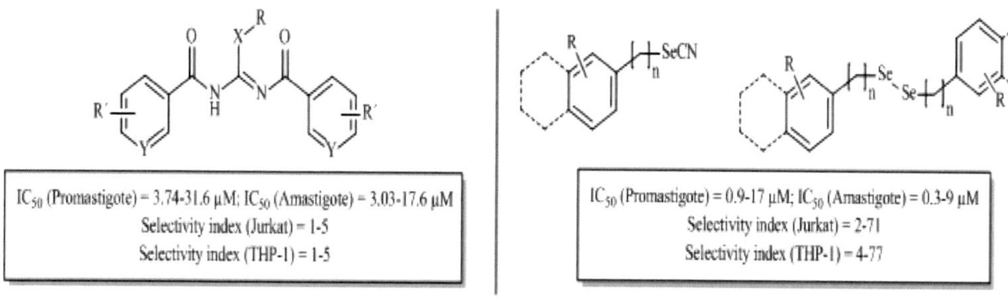

Figure 2. General structures for novel selenoderivatives with *in vitro* leishmanicidal activity against *Leishmania Infantum*.

Our research group suggests the use of selenocyanate and diselenide compounds as novel motifs for the development of new antileishmanial agents [57]. Seven compounds combined high potency and low cytotoxicity with selectivity index (SI) values higher than those obtained for miltefosine and edelfosine. Bis(4-aminophenyl)diselenide emerges as the most promising derivative (IC_{50} (promastigotes) = 0.96; IC_{50} (amastigotes) = 0.65; SI (Jurkat) = 32; SI (THP-1) = 24). This compound was assayed in amastigote-infected THP-1 cells. It is important to emphasize that treatment of the cells with this compound at 3 µM for 48 h is able to reduce infection rates by 74% while showing a negligible cytotoxic effect over the host cells. Its cytotoxic activity was also tested by microscopic visualization of promastigotes, amastigotes, Jurkat and THP-1 cells. The results presented in Figure 3 confirm the absence of cytotoxicity of bis(4-aminophenyl)diselenide at 1 µM and 3 µM in the human cell lines tested. Remarkably, this compound is able to kill 100% of the axenic amastigotes at 3 µM without almost any deleterious effect in the Jurkat or THP-1 cell lines.

3.4. Malaria

A study showed that sodium selenite had antimalarial activity against both the chloroquine-susceptible strain FCR-3 and chloroquine-resistant strain K-1 of *P. falciparum* [58]. The intracellular reduced glutathione level of parasitized red blood cells was decreased significantly by treatment with sodium selenite, and the decrease was consistent with their mortality. Moreover, sodium selenite presented a strong inhibitory effect on *in vitro* plasmodial development and was devoid of cytotoxicity towards human cells if there is co-treatment with $CuSO_4$. These results suggest that co-treatment with sodium selenite and $CuSO_4$ may be useful as a new antimalarial therapy.

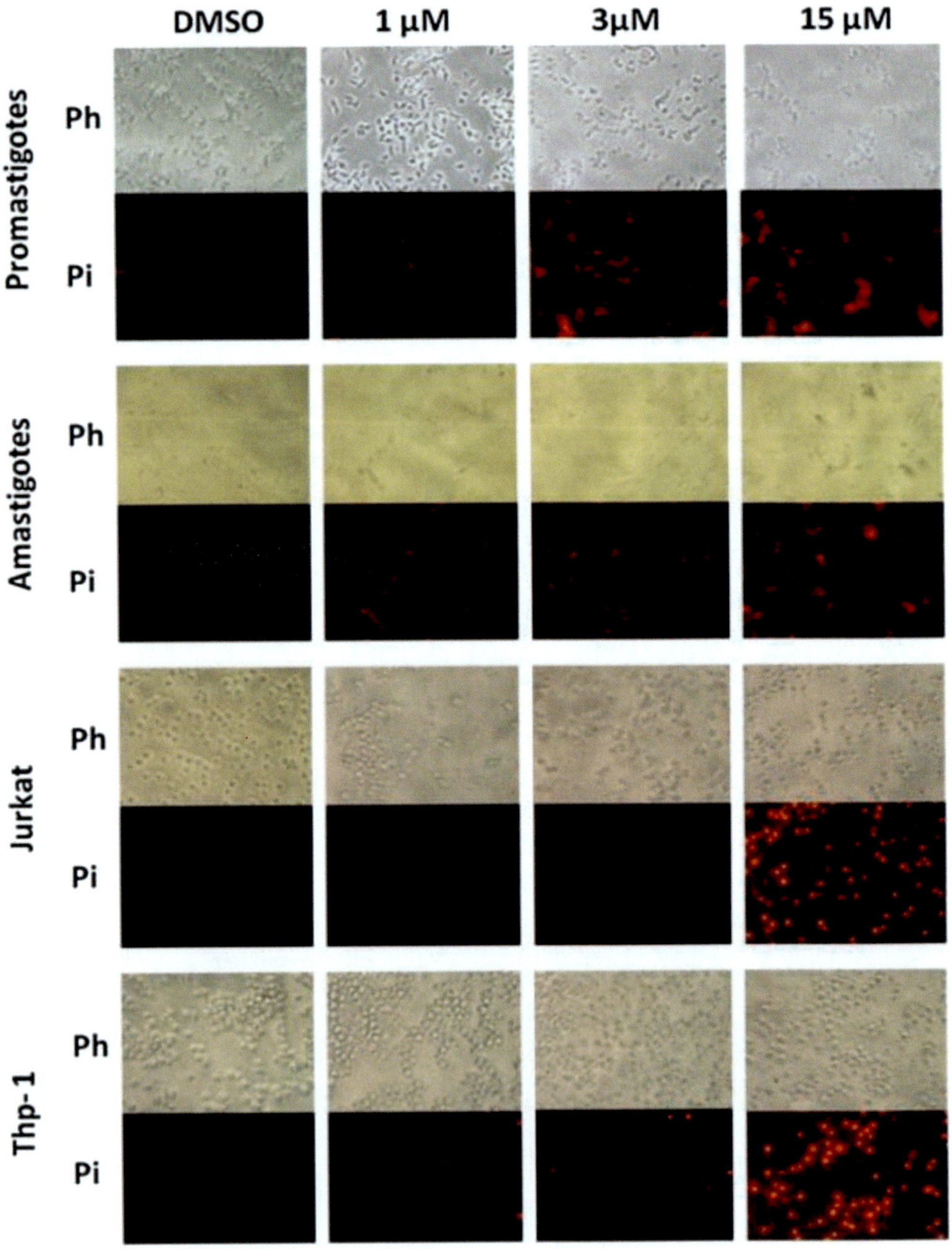

Figure 3. Microscopic visualization of *L. infantum* axenic promastigotes, *L. infantum* axenic amastigotes, Jurkat and THP-1 human cell lines. Cells were treated with bis(4-aminophenyl)diselenide at 1 µM, 3 µM and 15 µM. Non-treated control cells exposed to the maximum DMSO concentration of the assay are also shown. Propidium iodide was added to a final concentration of 5 µg/mL and the cells were immediately analyzed by phase contrast (Ph) and fluorescence (Pi) microscopy. Figure from Plano *et al.* [57] with permission.

A family of 5-selenated nucleotides has showed submicromolar inhibition of human and malarial orotate phosphoribosyltransferase (OPRT). OPRT is an essential enzyme for the *de novo* biosynthesis of pyrimidine nucleotides, promoting the attachment of orotic acid to phosphoribosylpyrophosphate [59]. This is the only pathway for pyrimidine nucleotide production in *P. falciparum* [60]. Considering that OPRT of the parasite (*Pf*OPRT) is a target for antimalarial drugs [60], these compounds could be pointed out as potential agents in the treatment of malaria.

3.5. Trypanosomiasis

Bosschaerts *et al.* have recently demonstrated the importance of selenoprotein P (SePP) in the African trypanosomiasis control [61]. SePP has two domains with respect to Se content and functions. The *N*-terminal, Se poor domain, encompassing two potential redox motifs, was predicted to have enzymatic (peroxidase) properties, while the C-terminal, Se rich domain plays a role in Se transport [62]. They suggested that the enzymatic activity of SePP, and not its Se transporter function, is important for African trypanosomiasis control. So, the antioxidant activity of SePP is important to avoid tissue damage and ensure survival.

3.6. Tuberculosis

Several studies have evidenced that TB patients have altered profile of trace elements in their sera [63-65]. So, the concentration of these trace elements -such as selenium, iron, zinc and cooper- in sera could be used as diagnostic parameters in monitoring responses to anti-TB chemotherapy. Various authors have reported beneficial effects for selenium and other micronutrients supplementation in TB patients [66-68].

We have been unable to find reports of any organic forms of Se in the treatment of TB. Nevertheless, in 2009 Miller *et al.* reported that Se-methylselenocysteine had a significant inhibitory effect in *Mycobacterium smegmatis* only when screening was performed under nutrient-limited culture conditions as opposed to nutrient-rich culture conditions [69].

Conclusion

This chapter includes a general revision of the implication of Se and selenocompounds in the development, prognosis and treatment of several tropical diseases. There are two facts that seem to pointed out a possible relationship between Se and parasites: a) several parasites require Se in their metabolism, since they own various selenoproteins; b) several clinical studies support a direct relationship between deficient status of Se in plama and a worst prognosis and more fatal consequences in various tropical diseases. In spite of this possible relationship, there are few studies in the literature about the effect of selenocompounds against tropical diseases.

So, due to the lack of studies and to the necessary research in the development of novel drugs against tropical diseases, our research group has started the evaluation of organoselenium compounds against *Leishmaniasis*. The obtained results showed a moderate anti-leishmanial activity for alkyl imidoselenocarbamate derivatives. However, the selenocyanate and diselenide derivatives exhibited promising leishmacidal effects, possessing more potent and selectivity effects than miltefosine and edelfosine.

Taking into account all facts exposed in this chapter and considering the obtained results by our research group, it can be concluded that: the diselenide function emerges as a novel scaffold against Leishmaniasis. Besides, we consider that the research in Se and selenocompounds could play a significant role in the development of novel drugs in the treatment of tropical diseases in a near future.

ACKNOWLEDGMENT

The authors wish to express their gratitude to the Ministerio de Educación y Ciencia, Spain (SAF 2009-07744) and to the Fundación para la Investigación Médica Aplicada for financial support.

REFERENCES

[1] Coura JR, de Castro SL. A Critical review on chagas disease chemotherapy. *Memorias Do Instituto Oswaldo Cruz.* 2002;97:3-24.

[2] Rassi A, Little WC, Xavier SS, Rassi SG, Rassi AG, Rassi GG, Hasslocher-Moreno A, Sousa AS, Scanavacca MI. Development and validation of a risk score for predicting death in chagas' heart disease. *New England Journal of Medicine*. 2006;355:799-808.

[3] http://www.filariasis.org/all_about_lf/epidemiology.html.

[4] Alvar J, Aparicio P, Aseffa A, Den Boer M, Canavate C, Dedet J-P, Gradoni L, Ter Horst R, Lopez-Velez R, Moreno J. The Relationship between Leishmaniasis and AIDS: the second 10 years. *Clinical Microbiology Reviews*. 2008;21:334-59.

[5] Le Pape P. Development of new antileishmanial drugs – current knowledge and future prospects. *Journal of Enzyme Inhibition and Medicinal Chemistry*. 2008;23:708-18.

[6] den Boer ML, Alvar J, Davidson RN, Ritmeijer K, Balasegaram M. Developments in the treatment of visceral leishmaniasis. *Expert Opinion on Emerging Drugs*. 2009;14:395-410.

[7] Palumbo E. Oral miltefosine treatment in children with visceral leishmaniasis: a brief review. *Brazilian Journal of Infectious Diseases*. 2008;12:2-4.

[8] Santa-Rita RM, Henriques-Pons A, Barbosa HS, de Castro SL. Effect of the lysophospholipid analogues edelfosine, ilmofosine and miltefosine against Leishmania amazonensis. *Journal of Antimicrobial Chemotherapy*. 2004;54:704-10.

[9] World Health Organization. *World malaria report* 2009;WHO [online], http://whqlibdoc.who.int/publications/2009/9789241563901_eng.pdf.

[10] Noedl H, Se Y, Schaecher K, Smith BL, Socheat D, Fukuda MM. Evidence of artemisinin-resistant malaria in western Cambodia. *New England Journal of Medicine*. 2008;359:2619-20.

[11] Noedl H, Socheat D, Satimai W. Artemisinin-resistant malaria in Asia. *New England Journal of Medicine.* 2009;361:540-1.

[12] Dondorp AM, Nosten F, Yi P, Das D, Phyo AP, Tarning J, Lwin KM, Ariey F, Hanpithakpong W, Lee SJ, Ringwald P, Silamut K, Imwong M, Chotivanich K, Lim P, Herdman T, An SS, Yeung S, Singhasivanon P, Day NPJ, Lindergardh N, Socheat D, White NJ. Artemisinin resistance in Plasmodium falciparum malaria. *New England Journal of Medicine.* 2009;361:455-67.

[13] Breman JG. Eradicating malaria. *Science Progress.* 2009;92:1-38.

[14] World Health Organisation. Multidrug and extensively drug-resistant TB (M/XDR-TB): *2010 global report on surveillance and response.* 2010;WHO [online], http://www.who.int/tb/publications/2010/9789245991 91/en/index.html.

[15] Corbett EL, Watt CJ, Walker N, Maher D, Williams BG, Raviglione MC, Dye C. The growing burden of tuberculosis: global trends and interactions with the HIV epidemic. *Archives of Internal Medicine.* 2003;163:1009-21.

[16] US Centers for Disease Control. *Multidrug-resistant tuberculosis* (MDR TB) fact sheet. 2010;CDC [online], http://www.cdc.gov/tb/publications/factsheets/drtb/mdrtb.htm.

[17] US Centers for Disease Control. 2007;*United States Centers for Disease Control,* Atlanta.

[18] Mitchison DA, Nunn AJ. Influence of initial drug resistance on the response to short-course chemotherapy of pulmonary tuberculosis. *The American Review of Respiratory Disease.* 1986;133:420-30.

[19] Guo H, Seet Q, Denkin S, Parsons L, Zhang Y. Molecular characterization of isoniazid-resistant clinical isolates of Mycobacterium tuberculosis from the USA. *Journal of Medical Microbiology.* 2006;55:1527-31.

[20] Scorpio A, Zhang Y. Mutations in pncA, a gene encoding pyrazinamidase /nicotinamidase, cause resistance to the antituberculous drug pyrazinamide in tubercle bacillus. *Nature Medicine.* 1996;2:662-7.

[21] Zhang Y, Wade MM, Scorpio A, Zhang H, Sun Z. Mode of action of pyrazinamide: disruption of Mycobacterium tuberculosis membrane transport and energetics by pyrazinoic acid. *Journal of Antimicrobial Chemotherapy.* 2003;52:790-5.

[22] Sreevatsan S, Stockbauer K, Pan X, Kreiswirth B, Moghazeh S, Jacobs W, Jr, Teleneti A, Musser JM. Ethambutol resistance in Mycobacterium tuberculosis: critical role of embB mutations. *Antimicrobial Agents and Chemotherapy.* 1997;41:1677-81.

[23] DeBarber AE, Mdluli K, Bosman M, Bekker L-G, Barry CE. Ethionamide activation and sensitivity in multidrug-resistant Mycobacterium tuberculosis. *Proceedings of the National Academy of Sciences.* 2000;97:9677-82.

[24] Sreevatsan S, Pan X, Stockbauer K, Williams D, Kreiswirth B, Musser J. Characterization of rpsL and rrs mutations in streptomycin-resistant Mycobacterium tuberculosis isolates from diverse geographic localities. *Antimicrobial Agents and Chemotherapy.* 1996;40:1024-6.

[25] Lobanov AV, Delgado C, Rahlfs S, Novoselov SV, Kryukov GV, Gromer S, Hatfield DL, Becker K, Gladyshev VN. The Plasmodium selenoproteome. *Nucleic Acids Research.* 2006;34:496-505.

[26] Bauer H, Gromer S, Urbani A, Schnolzer M, Schirmer RH, Muller HM. Thioredoxin reductase from the malaria mosquito Anopheles gambiae - Comparisons with the

orthologous enzymes of Plasmodium falciparum and the human host. *European Journal of Biochemistry.* 2003;270:4272-81.

[27] Becker K, Gromer S, Schirmer RH, Muller S. Thioredoxin reductase as a pathophysiological factor and drug target. *European Journal of Biochemistry.* 2000;267:6118-25.

[28] Cassago A, Rodrigues EM, Prieto EL, Gaston KW, Alfonzo JD, Iribar MP, Berry MJ, Cruz AK, Thiemann OH. Identification of Leishmania selenoproteins and SECIS element. *Molecular and Biochemical Parasitology.* 2006;149:128-34.

[29] Sculaccio SA, Rodrigues EM, Cordeiro AT, Magalhaes A, Braga AL, Alberto EE, Thiemann OH. Selenocysteine incorporation in Kinetoplastid: Selenophosphate synthetase (SELD) from Leishmania major and Trypanosoma brucei. *Molecular and Biochemical Parasitology.* 2008;162:165-71.

[30] Lobanov AV, Gromer S, Salinas G, Gladyshev VN. Selenium metabolism in Trypanosoma: characterization of selenoproteomes and identification of a Kinetoplastida-specific selenoprotein. *Nucleic Acids Research.* 2006;34:4012-24.

[31] Salinas G, Selkirk ME, Chalar C, Maizels RM, Fernández C. Linked thioredoxin-glutathione systems in platyhelminths. *Trends in Parasitology.* 2004;20:340-6.

[32] Kuntz AN, Davioud-Charvet E, Sayed AA, Califf LL, Dessolin J, Arnér ESJ, Williams DL. Thioredoxin glutathione reductase from *Schistosoma mansoni*: an essential parasite enzyme and a key drug target. *PLoS Medicine.* 2007;4:e206.

[33] Rivera MT, De Souza AP, Moreno AHM, Xavier SS, Gomes JAS, Rocha MOC, Correa-Oliveira R, Neve J, Vanderpas J, Araujo-Jorge TC. Progressive Chagas' cardiomyopathy is associated with low selenium levels. *American Journal of Tropical Medicine and Hygiene.* 2002;66:706-12.

[34] de Souza AP, de Oliveira GM, Neve J, Vanderpas J, Pirmez C, de Castro SL, Araujo-Jorge TC, Rivera MT. Trypanosoma cruzi: host selenium deficiency leads to higher mortality but similar parasitemia in mice. *Experimental Parasitology.* 2002;101:193-9.

[35] Gomez RM, Solana ME, Levander OA. Host selenium deficiency increases the severity of chronic inflammatory myopathy in Trypanosoma cruzi-inoculated mice. *Journal of Parasitology.* 2002;88:541-7.

[36] de Souza AP, de Oliveira GM, Vanderpas J, de Castro SL, Rivera MT, Araujo-Jorge TC. Selenium supplementation at low doses contributes to the decrease in heart damage in experimental Trypanosoma cruzi infection. *Parasitology Research.* 2003;91:51-4.

[37] Davis CD, Brooks L, Calisi C, Bennett BJ, McElroy DM. Beneficial effect of selenium supplementation during murine infection with Trypanosoma cruzi. *Journal of Parasitology.* 1998;84:1274-7.

[38] de Souza AP, Tang B, Tanowitz HB, Araújo-Jorge TC, Jelicks LA. Magnetic resonance imaging in experimental Chagas disease: a brief review of the utility of the method for monitoring right ventricular chamber dilatation. *Parasitology Research.* 2005;97:87-90.

[39] de Souza AP, Jelicks LA, Tanowitz HB, Olivieri BP, Medeiros MM, Oliveira GM, Pires ARC, dos Santos AM, Araujo-Jorge TC. The benefits of using selenium in the treatment of Chagas disease: prevention of right ventricle chamber dilatation and reversion of Trypanosoma cruzi-induced acute and chronic cardiomyopathy in mice. *Memorias Do Instituto Oswaldo Cruz.* 2010;105:746-51.

[40] de Souza AP, Sieberg R, Li H, Cahill HR, Zhao DZ, Araujo-Jorge TC, Tanowitz HB, Jelicks LA. The role of selenium in intestinal motility and morphology in a murine model of Typanosoma cruzi infection. *Parasitology Research.* 2010;106:1293-8.

[41] Docampo R. Recent Developments in the chemotherapy of chagas disease. *Current Pharmaceutical Design.* 2001;7:1157-64.

[42] Rodriguez JB. Specific molecular targets to control tropical diseases. *Current Pharmaceutical Design.* 2001;7:1105-16.

[43] Cazzulo J, Stoka V, Turk V. The major cysteine proteinase of Trypanosoma cruzi: A valid target for chemotherapy of chagas disease. *Current Pharmaceutical Design.* 2001;7:1143-56.

[44] McKerrow JH. Development of cysteine protease inhibitors as chemotherapy for parasitic diseases: insights on safety, target validation, and mechanism of action. *International Journal for Parasitology.* 1999;29:833-7.

[45] Piovan L, Alves MFM, Juliano L, Brömme D, Cunha RLOR, Andrade LH. Chemoenzymatic synthesis of organoselenium(IV) compounds and their evaluation as cysteine protease inhibitors. *Journal of the Brazilian Chemical Society.* 2010;21:2108-18.

[46] Singh A, Kamal S, Rathaur S. Filarial selenium glutathione peroxidase: a probable immunodiagnostic marker for lymphatic filariasis. *Transactions of the Royal Society of Tropical Medicine and Hygiene.* 2010;104:524-8.

[47] Culha G, Yalin E, Sangun K. Alterations in serum levels of trace elements in cutaneous leishmaniasis patients in endemic region of Hatay (Antioch). *Asian Journal of Chemistry.* 2008;20:3104-8.

[48] Araujo AP, Rocha OGF, Mayrink W, Machado-Coelho GLL. The influence of copper, selenium and zinc on the response to the Montenegro skin test in subjects vaccinated against American cutaneous leishmaniasis. *Transactions of the Royal Society of Tropical Medicine and Hygiene.* 2008;102:64-9.

[49] Burk RF, Hill KE. Selenoprotein P - Expression, functions, and roles in mammals. Biochimica et Biophysica Acta (BBA) - *General Subjects.* 2009;1790:1441-7.

[50] Mukhopadhyay R, Madhubala R. Effect of antioxidants on the growth and polyamine levels of Leishmania donivani. *Biochemical Pharmacology.* 1994;47:611-5.

[51] Krauth-Siegel RL, Meiering SK, Schmidt H. The parasite-specific trypanothione metabolism of Trypanosoma and Leishmania. *Biological Chemistry.* 2003;384(4):539-49.

[52] Stump B, Kaiser M, Brun R, Krauth-Siegel RL, Diederich F. Betraying the parasite's redox system: Diaryl sulfide-based inhibitors of rrypanothione reductase: subversive substrates and antitrypanosomal properties. *ChemMedChem.* 2007;2:1708-12.

[53] Perez JM, Fuertes MA, Nguewa PA, Castilla J, Alonso C. Anticancer compounds as leishmanicidal drugs: challenges in chemotherapy and future perspectives. *Current Medicinal Chemistry.* 2008;15:433-9.

[54] da Silva DB, Tulli ECO, Militão GCG, Costa-Lotufo LV, Pessoa C, de Moraes MO, Alburquerque S, de Siqueira JM. The antitumoral, trypanocidal and antileishmanial activities of extract and alkaloids isolated from Duguetia furfuracea. *Phytomedicine.* 2009;16:1059-63.

[55] Sanmartin C, Font M, Palop JA. Molecular symmetry: A structural property frequently present in new cytotoxic and proapoptotic drugs. *Mini-Reviews in Medicinal Chemistry.* 2006;6:639-50.

[56] Moreno D, Plano D, Baquedano Y, Jiménez-Ruiz A, Antonio Palop J, Sanmartín C. Antileishmanial activity of imidothiocarbamates and imidoselenocarbamates. *Parasitology Research.* 2011;108:233-9.

[57] Plano D, Baquedano Y, Moreno-Mateos D, Font M, Jimenez-Ruiz A, Antonio Palop J, et al. Selenocyanates and diselenides: A new class of potent antileishmanial agents. *European Journal of Medicinal Chemistry.* 2011;46:3315-23.

[58] Taguchi N, Hatabu T, Yamaguchi H, Suzuki M, Sato K, Kano S. Plasmodium falciparum: selenium-induced cytotoxicity to P. falciparum. *Experimental Parasitology.* 2004;106:50-5.

[59] Zhang Y, Luo M, Schramm VL. Transition states of Plasmodium falciparum and human orotate phosphoribosyltransferases. *Journal of the American Chemical Society.* 2009;131:4685-94.

[60] Krungkrai SR, Aoki S, Palacpac NMQ, Sato D, Mitamura T, Krungkrai J, et al. Human malaria parasite orotate phosphoribosyltransferase: functional expression, characterization of kinetic reaction mechanism and inhibition profile. *Molecular and Biochemical Parasitology.* 2004;134:245-55.

[61] Bosschaerts T, Guilliams M, Noel W, Herin M, Burk RF, Hill KE, Brys L, Raes G, Ghassabeh GH, de Baetselier P, Beschin A. Alternatively activated myeloid cells limit pathogenicity associated with African trypanosomiasis through the IL-10 inducible gene selenoprotein P. *Journal of Immunology.* 2008;180:6168-75.

[62] Hill KE, Zhou J, Austin LM, Motley AK, Ham A-JL, Olson GE, Atkins JF, Gesteland RF, Burk RF. The selenium-rich C-terminal domain of mouse selenoprotein P is necessary for the supply of selenium to brain and testis but not for the maintenance of whole body selenium. *Journal of Biological Chemistry.* 2007;282:10972-80.

[63] Kassu A, Yabutani T, Mahmud ZH, Mohammad A, Nguyen N, Huong BTM, Hailemariam G, Diro E, Ayele B, Wondmikun Y, Motonaka J, Ota F. Alterations in serum levels of trace elements in tuberculosis and HIV infections. *European Journal of Clinical Nutrition.* 2006;60:580-6.

[64] Ramakrishnan K, Sharma SP, Shenbagarathai R, Kavitha K, Thirumalaikolundusubramanian P. Serum selenium levels in pulmonary tuberculosis levels with and without HIV/AIDS. *Retrovirology.* 2009;6:P76.

[65] van Lettow M, West CE, van der Meer JWM, Wieringa FT, Semba RD. Low plasma selenium concentrations, high plasma human immunodeficiency virus load and high interleukin-6 concentrations are risk factors associated with anemia in adults presenting with pulmonary tuberculosis in Zomba district, Malawi. *European Journal of Clinical Nutrition.* 2005;59:526-32.

[66] Range N, Changalucha J, Krarup H, Magnussen P, Andersen AB, Friis H. The effect of multi-vitamin/mineral supplementation on mortality during treatment of pulmonary tuberculosis: a randomised two-by-two factorial trial in Mwanza, Tanzania. *British Journal of Nutrition.* 2006;95:762-70.

[67] Seyedrezazadeh E, Ostadrahimi A, Mahboob S, Assadi Y, Ghaemmagami J, Pourmogaddam M. Effect of vitamin E and selenium supplementation on oxidative stress status in pulmonary tuberculosis patients. *Respirology.* 2008;13:294-8.

[68] Villamor E, Mugusi F, Urassa W, Bosch RJ, Saathoff E, Matsumoto K, Meydani SN, Fawzi WW. A trial of the effect of micronutrient supplementation on treatment outcome, T cell counts, morbidity, and mortality in adults with pulmonary tuberculosis. *Journal of Infectious Diseases.* 2008;197:1499-505.

[69] Miller CH, Nisa S, Dempsey S, Jack C, O'Toole R. Modifying culture conditions in chemical library screening identifies alternative inhibitors of Mycobacteria. *Antimicrobial Agents and Chemotherapy.* 2009;53:5279-83.

In: Selenium: Sources, Functions and Health Effects
Editors: Chinatsu Aomori and Megumi Hokkaido
ISBN: 978-1-61942-061-8

Chapter 8

PRODUCTS OF ANIMAL ORIGIN AS A SOURCE OF SE IN HUMAN DIET

Bogumiła Pilarczyk,[1,*] Agnieszka Tomza-Marciniak[1] and Andrzej Marciniak[2]

[1]Department of Animal Reproduction Biotechnology and Environmental Hygiene, West Pomeranian University of Technology in Szczecin, Szczecin, Poland

[2]West Pomeranian Research Centre in Szczecin, Institute of Technology and Life Sciences, Szczecin, Poland

Selenium consumption in European countries has declined considerably over the last few decades [Murphy and Cashman 2001], mainly as a result of eating low-selenium foods. Geographical location is a factor that has a significant effect on Se concentration in products of plant and animal origin. Because selenium content in the environment is low in most European countries, the majority of food products coming from this area contains small amounts of this trace element. Human Se intake depends on the amount and type of ingested food.

Admittedly, selenium is present in almost every commercially available food product but not all of these products are a good and sufficient source of this element. Research has shown that a valuable source of Se are high-protein food products of animal origin such as meat, offal, fish and eggs [Van der Torre et al. 1991, Wen et al. 1997, Klapec et al. 1998, Golubkina and Papazyan 2006], especially since these products mainly contain the readily available forms selenomethionine and selenocysteine. Finley et al. [2004] reported that Se from pork or beef is highly bioavailable for selenoprotein synthesis.

Recent research has proven that bioavailability of Se from meat is as high as that from wheat [Van der Torre et al. 1991]. Nutrient bioavailability defines the proportion of substances ingested with food that are digested, absorbed and metabolized. According to Kuczyńska and Biziuk [2007], selenium bioavailability from food is about 55%, and it is estimated to be 20-50% from fish, 15% from meat, and about 2% from dairy products.

* E-mail: Bogumila.Pilarczyk@zut.edu.pl

The authors report that selenium absorption is enhanced by low-molecular proteins and vitamins E and A. Selenium absorption is adversely affected by a high intake of vitamin C, sweets, and elevated concentrations of heavy metals in the diet. Researchers differ on selenium supplementation.

Some believe that it is indispensable in deficient areas, others consider it unnecessary. Bjelakovic et al. [2008] hold the view that selenium supplementation does not increase selenoprotein synthesis. The authors report that the use of selenium from sources other than food does not produce desired results. Adequate amounts of selenium can therefore be ensured by a varied diet that includes selenium-rich products.

Concentration of Selenium in Dairy Products (Milk, Yoghurt, Cheese, Kefir)

Due to its nutritional value, cow's milk is very important to human nutrition. It is an indispensable source of vitamins and minerals, including selenium. As reported by Puls (1988), the optimum concentration of Se in cow's milk exceeds 0.03 μg/ml, with levels below 0.009 μg/ml considered deficient and those ranging from 0.011 to 0.029 μg/ml considered marginal.

Selenium content of milk products varies from country to country (Tab. 1), largely due to geographical location and the geochemical conditions prevailing in a given area. Selenium content of milk and milk products is dependent on many factors such as the concentration of this element in the plants and feeds ingested by dairy cattle.

This, in turn, depends on soil Se content, its chemical form, soil pH and redox potential, the presence of organic substances and heavy metals, as well as climatic factors and the activity of soil microorganisms. In general, if a herd receives no Se supplements and the concentration of this element in soil and plants is low, milk products from these animals will also be poor in selenium.

Selenium content of milk also depends on the cow feeding system. Particular attention should be given to Se concentration in dairy cows from organic and conventional farms with a traditional feeding system, especially in selenium-deficient areas. Pilarczyk et al. [2011] showed that the mean Se concentration in the milk of cows from an organic farm and from conventional farms, in which animals were traditionally fed, was markedly lower (0.016 μg/ml and 0.005 μg/ml, respectively) compared to the milk of cows on TMR (total mixed ration) diets (0.042 μg/ml).

It is also worth noting the effect of season on the amount of Se in milk. It appears that during the winter period, the milk of cows and goats may contain twice as much selenium as the milk obtained during the summer period. This is related to the winter feeding of these animals with ground grains (wheat, oats), which are an excellent source of selenium.

Selenium concentration in milk products depends on its concentration in milk. When milk contains marginal concentrations of selenium, milk products will be also poor in this element [Pilarczyk et al. 2009a, 2010b]. Among milk products, the highest Se content was characteristic of cheese. The selenium content of the cheese was about 2-4 times that of milk.

Table 1. Se content of cow's milk and milk products in different countries

Product	Country	Average	Range	References
milk ($\mu g \cdot ml^{-1}$)	Croatia	0.029	0.019 - 0.041	Klapec et al. (2004)
	Australia	n/d	0.023 - 0.026	McNaughton and Marks (2002)
	Pakistan	0.170	n/d	Iqbal et al. (2008)
	Slovakia	0.007	n/d	Kadrabova et al. (1997)
	Israel	0.073	n/d	Lavi and Alfassi (1990)
	Poland	0.020	0.009 - 0.032	Pilarczyk et al. (2010b)
	Turkey	15.24*	5.78 - 32.02*	Yanardağ and Orak (1999)
yoghurt ($\mu g \cdot ml^{-1}$)	Croatia	0.029	n/d	Klapec et al. (2004)
	UK	0.014	n/d	Barclay et al. (1995)
	Greece	0.017	n/d	Baratakos et al. (1987)
	Slovakia	0.006	0.004 - 0.008	Kadrabova et al. (1997)
	Poland	0.010	n/d	Pilarczyk et al. (2010)
	Australia	0.020	n/d	McNaughton and Marks (2002)
	Pakistan	0.030	n/d	Iqbal et al. (2008)
kefir ($\mu g \cdot ml^{-1}$)	Poland	0.012	nd - 0.035	Pilarczyk et al. 2010b
	Turkey	24.47*	13.92 – 39.51*	Yanardağ and Orak (1999)
cottage cheese ($\mu g \cdot g^{-1}$ w.w.)	Slovakia	0.028	n/d	Kadrabova et al. (1997)
	England	0.035	n/d	Barclay et al. (1995)
	Croatia	0.039	n/d	Klapec et al. (2004)
	Mexico	0.080	n/d	Wyatt et al. (1996)
	Poland	0.072	0.014 - 0.142	Pilarczyk et al. (2010b)
edam cheese ($\mu g \cdot g^{-1}$ w.w.)	Croatia	0.073	n/d	Klapec et al. (2004)
	England	0.064	n/d	Barclay et al. (1995)
	Slovakia	0.040	n/d	Kadrabova et al. (1997)
	Poland	0.088	0.047 - 0.113	Pilarczyk et al. (2010b)
gouda cheese ($\mu g \cdot g^{-1}$ w.w.)	Greece	0.085	n/d	Pappa et al. (2006)

*ng/g
n/d – no data, nd – not detected

CONCENTRATION OF SELENIUM IN MEAT

Meat and meat preparations have many nutritional and functional properties. Animal protein is the main source of dietary selenium, accounting for about 40-60% of the total selenium intake. Most Se in meat (beef, pork) is in the form of selenomethionine and selenocysteine, although this might depend on the feedstuffs consumed by the animal. A consequence of the insufficient content of this element in the nutrition of animals whose meat is used for consumption is the increased risk of Se deficiency in humans. From the consumer's point of view, it is appropriate to enrich pork, beef and poultry meat with selenium, particularly its organic forms (selenomethionine and selenocysteine), especially in the countries with selenium deficiency. Se from beef shows an especially high bioavailability for humans [Van der Torre et al. 1991]. This availability may be lower for pork. Bügel et al. [2004] report that although Se absorption and retention from pig meat in the bodies of volunteers participating in the study was high, which may be indicative of high bioavailability, the availability of Se for blood Se protein appears to be low. The authors believe that Se in pork occurs not as selenomethionine but in a different form.

Table 2. Se content ($\mu g \cdot g^{-1}$ w.w.) of meat in different countries

	Country	Average	Range	References
beef	UK	0.076	n/d	Barclay et al. (1995)
	Slovakia	0.057	0.030 - 0.067	Kadrabova et al. (1997)
	Croatia	0.131	n/d	Klapec et al. (2004)
	Ireland	0.081	0.061 - 0.105	Murphy and Cashman (2001)
	Poland	0.064	0.030 - 0.087	Pilarczyk et al. (2010b)
	France	0.099	n/d	Simonoff et al. (1988)
	Austria	0.060	n/d	Gamerith et al. (1991)
	Germany	0.058	n/d	Oster and Prellwitz (1989)
	Greece	0.084	n/d	Bratakos et al. 1987)
	USA	0.190	n/d	Pennington et al. (1995)
	Australia	n/d	0.072 - 0.121	McNaughton and Marks (2002)
	Mexico	0.080	n/d	Wyatt et al. (1996)
pork	UK	0.140	n/d	Barclay et al. (1995)
	Ireland	0.104	0.082 - 0.129	Murphy and Cashman (2001)
	USA	0.265	n/d	Pennington et al. (1995)
	Poland	0.078	0.067 - 0.089	Pilarczyk et al. (2010b)
	Russia	0.129	n/d	Golubkina and Khotimchencko (1994)
chicken	UK	0.070	n/d	Holland et al. (1991)
	Ireland	0.115	0.086 - 0.147	Murphy and Cashman (2001)
	Poland	0.094	0.076 - 0.139	Pilarczyk et al. (2010b)
	Australia	n/d	0.116 - 0.280	McNaughton and Marks (2002)
	USA	0.130	n/d	Pennington et al. (1995)
	Spain	0.073	0.058 - 0.084	Díaz-Alarcón et al. (1996)

n/d – no data

Selenium concentration in meat ranges very widely from several hundredth to several tenth of μg per g of the product (Tab. 2). Its content depends primarily on the feeding method and the type and quality of feeds given to animals. Adding Se to livestock feeds is a way of increasing the concentration of this trace element in animal tissues (in meat). However, when supplementing selenium, account must be taken of the fact that the availability of different Se forms varies according to animal species. In pigs, as much as 77% of Se in the form of selenite is resorbed from the digestive tract compared to only 29% absorbed in ruminants [Grela and Sembratowicz 1997]. For the supplementation to be effective, it is important that the most readily available form of selenium for a given animal species be administered. This is the precondition for increasing the retention of this element, and thus for increasing Se concentration in meat; there is also a lower risk of excessive fecal selenium excretion and environmental contamination through the transfer of excessive Se loads into the soil (fertilization) and then into surface waters. In addition, Se supplementation has a positive effect on the quality and technological properties of meat. Selenium inhibits the oxidation of cell membranes, which is the reason for increased exudation of meat juices, and improves the colour and shelf life of meat, thus allowing it to be stored for extended periods of time.

It is worthy of note that during the heat treatment of meat and other products, selenium loss may exceed 40%. Cooking generates the smallest loss, and frying and roasting the greatest loss. On the whole, the harsher the cooking conditions the more selenium is lost. Bratakos et al. [2007] observed that fish lost 36–46%, meats 13–41%, cereals 20–30%, vegetables 12 – 37%, and pulses and cereal products 5–10% selenium.

Concentration of Selenium In Offal

When examining the proportion of selenium in different products, it can be stated that an important source of this element is offal, mainly liver and kidneys. In general, selenium content of offal is several times that of the meat of the same animals. The liver of most animal species contains about 4 times more Se than skeletal muscles, and the kidneys of cattle, pigs and sheep may accumulate 10 to 16 times as much selenium as muscles [Combs and Combs 1986].

The ability to accumulate Se results mainly from the physiological function of these organs, especially the liver. Selenium is absorbed in the digestive tract, where organic and inorganic selenium compounds are transported from the intestine to the liver. In this organ they are reduced to selenides, which are used for synthesis of different selenoproteins.

Table 3. Content of Se ($\mu g \cdot g^{-1}$ w.w.) in offal in different countries

	Country	Average	Range	References
chicken liver	Croatia	0.298	n/d	Klapec et al. (2004)
	Greece	0.398	n/d	Baratakos et al. (1987)
	Slovakia	0.398	n/d	Kadrabova et al. (1997)
	Spain	n/d	0.280 - 1.420	Díaz-Alarcón et al. (1996)
	Poland	0.401	0.309 - 0.643	Pilarczyk et al. (2010b)
	Mexico	0.020	n/d	Wyatt et al. (1996)
beef liver	Slovakia	0.096	0.058 - 0.154	Kadrabova et al. (1997)
pig's liver	Croatia	0.285	n/d	Klapec et al. (2004)
	Greece	0.277	n/d	Baratakos et al. (1987)
	Slovakia	0.230	n/d	Kadrabova et al. (1997)
	Spain	n/d	0.256 - 0.800	Díaz-Alarcón et al. (1996)
	Poland	0.307	0.020 - 0.413	Pilarczyk et al. (2010b)
	Mexico	0.290	n/d	Wyatt et al. (1996)
wild boar's liver	Poland	0.189	0.099 - 0.254	Pilarczyk et al. (2010b)
roe deer's liver	Poland	0.385	0.182 - 0.598	Pilarczyk et al. (2010b)
red deer's liver	Poland	0.315	0.166 - 0.645	Pilarczyk et al. (2010b)
	USA	0.676	0.04 - 1.30	McDowell et al. (1995)
	Norway	0.090	0.04 - 1.00	Vikøren et al. (2005)
pork kidneys	Spain	1.196	n/d	Díaz-Alarcón et al. (1996)
rabbit kidneys	Spain	1.165	n/d	Díaz-Alarcón et al. (1996)
fallow deer's liver	Poland	0.071	0.048 - 0.098	Pilarczyk et al. (2010b)
wild boar's kidneys	Poland	1.20	0.143 - 2.033	Pilarczyk et al. (2010b)
roe deer's kidneys	Poland	1.20	0.143 - 2.033	Pilarczyk et al. (2009a)
red deer's kidneys	Poland	2.72	n/d	Pilarczyk et al. (2009a)

n/d – no data

The remaining part is transported with blood to other organs [Kuczyńska and Biziuk 2007]. Tissue distribution of selenium depends largely on its dietary content. When selenium is present in adequate amounts, then the highest quantities of this element are found in the liver and kidneys. When dietary Se content is low, the brain, endocrine glands and reproductive organs are supplied with selenium in the first place, before the liver, heart and skeletal muscles [Behne et al. 1995]. In the case of offal from domestic animals, selenium concentration is strictly dependent on the composition of the feed and the use of mineral supplements that contain Se. For free-living animals, the most important factors are geographical origin and availability of selenium-rich foods in a given area. The selenium content of the offal from different farm and free-living animals in different countries is shown in Table 3.

CONCENTRATION OF SELENIUM IN EGGS

Eggs are a highly valuable food product. From the nutritional point of view, they are a very good source of basic nutrients. They contain high-value and easily assimilated protein, and are rich in major and trace elements and vitamins of the B group. Eggs are also a good source of antioxidants (vitamins E and A, beta-carotene, selenium) that protect us from the harmful effects of free radicals.

Table 4. Content of Se ($\mu g \cdot g^{-1}$ w.w.) in hen eggs in different countries

	Country	Average	Range	References
whole eggs	France	0.150	n/d	Simonoff et al. (1988)
	UK	n/d	0.009 - 0.012	Holland et al. (1991)
	Croatia	0.177	n/d	Klapec et al. (2004)
	Greece	0.087	n/d	Baratakos et al. (1987)
	Poland	0.159	0.010 - 0.396	Pilarczyk et al. (2010)
	Poland	0.160	0.148 - 0.385	Jabłońska. (2007)
	Slovakia	0.215	0.189 - 0.233	Kadrabova et al. (1997)
	New Zealand	0.240	n/d	Thomson and Robinson (1980)
	Australia	n/d	0.19 - 0.414	McNaughton and Marks (2002)
	USA	n/d	0.225 - 0.308	USDA* (1999)
yolk	France	0.237	n/d	Simonoff et al. (1988)
	UK	0.200	n/d	Holland et al. (1991)
	Slovakia	0.342	0.281 - 0.443	Kadrabova et al. (1997)
	Ireland	0.241	n/d	Murphy and Cashman (2001)
	Russia	0.102	n/d	Golubkina and Papazyan (2006)
	Australia	0.260	n/d	McNaughton and Marks (2002)
	USA	n/d	0.225 - 0.308	USDA* (1999)
white	France	0.063	n/d	Simonoff et al. (1988)
	UK	0.060	n/d	Holland et al. (1991)
	Slovakia	0.086	0.082 - 0.109	Kadrabova et al. (1997)
	Ireland	0.064	n/d	Murphy and Cashman (2001)
	Russia	0.261	n/d	Golubkina and Papazyan (2006)
	Australia	0.090	n/d	McNaughton and Marks (2002)
	USA	0.176	n/d	USDA* (1999)

n/d – no data

* United States Department of Agriculture

Eggs are potent accumulators of selenium. Selenium in eggs is characterized by very high bioavailability. The concentration of this trace element in eggs depends on the hen feeding method, dietary Se concentration and the form of Se. As a rule, eggs from organically and conventionally raised hens have lower concentrations of selenium compared to eggs obtained under commercial conditions. The distribution of Se in eggs is unequal and depends, to a certain extent, on Se forms used in animal nutrition. The deposition of organic selenium is most effective in egg yolk. It also turns out that selenocysteine is deposited mainly in yolk, and selenomethionine in albumen. In general, egg yolk is a richer source of selenium than egg albumen. It may contain 4 times as much selenium as albumen. Selenium content decreases in the order: yolk>chalazae>internal viscous albumen>external liquid egg white [Golubkina and Papazyan 2006]. A new product on the market are bio-eggs, which are deliberately enriched with certain vitamins and/or trace elements, including selenium.

Table 5. Content of Se ($\mu g \cdot g^{-1}$ w.w.) in fish in different countries

Fish	Country	Average	Range	References
carp	Slovakia	0.243	0.204 - 0.273	Kadrabova et al. 1997
trout	Slovakia	n/d	0.196 - 0.207	Kadrabova et al. 1997
plaice	Ireland	0.282	0.268 - 0.298	Murphy and Cashman 2001
cod	Ireland	0.265	0.233 - 0.299	Murphy and Cashman 2001
mackerel	Poland	0.320	n/d	Kuczyńska and Biziuk (2007)
flounder	Poland	0.136	0.095 - 0.154	Pilarczyk et al. (2010b)
sprat	Poland	0.175	0.144 - 0.310	Pilarczyk et al. (2010b)
salmon	Poland	0.171	0.068 - 0.244	Pilarczyk et al. (2010b)
herring	Sweden	0.347	n/d	Önning (2000)
mackerel	Sweden	0.117	n/d	Önning (2000)
plaice	Sweden	0.322	n/d	Önning (2000)
turbot	Sweden	0.473	n/d	Önning (2000)
flounder	Sweden	0.101	n/d	Önning (2000)
dab	Sweden	0.068	n/d	Önning (2000)
salmon	Norway	0.195	n/d	Plessi et al. (2001)
cod	Norway	0.188	n/d	Plessi et al. (2001)
pilchard	Italy	0.678	n/d	Plessi et al. (2001)
perch	Italy	0.073	n/d	Plessi et al. (2001)
mackerel	Italy	0.356	n/d	Plessi et al. (2001)
sea bass	Turkey	0.294	n/d	Özden and Erkan (2008)
sea bream	Turkey	0.240	n/d	Özden and Erkan (2008)
common dentex	Turkey	0.285	n/d	Özden and Erkan (2008)
sea bass[A]	Croatia	0.397	n/d	Satovič and Dubravka (2004)
sea bass[B]	Croatia	n/d	0.235 - 0.287	Satovič and Dubravka (2004)
pilchard	Croatia	0.580	n/d	Satovič et al. (2003)
salmon	Taiwan	2.01	n/d	Chien et al. (2003)
bonneville whitefish	USA/Idaho	1.01	n/d	Essig and Kosterman (2008)
rainbow trout	USA/Idaho	n/d	0.06 - 0.71	Essig and Kosterman (2008)
lake trout	USA/Idaho	n/d	0.11 - 0.85	Essig and Kosterman (2008)
channel catfish	USA/Idaho	n/d	0.11 - 0.19	Essig and Kosterman (2008)
common carp	USA/Idaho	0.49	n/d	Essig and Kosterman (2008)
yellow perch	USA/Idaho	n/d	0.05 - 0.53	Essig and Kosterman (2008)

n/d – no data, [A] – farmed fish, [B] – free-living fish

Concentration of Selenium in Fish

Selenium concentration in fish varies widely according to species, site of occurrence and trophic level of fish. Because of their ability to accumulate considerable amounts of selenium, fish are regarded as a good dietary source of this trace element for humans and animals. The content of selenium in fish and fish products ranges from 0.2 to 0.9 μg Se/g (Food Standards Agency 2002). Selenium bioavailability from fish is also high. Many studies have shown that Se intake from fish is closely correlated to the increased activity of glutathione peroxidase (GSH-Px) and concentration of selenoprotein P [Bergmann et al. 1998, Hagmar et al. 1998]. Wen et al. [1997] observed that out of veal, chicken, beef, pork, lamb, flounder, tuna, selenomethionine and sodium selenite, Se from flounder was the most efficient in restoring proper Se concentrations in the liver and skeletal muscles of Se-deficient rats. Furthermore, Fox et al. [2004] reported that selenium retention from fish was significantly higher than from selenate and yeast. Hagmar et al. [1998] observed a positive correlation between the number of fish consumed and plasma Se concentration, which in the group of humans eating more fish was higher by 81%. However, some studies indicate that bioavailability from tuna and other seafood is low [Mutanen 1986, Meltzer et al. 1993]. It is difficult to explain the differences in research results. The probable reasons are differences specific to individual populations and the use of different methods for evaluation of Se bioavailability. The low availability of Se from fish may also result from the tissue accumulation of heavy metals, which significantly inhibit the absorption of selenium [Cappon and Smith 1982]. Selenium concentration increases in the order of bivalve molluscs<crustacean<fish.

Meeting the Daily Selenium Requirement in Different Countries of Europe

Food is the main source of selenium for humans. Rayman [2000] reports that daily selenium requirement depends on age and state of the body and ranges from 20 μg (children) to 75 μg (adults). A safe daily dose of selenium should not exceed 400 mg. Daily Se intake varies by country and reflects the level of this trace element in food products. Eating habits play a considerable role in this respect. Bioavailability of selenium from food depends on its chemical form and other factors (e.g. protein content, amount of fats, content of heavy metals). Selenium content of food also depends on the temperature and duration of heat treatment of food. Se content in processed foods, e.g. tinned food and concentrates is half that of the original raw materials. Organic selenium compounds are the most readily available form for humans. Inorganic selenium compounds are absorbed through passive transport (diffusion in the presence of sodium ions) in the small intestine, whereas organic selenium forms are absorbed by active transport in the intestine [Kuczyńska and Biziuk 2007]. Recommended selenium intake for different age groups is given in Table 6. In Europe, selenium consumption has decreased over the recent years. The Scientific Committee on Food [2000] and FAO/WHO [2002] report that daily intake of this element averaged 75 μg in the 1970s and is half that today. The reason is the increased consumption by Europeans of local foods that replaced products imported from Canada and the USA, which contain more selenium than their counterparts of European origin. Daily Se intake in European countries shows considerable differences (Tab. 7). In Croatia the mean daily intake of Se is 27.3 μg/day [Klapec et al. 1998], which is one-half of the daily requirement. Relatively low intake of this trace element is also found in Slovakia, Poland, Germany, Spain and Sweden (32-38 μg/day). In Finland, daily selenium intake exceeds 100 μg/day, which is associated with the use of selenium-containing fertilizers in plant production. Animal protein is the main source of dietary selenium.

Table 6. Recommended dietary allowances of selenium for humans [Dietary Reference Intakes 2000]

Population groups	Recommended dietary allowances (μg/day)	Safe maximum dose with no risk of side effects (μg /day)
Children 1-8 years	20-30	90-150
Children 9-13 years	40	280
Teenagers 14-18 years	55	400
Adults	55	400
Pregnant women	60	400
Nursing mothers	70	400

Table 7. Daily intake of Se (μg/day) in selected countries of Europe

	Daily intake of Se with products of animal origin	% of recommended dose	Total daily intake of Se
Belgium	27	49	28-61[1]
France	38	69	29-43[2]
Germany	30	55	35[3]
Greece	28	51	39.3-110[4]
Denmark	37	67	38-47[5]
Italy	36	66	n/d
Netherlands	27	49	40-54[5]
Spain	35	63	44,6-50,4[6]
Sweden	35	64	38[7]
UK	32	58	63[5]
Ireland	36	65	50[8]
Slovakia	20	36	38[9]
Poland	21	37	30-40[10]
Finland	45	82	100-110[5]
Switzerland	34	62	70[7]

[1]Robberecht et al. 1994,
[2]Lamand et al. 1994,
[3]Alfthan and Nève 1996,
[4]Bratakos and Ioannou (1991),
[5]Scientific Committee on Food (2000),
[6]Navarro et al. 1995,
[7]Kumpulainen (1993),
[8]Murphy et al. (2002),
[9]Kadrabova et al. 1998,
[10]Wąsowicz et al. (2003), Pilarczyk et al. (2010b).

In European countries, the proportion of animal products in meeting the total Se requirement is large and usually ranges from 50 to 60%. This proportion varies widely in different countries, ranging from 37 to 82% according to eating habits of a population and Se concentration in food. These data indicate that food products of animal origin are a valuable source of selenium. Proper nutrition that includes essential trace elements, in particular selenium, is the best, easiest and cheapest way of preventing many diseases in humans. Therefore it seems necessary to develop appropriate prophylactic programmes to prevent selenium deficiency and the resulting selenium-dependent diseases.

The most effective method of preventing deficiency and ensuring increased intake by humans is to use selenium-enriched feeds in animal husbandry and breeding. It should be remembered that compared to plant feeds, feedstuffs of animal origin contain more selenium that is less readily absorbed by the body than that from plant feeds (bioavailability of 20% vs. 70% on average). People need no encouragement to take medication; it is more difficult to convince them that a well-balanced diet may be an excellent medicine.

References

Alfthan, G. and Nève, J. (1996). Selenium intakes and plasma selenium levels in various populations. In J. Kumpulainen and J. Salonen (Eds.), Natural Antioxidants and Food Quality in Atherosclerosis and Cancer Prevention, pp. 161-167. Cambridge, Royal Society of Chemistry.

Behne, D., Weiss-Nowak, C., Kalcklosch, M., Westphal, C., Gessner, H. and Kyriakopoulos A. (1995). Studies on the distribution and characteristics of new mammalian selenium-containing proteins. *Analyst* 120, 823-825.

Bratakos, M. and Ioannou, P. (1991). Selenium in human milk and dietary selenium intake by Greeks. *The Science of The Total Environment* 105, 101-107.

Bratakos, M., Zafiropolus, T., Siskos, P. and Ioannou, P. (1987). Selenium in foods produced and consumed in Greece. *Journal of Food Science* 52, 817-822.

Bratakos, M., Zafiropoulos, T., Siskos, P. and Ioannou P. (2007). Selenium losses on cooking Greek foods. *International Journal of Food Science and Technology* 23, 6, 585-590.

Barclay, M.N.I., McPherson, A., and Dixon, J. (1995). Selenium content of a range of foods. *Journal of Food Composition Analysis* 8, 307-318.

Bergmann, S., Neumeister, V., Siekmeier, R., Mix, C., Wahrburg, V. and Jaross W. (1998). Food supply abundant increase of serum selenium concentrations in middle-aged Dresden women between 1990 and 1996. DRECAN-Team. Dresden Cardiovascular Risk and nutrition. *Toxicology Letters* 18, 196-197.

Bjelakovic, G., Nikolova, D., Gluud, L.L., Simonetti, R.G. and Gluud, C. (2008). Antioxidant supplements for prevention of mortality in healthy participants and patients with various diseases. Cochrane Database of Systematic Reviews 2008. http://www.cfah.org/hbns /archives/viewSupportDoc.cfm?supportingDocID=578

Bügel, S., Sandström, B. and Skibsted, L. (2004). Pork meat: A good source of selenium? *Journal of Trace Elements in Medicine and Biology* 17, 4, 307-311.

Cappon, C. and Smith, J. (1982). Chemical form and distribution of mercury and selenium in edible seafood. *Journal of Analytical Toxicology* 6, 1, 10-21.

Chien, L.C., Yeh, C.Y., Huang, S.Y., Shieh, M.J. and Han, B.C., (2003). Pharmacokinetic model of daily selenium intake from contaminated seafood in Taiwan. *The Science of The Total Environment* 311, 1-3, 57-64.

Combs, G.F. Jr.and Combs, S.B (1986). Selenium in foods and feeds. In: Role of selenium in nutrition, pp. 41-126. New Jork, NY Academic Press.

Díaz-Alarcón., J.P., Navarro-Alarcón, M., López-Martínez, M. and López-García de la Serrana, H., (1996). Determination of Selenium in Meat Products by Hydride Generation Atomic Absorption Spectrometry - Selenium Levels in Meat, Organ Meats, and Sausages in Spain. *Journal of Agricultural and Food Chemistry* 44, 1494-1497.

Dietary Reference Intakes For Vitamin C, Vitamin E, Selenium, and Carotenoids. A Report of the Panel on Dietary Antioxidants and Related Compounds, Subcommittees on Upper Reference Levels of Nutrients and Interpretation and Uses of Dietary Reference Intakes, and the Standing Committee on the Scientific Evaluation of Dietary Reference Intakes. NATIONAL ACADEMY PRESS Washington, D.C., 2000.

Essig, D. and Kosterman, M., (2008). Arsenic, Mercury, and Selenium in Fish Tissue from Idaho Lakes and Reservoirs: A Statewide Assessment. Idaho department of Environmental Quality. http://www.deq.idaho.gov/ media/639760-arsenic_mercury_fish_tissue_report_0508.pdf

FAO/WHO (2002). Human Vitamin and Mineral Requirements. Report of an expert consultation, Bangkok, Thailand. Chapter 15 Selenium. Available at: http://www.fao.org/DOCREP/004/Y2809E/y2809e0l.htm#bm21.1

Finley, J., Grusak, M., Keck, A. and Gregoire, B. (2004). Bioavailability of selenium from meat and broccoli as determined by retention and distribution of ^{75}Se. *Biological Trace Element Research* 99, 191-209.

Food Standards Agency (2002): Mc Lance and Widdowsen's. The Composition of Foods, Sixth Summary edition. Cambridge, Royal Society of Chemistry.

Fox, T.E., Van den Heuvel, E.G., Atherton, C.A., Dainty, J.R., Lewis, D.J., Langford, N.J., Crews, H.M., Luten, J.B., Lorentzen, M., Sieling, F.W., van Aken-Schneyder, P., Hoek, M., Kotterman, M.J., van Dael, P. and Fairweather-Tait, S.J. (2004). Bioavailability of selenium from fish, yeast and selenate: a comparative study in humans using stable isotopes. *European Journal of Clinical Nutrition* 58, 343-349.

Gamerith, W., Lichtenegger, F., Schindler, E. and Steindl, W. (1991). *Selengehalt in Fleisch* und Nieren von Rindern und Schweinen in der Steiermark. *Fleischwirtschaft* 71, 1343-1345.

Golubkina, N.A. and Khotimchencko, S.A. (1994). Selenium in the food products of Ural economic region. Gigiena i Sanitariia 7, 12- 14.

Golubkina, N.A. and Papazyan T.T. (2006) Selenium distribution in eggs of avian species. Comparative Biochemistry and Physiology Part B: *Biochemistry and Molecular Biology* 145, 3-4, 84-8.

Grela, E.R. and Sembratowicz, I. (1997). Organiczne związki selenu w żywieniu zwierząt. *Medycyna Weterynaryjna* 53, 7, 385-386.

Dagmar, L., Persson-Moschos, M., Akkesson, B. and Shutz, A. (1998). Plasma levels of selenium, selenoprotein P and glutathione peroxidase and their correlations to fish intake and serum levels of thyrotropin and thyroid hormones: a study on Latvian fish consumers. *European Journal of Clinical Nutrition* 52, 796-800.

Holland, B., Welch, A., Unwin, I., Buss, D., Paul, A. and Southgate, D. (1991). McCance and Widdowson's the composition of foods. Cambridge, Royal Society of Chemistry/Ministry of Agriculture, Fisheries and Food.

Iqbal, S., Kazi, T.G., Bhanger, M.I., Akhtar, M., and Sarfraz, R.A. (2008). Determination of selenium content in selected Pakistani foods. *International Journal of Food Science and Technology*, 43, 339-345.

Jabłoński, E., Gromadzińska, J., Bertrandt, J., Kłos, A., Dargo, A. and Wąsowicz, W. (2007). Zawartość selenu w wybranych artykułach żywnościowych pochodzących z Polski centralnej. Żywienie Człowieka i Metabolizm, 34, 1440-1445

Kadrabova, J., Madaric, A., and Ginter, E. (1997). The selenium content of selected food from the Slovak Republic. *Food Chemistry*, 58, 1-2, 29-32.

Klapec, T., Mandić, M. L., Grgić, J., Primorac, L., Ikić, M., Lovrić, T., Grgić, Z. and Herceg, Z. (1998). Daily dietary intake of selenium in eastern Croatia. *The Science of the Total Environment*, 217, (1-2), 127-136.

Klapec, T., Mandić, M.L., Grgić, J., Primorac, L., Perl, A., and Krstanovic, V. (2004). Selenium in selected foods grown or purchased in eastern Croatia. *Food Chemistry*, 85, 445-452.

Kuczyńska, J. and Biziuk, M. (2007). Biogeochemia selenu i jego monitoring w materiałach biologicznych pochodzenia ludzkiego. *Ecological Chemistry and Engineering* 14, S1, 47-65.

Kumpulainen, J.T. (1993). Selenium in foods and diets of selected countries. *Journal of Trace Elements and Electrolytes in Health and Diseases* 7, 107-108.

Lamand, M., Tressol, J.C. and Bellanger, J. (1994). The mineral and trace element composition in French food items and intake levels in France. *Journal of Trace Elements and Electrolytes in Health and Diseases* 8, 195-202.

Lavi, L. and Alfassi, Z.B. (1990). Determination of trace amounts of cadmium, cobalt, chromium, iron, molybdenum, nickel, selenium, titanium, vanadium and zinc in blood and milk by neutron activation analysis. *Analyst*, 115, 817-822.

McDowell, L.R., Forrester, D.J., Linda, S.B., Wright, S.D. and Wilkinson, N.S. (1995). Selenium status of white-tailed deer in southern Florida. *Journal of Wildlife Disease* 31, 205-211.

McNaughton, S.A. and Marks G.C. (2002). Selenium Content of Australian Foods: A Review of Literature Values. *Journal of Food Composition and Analysis* 15, 169-182.

Meltzer, H.M., Bibow, K., Paulsen, I.T., Mundal, H.H., Norheim, G. and Holm. H, (1993). Different bioavailability in humans of wheat and fish selenium as measured by blood-platelet response to increased dietary Se. *Biological Trace Element Research* 36, 229-241.

Murphy, J. and Cashman, K.D. (2001). Selenium content of a range of Irish foods. *Food Chemistry* 74, 493-498.

Murphy, J., Hannon, E.M., Kiely, M., Flynn, A. and Cashman, K.D. (2002). Selenium intakes in 18 – 64-y-old Irish adults. *European Journal of Clinical Nutrition* 56, 402-408.

Mutanen, M. (1986). Bioavailability of selenium. *Annual Clinical Research* 18, 48-54.

Navarro, M., López, H., Ruiz, M. L., González, S., Pérez, V., and López, M. C. (1995). Determination of selenium in serum by hydride generation atomic absorption spectrometry for calculation of daily dietary intake. *The Science of the Total Environm*ent 175, 3, 245-252.

Oster, O. and Prellwitz, W. (1989). The daily dietary selenium intake of West Germany adults. *Biological Trace Element Research* 20, 1-14.

Önning, G. (2000). Separation of soluble selenium compounds in different fish species. *Food Chemistry*, 68, 2, 133-139.

Özden, Ö. and Erkan, N. (2008). Comparison of biochemical composition of three aqua cultured fishes (*Dicentrarchus labrax, Sparus aurata, Dentex dentex*). *International Journal of Food Sciences and Nutrition* 59, 7-8, 545-557.

Pappa, E., Pappas, A. and Surai, P. (2006). Selenium content in selected foods from the Greek market and estimation of the daily intake. *The Science of the Total Environment* 372, 100-108.

Pennington, J., Schoen, S., Salmon, G., Young, B., Johnson, R. and Marst, R. (1995). Composition of core foods of the US food supply, 1982-1991. III. Copper, manganese, selenium and iodine. *Journal of Food Composition and Analysis* 8, 171-217.

Pilarczyk, B., Balicka-Ramisz, A., Mituniewicz-Małek, A., Ramisz, A., Sablik, P. and Dmytrów, I. (2009). Selengehalte im Blutserum, in der Kuhmilch und in ihren Produkten in Westpommern. *Tierärztliche Umschau* 64, 327-331.

Pilarczyk, B., Balicka-Ramisz, A., Ramisz, A., Adamowicz, E., Bujak, T., Tomza-Marciniak, A., Bąkowska, M. and Wieczorek-Dąbrowska, M. (2008). Selenium concentration in roe deer from the Western Pomerania, Poland. *The Bulletin of the Veterinary Institute in Pulawy* 52, 631-633.

Pilarczyk, B., Balicka-Ramisz, A., Ramisz, A., Adamowicz, E., Pilarczyk, R., Tomza-Marciniak, A., Bąkowska, M. (2009). Selenium concentration in liver and kidney of free living animals (roe and red deer) from West Pomerania (Poland). *European Journal of Wildlife Research* 55, 279-283.

Pilarczyk, B., Hendzel, D., Pilarczyk, R., Tomza-Marciniak, A., Błaszczyk, B., Dąbrowska-Wieczorek, M., Bąkowska, M., Adamowicz, E. and Bujak, T. (2010b). Liver and kidney concentrations of selenium in wild boars (*Sus scrofa*) from northwestern Poland. *European Journal of Wildlife Research* 55, 5, 797-802.

Pilarczyk, B., Pilarczyk, R., Tomza-Marciniak, A., Kowieska, A., Wójcik, J., Sablik, P., Tylkowska, A. and Hendzel, D. (2011). Selenkonzentration im Blutserum und in der Milch von Kuhen aus einem ökologischen Betrieb und aus konventionellen Betrieben in einem Selenmangelgebiet. *Tierarztliche Umschau* 66, 6, 248-252.

Pilarczyk B., Tomza-Marciniak, A., Mituniewicz-Małek, A., Wieczorek, M., Pilarczyk, R., Wójcik, J., Balicka-Ramisz, A., Bąkowska, M. and Dmytrów, I. (2010). Selenium content in selected products of animal origin and estimation of the degree of cover daily Se requirement in Poland. *International Journal of Food Science and Technology* 45, 186-191.

Plessi, M., Bertelli, D. and Monzani, A. (2001). Mercury and selenium content in selected seafood. *Journal of Food Composition and Analysis* 14, 461-467.

Puls, R. (1988). Mineral Levels in Animal Health. *Sherpa International, Clearbrook,* B.C., Canada.

Rayman, M.P. (2000). The importance of selenium to human health. *Lancet* 356, 233-241.

Robberecht, H.J., Hendrix, P., Van Cauenbergh, R. and Deelstra, H.A. (1994). Actual daily intake of selenium in Belgium using duplicate portion sampling. Zeitschrift für Lebensmittel-Untersuchung und -Forschung A (Berlin) 199, 251-254.

Satovič, V. and Beker, D. (2004). Selenium content in sea bass of the Adriatic Sea. European Food Research and Technology, part A, 218, 2, 111-113.

Satovič, V., Beker, D. and Gumhalter-Karolyi, L. (2003). Selenium content in the pilchard of the Adriatic Sea. European Food Research and Technology. 217, 2, 154-156.

Scientific Committee on Food (2000). Opinion of the Scientific Committee on Food on the Tolerable Upper Intake Level of Selenium. Available from http://ec.europa.eu/food/fs/sc /scf/out80g_en.pdf

Simonoff, M., Hamon, C., Moretto, P., Llabador, Y. and Simonoff, G. (1988). *Selenium in foods in France*. Journal of Food Composition and *Analysis* 1, 295-302.

Thomson, C.D. and Robinson, M.F. (1980). Selenium in human health and disease with emphasis on those aspects peculiar to New Zealand. *American Journal of Clinical Nutrition 33*, 2, 303-23.

United States Department of Agriculture (1999). USDA Nutrient database for standard reference, Release 13. http://www.nal.usda.gov/fnic/foodcomp /Data/SR13/sr13.html.

Van der Torre, H., Van Dokkum, W., Schaafsma, G., Wedel, M. and Ockhuizen, T. (1991). Effect of various levels of selenium in wheat and meat on blood Se status indices and on Se balance in Dutch men. *British Journal of Nutrition* 65, 69-80.

Vikøren, T., Bernhoft, A., Waler, T. and Handeland, K. (2005). Liver concentrations of copper, cobalt, and selenium in wild Norwegian red deer (*Cervus elaphus*). *Journal of Wildlife Disease* 41, 569-579.

Wąsowicz, W., Gromadzińska, J., Rydzyński, K. and Tomczak, J. (2003). Selenium status of low-selenium area residents: Polish experience. *Toxicology Letters* 137, 95-101.

Wen, H.Y., Davis, R.L., Shi, B., Chen, J.J., Chen, L., Boylan, M. and Spallholz, J.E. (1997). Bioavailability of selenium from veal, chicken, beef, pork, lamb, flounder, tuna, selenomethionine, and sodium selenite assessed in selenium-deficient rats. *Biological Trace Element Research* 58, 43-53.

Wyatt, C.J., Meléndez, J.M., Acuoa, N., and Rascon, A. (1996). Selenium (Se) in foods in northern Mexico, their contribution to the daily Se intake and the relationship of Se plasma levels and glutathione peroxidase activity. *Nutrition Research*, 16, 949-960.

Yanardağ, R. and Orak, H. (1999). Selenium content of milk and milk products of Turkey. II. *Biological Trace Element Research* 68, 1, 79-95.

In: Selenium: Sources, Functions and Health Effects
Editors: Chinatsu Aomori and Megumi Hokkaido
ISBN: 978-1-61942-061-8

Chapter 9

STRESS SIGNALING AND PROAPOPTOTIC PATHWAYS INDUCED BY SELENITE IN MALIGNANT COLONOCYTES

Věra Králová*[*]*, Soňa Benešová and Emil Rudolf
Department of Medical Biology and Genetics,
Charles University in Prague, Faculty of Medicine in Hradec Králové,
Šimkova, Hradec Králové, Czech Republic

ABSTRACT

Sodium selenite (NaSe) continues to be studied as a prospective chemopreventive agent against several types of malignancies including colon cancer. Recent studies demonstrated that in malignant or premalignant cells of digestive system NaSe induces a wide range of effects which may in the end either stabilize or damage these cells. Due to a multifaceted nature of NaSe-dependent mechanisms and their natural crosstalk, their timing and contribution for final observed outcome in the studied model is of paramount importance. Thus the purpose of this work was to investigate the kinetics of selected mechanisms of NaSe in human colon cancer cells HCT-116 during 24 h of exposure. Our results indicate that NaSe has moderate genotoxic effects with subsequent activation of DNA-damage pathway in HCT-116 cells. Furthermore, in thus exposed colonocytes mitogen stress kinase signaling (in particular p38) occurs as well as an early direct mitochondrial changes with resulting cellular degeneration bearing features of apoptosis. These results suggest that in HCT-116 cells NaSe mediates a series of changes of which DNA damage response and direct mitochondrial effects seem to complement each other and to contribute to final observed cell death.

Keywords: Sodium selenite, HCT-116; DNA damage, stress response, cell death

---

[*] Emil Rudolf, Ph.D. Department of Medical Biology and Genetics, Charles University in Prague, Faculty of Medicine in Hradec Králové, Šimkova 870, 500 38 Hradec Králové Czech Republic, Tel: +420 495816393; Fax: +420 495816495 E-mail: rudolf@lfhk.cuni.cz

1. Introduction

Trace element selenium is beneficial for maintenance of homeostasis of mammalian cells, tissues and organs in a number of ways. In particular, growing awareness of its importance in prevention of malignant diseases was repeatedly supported by many experimental, clinical as well as epidemiological studies [1]. Here, it has often been argued that organic selenium forms are superior to inorganic ones, namely because of their lesser toxicity, supposed better bioavailability and modes of action [2]. Nevertheless, after premature termination of the Selenium and Vitamin E Cancer Prevention Trial (SELECT) which unsuccessfully evaluated an ability of an organic selenium to prevent prostate cancer and secondary cancers [3], inorganic sodium selenite (NaSe) came again into scientific spotlight. Since the first reports on NaSe biological activity, it continues to attract scientific as well as clinical attention due to its reported range of antineoplastic effects against several types of malignancies [4]. In particular, NaSe appears to hold a promising chemopreventive potential towards colorectal carcinogenesis where it has shown a number of specific as well as non-specific effects [5]. The efficacy of NaSe in colon cancer is in part exaplained by its natural administration, accumulation and retention in digestive system tissues. In addition, when diffused into cells NaSe has a straightforward intracellular metabolism, i.e. in the presence of thioredoxin reductase, it is reduced into selenide which continously cycles to generate reactive oxygen species (ROS) and oxygen [6]. Thus when in the cytoplasm, NaSe may perturb the intracellular redox balance by reacting with intracellular thiols while mounting significant levels of oxidative stress. Alternatively, NaSe has been reported to directly target specific intracellular molecules or signaling pathways including Wnt/β-catenin, MAP kinases or proapoptotic Bax [6, 7].

In our previous experiments we demonstrated that *in vitro* NaSe inhibits growth of colon cancer cells with differing status of p53 [8, 9]. Moreover, in thus treated cells NaSe altered cell cycle distribution and stimulated stress signaling, resulting in activation of cell death – apoptosis. NaSe-induced apoptosis in colon cancer cells is by far not entirely understood despite recent advancements in our knowledge in various cancer models [6, 10, 11]. Thus the aim of the present work was to to investigate the kinetics of selected mechanisms of NaSe in human colon cancer cells HCT-116 during 24 h of exposure. Our results suggest that in HCT-116 cells NaSe mediates a series of changes of which DNA damage response and direct mitochondrial effects seem to complement each other and to contribute to final observed cell death.

2. Materials and Methods

2.1. Cell Line

Human colorectal carcinoma cells HCT-116 were obtained from the American Type Culture Collection (ATCC number CCL-247). Cell line was maintained in Dulbecco´s Modified Eagle´s Medium (DMEM) supplemented with 10% fetal bovine serum, 100 U/ml penicillin and 100 μg/ml streptomycin at 37 °C in a humidified atmosphere with 5% CO_2.

Cells were passaged upon reaching 80% confluence approximately twice a week. Cultures were regularly checked for the presence of mycoplasmas.

2.2. Treatments

Sodium selenite (Sigma-Aldrich, Prague, Czech Republic) was dissolved in distilled water as stock concentration of 1.5 mM. Stock solutions of antioxidant NAC (Sigma-Aldrich, Prague, Czech Republic) was prepared in serum free DMEM. The working concentration of NAC was achieved by diluting its stock solution in treatment medium: NAC (N-acetyl cystein 250 μM – added to cells 24 h prior to NaSe). The concentrations of the employed pharmacological inhibitors were as follows: Pifithrin – specific p53 inhibitor (30 μM – added to cells 1 h before NaSe exposure, Sigma-Aldrich, Prague, Czech Republic), p38 and JNK inhibitors SB203580 and SP600125 (10 μM, – both added to cells 30 min prior to NaSe; Calbiochem, EMD Biosciences, Inc., La Jolla, Ca, USA).

2.3. Cell Death Assays

Phosphatidyl serine translocation NaSe-treated and control HCT-116 cultures grown on slides were rinsed with phosphate saline buffer, 1 ml of Annexin-binding buffer with 20 μl of Annexin V-FITC (A) and 20 μl of propidium iodide (PI) was added, and slides were left at dark for 15 min. Next, labeling medium was aspirated and 1 ml of Annexin-binding buffer was added for 1 min. The buffer was aspirated and slides were mounted into SlowFade® medium (Molecular Probes, Inc. Eugene, U.S.A.) and examined under a fluorescence microscope Nikon Eclipse E 400 (Nikon, Prague, Czech Republic) with the digital color matrix camera COOL 1300 (VDS, Vosskühler, Germany). Annexin and PI positivity were evaluated by the software NIS Elements AR 3.20 in at least 1,000 cells.

Morphology analysis by time-lapse videomicroscopy HCT-116 cells were seeded into plastic tissue-culture dishes with glass bottom and left for 24 h in an incubator with 5% CO2 at 37°C. Next day the growth medium was replaced with a medium containing 10 μM NaSe. The tissue-culture dishes were transferred into a time-lapse imaging system BioStation IM (Nikon, Prague, Czech Republic) combining an incubator, a motorized microscope and a cooled CCD camera. Recording was carried out in a multipoint and multichannel manner employing various time-lapse modes and upon small as well as high magnifications to allow global as well as detailed view of changes in behavior of treated cell populations. Recorded sequences were subsequently semi automatically analyzed with the software NIS Elements AR 3.20 (Nikon, Prague, Czech Republic).

2.4. Caspase Activity

NaSe-treated and control cultures at 6, 12, 18 and 24 h were harvested by centrifugation (600 g, 5 min, JOUAN MR 22, Trigon, Prague, Czech Republic) and lysed on ice for 20 min in a lysis buffer containing 50 mM HEPES, 5 mM CHAPS and 5 mM DTT. The lysates were centrifuged at 14,000 g, 10 min, 4°C, and the supernatants were collected and stored at -80°C. The enzyme activity was measured in a 96-well microplate using a kinetic fluorometric method based on the hydrolysis of the fluorogenic caspase-specific substrate (DEVD-AFC for caspase-3, Ac-LEHD-AFC for caspase-9 and IETD-AFC for caspase-8, 37°C, 1 h) by individual caspases. Specific inhibitors of caspase-9, -8 and -3 were used to confirm the specificity of the cleavage reaction. Fluorescence was recorded at 460/40 nm after excitation at 360/40 nm using TECAN SpectraFluor Plus (TECAN Austria GmbH, Grödig, Austria). Results are shown as fold increase in activity relative to untreated cells.

2.5. DNA Damage – Comet Assay

DNA damage in NaSe-treated cells was determined by alkali single-cell gel electrophoresis (comet assay). Control and treated HCt-116 cells were at particular time intervals harvested by 0.25% trypsin, centrifuged for 5 min at 1,500 rpm at 4 °C, suspended in 0.6% agarose and mounted to microscopic slides. Next, cells were lysed in cooled lysis buffer (2.5 M NaCl, 100 mM EDTA, 10 mM Trizma, 1 % Triton X-100 and 10 % DMSO) for 1.5 h at 4 °C. After rinsing in TRIS, cells were allowed to unwind DNA in alkali buffer for 30 min. Electrophoresis was performed at 25 V and 300 mA for 30 min. After neutralizing slides and draining them, cells were stained by 100 μl ethidium bromide (Sigma-Aldrich, Prague, Czech Republic). One hundred cells from three independent samples were scored for tail migration intensity.

2.6. Oxidative Stress

Generation of hydrogen peroxide and/or hydroxyl radical was monitored by intracellular conversion of 2´-7´- dichlorodihydrofluorescein diacetate (DFCH/DA – Sigma-Aldrich, Prague, Czech Republic) into a fluorescent product dichlorofluorescein (DCF). Colonic cells were seeded into cultivation flasks and cultivated to 75% confluence at 37 °C and 5 % CO_2. After exposure to NaSe control and treated cells were detached by a cell scraper and collected by centrifugation (50 x g, 5 min, 4 °C – JOUAN M21, Trigon, Prague, Czech Republic). The cells were resuspended in DMEM (pH adjusted to 7.2) and 5 μM DFCH/DA was added (5 min, 37 °C). Changes in the fluorescence intensity (485 nm excitation; 538 nm emission) were measured by Shimadzu UV - Visible Spectrophotometer UV – 1601 (Shimadzu Deutschland GmbH, Duisburg, Germany). The data were expressed as a percentage of fluorescence intensity increase per 10^6 cells.

2.7. Expression of p53, Bax, Bcl-2 and Bcl-XL proteins

NaSe-treated and control cells were at individual time intervals collected by trypsinization, fixed in 4% paraformaldehyde and permeabilized with cold methanol/Triton-X in 5% BSA. For detection of individual proteins, the cells were then incubated with the following primary antibodies: anti-Bax 1:100; anti-Bcl-2 1:100 and anti-Bcl-XL 1:100 at 4°C for 1 h. After washing with cold PBS (5 min, 25°C), FITC-conjugated goat anti-rabbit or anti-mouse (1:10) was added for 1 h at 4°C. Samples were then rinsed in PBS and the fluorescence of 30,000 cells was analyzed by a flow cytometer Cell Lab Quanta™ SC (Beckman Coulter Inc.).

2.8. MAPK Activities

NaSe-treated and control cells were harvested and collected by centrifugation. Whole cell extracts were prepared by lysis in cell extraction buffer (10 mM Tris, 100 mM NaCl, 1 mM EDTA, 1 mM EGTA, 1 mM NaF, 20 mM $Na_4P_2O_7$, 2 mM Na_3VO_4, 1% Triton X-100, 10% glycerol, 0.1% SDS and 0.5% deoxycholate with 1 mM protease inhibitor cocktail) for 30 min, on ice, with vortexing at 10 min intervals. ERK, p38 and c-Jun N-terminal kinase (JNK) activities were measured using Elisa kits (Sigma-Aldrich, St. Luis, MO, USA and Calbiochem, San Diego, CA, USA) specific for total ERK and phospho-ERK ($pTpY^{185/187}$), total and phospho-p38 ($pTpY^{180/182}$) and total and phospho-JNK ($pTpY^{183/183}$) according to manufacturer's instructions.

The assays were performed in 96-well plate format and samples were read against standard curves obtained from ERK and phospho-ERK, p38 and phospho-p38, JNK and phospho-JNK standards. Results were normalized to micrograms of protein in the cell extract and expressed as the ratio of phospho to total kinase in the same sample. The results of NaSe treatment were expressed as percentage of control values.

2.9. Mitochondrial Membrane Potential

HCT-116 cells were seeded to 96-well plates with black bottom and allowed to grow overnight at 37°C and 5% CO_2. At specified time intervals following the treatment with NaSe, cells were rinsed in PBS and stained with cationic JC-1 dye (10 µg/ml) for 15 min at 37°C.

Following the rinsing with warm cultivation medium, changes in mitochondrial transmembrane potential in at least 50,000 target cells for total assay were assessed by a multiplate reader TECAN SpectraFluo Plus (TECAN Austria GmbH) at 485/520 nm (for the monomer form) and 520/590 nm (for the aggregate form) filter combinations, respectively. Results were expressed as changes in ratio of (J aggregates/monomer) for each different treatment interval or treatment group.

Statistics

Statistical analysis was carried out with a statistical program GraphPad Prism, using one-way Anova test with Dunnet's post test for multiple comparisons. Results were compared with control samples, and means were considered significant if $P<0.05$.

3. Results

3.1. Apoptosis

The proapoptotic effects of NaSe at a previously described concentration of 10 μM in human colon cancer cell line HCT-116 were examined using various markers of early and advanced stages of cell death – apoptosis. In particular, kinetic analyses of Annexin V positivity, membrane blebbing and caspase-8, caspase-9 and caspase-3 activities were carried out.

Data shown in Figure 1A demonstrate that phosphatidylserine translocation occurring at early stages of apoptosis was significantly increased in NaSe-exposed cells in the treatment interval of 8 to 16 h and reached maximum at 24 h of experiment. Similarly, membrane blebbing followed the indicated time course too (Figure 2).

On the other hand, kinetic analyses of activities of selected caspases revealed that caspase-8 was not involved in this model while caspase-9 activity significantly increased as early as at 8 h of treatment. Caspase-3 activity increased too but in a slower manner (Figure 3).

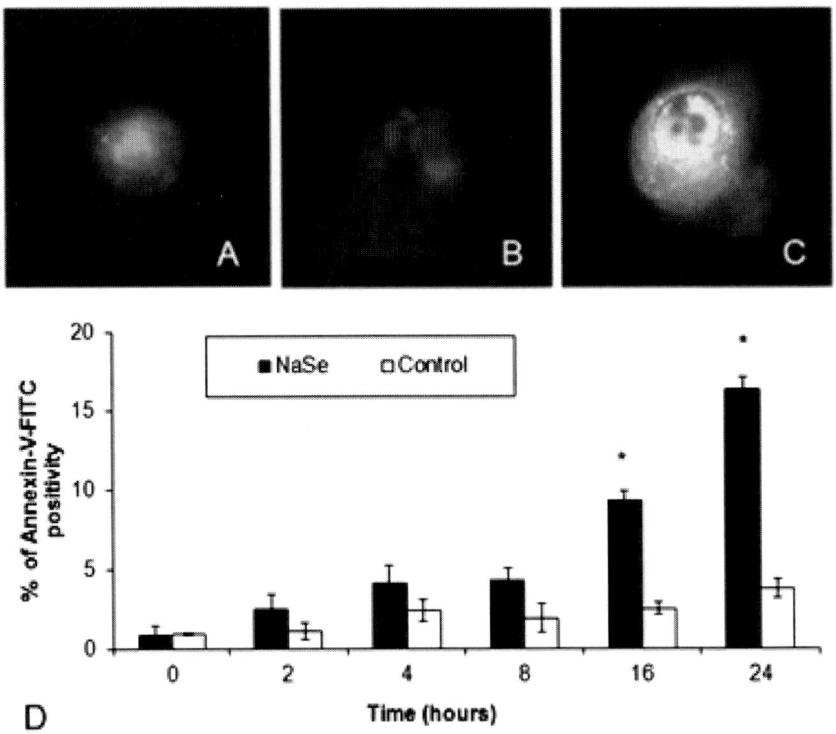

Figure 1. Quantitation of Annexin V positivity in human colon cancer cells HCT-116 exposed to 10 μM sodium selenite (NaSe) by fluorescent microscopy. Early apoptotic cells with translocated phosphatidylserine were of green color (A), late apoptotic cells that were positive for propidium iodide labeled chromatin were of red color (B) and swollen necrotic cells showed early positivity for both green and red (C). Data shown in part D of the figure summarize early apoptotic cells only. Values represent means ± SD of at least three experiments *P<0.05 with one way-Anova test and Dunnett's post test for multiple comparisons.

3.2. Oxidative Stress and DNA Damage

Generation of oxidative stress in treated and control colon cancer cells HCT-116 during 24 h of exposure to 10 μM NaSe was determined spectrophometrically. First elevation in ROS levels in treated HCT-116 cells was observed at 2 h of exposure to NaSe and this increase became significant at 4 h of treatment (Figure 4). Generally, during the entire experiment ROS levels in treated HCT-116 cells continuously grew.

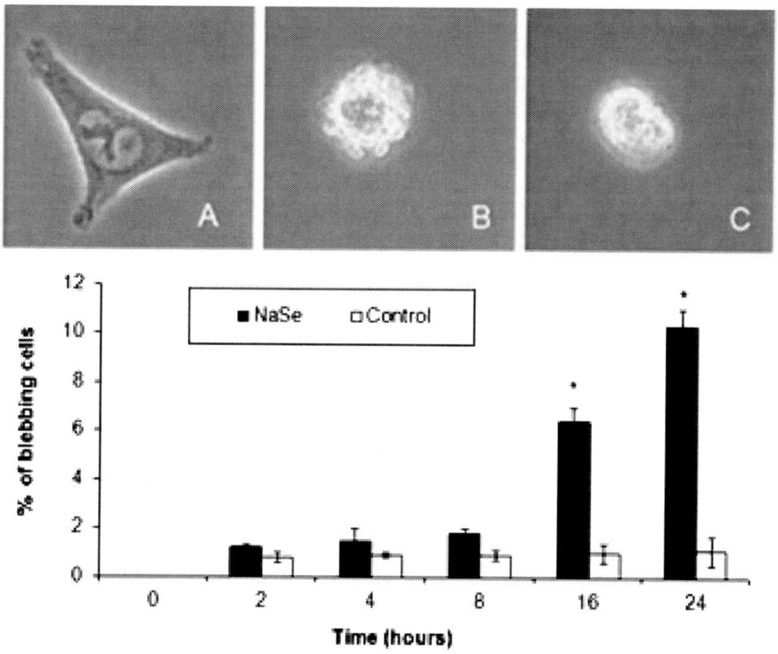

Figure 2. Quantitation of membrane blebbing in human colon cancer cells HCT-116 exposed to 10 μM sodium selenite (NaSe) by time-lapse videomicroscopy. Typical morphology of normal cell (A), blebbing cell (B) and late apoptotic cell (C). Data shown in part D of the figure summarize blebbing cells only. Values represent means ± SD of at least three experiments *P<0.05 with one way-Anova test and Dunnett's post test for multiple comparisons.

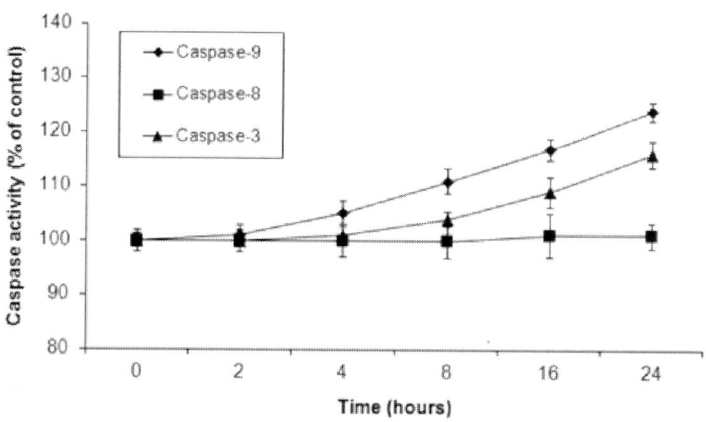

Figure 3. Changes in caspase-8, caspase-9 and caspase-3 activities during 24 h of treatment of colon cancer cells HCT-116 with 10 μM of sodium selenite (NaSe). Values represent means ± SD of at least three experiments.

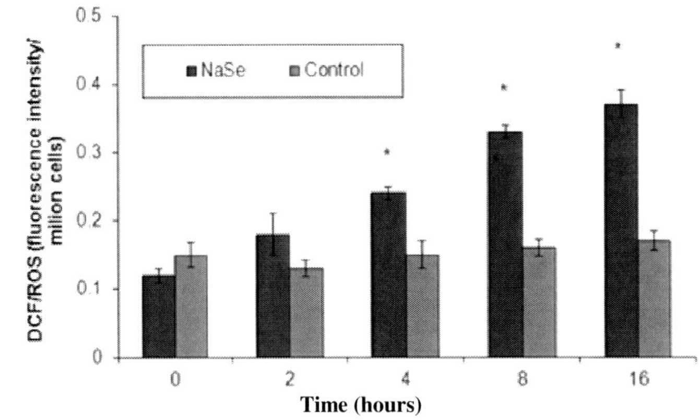

Figure 4. Determination of oxidative stress by spectrophotometric detection of conversion of 2´-7´-dichlorodihydrofluorescein diacetate into a fluorescent product dichlorofluorescein in human colon cancer cells HCT-116 exposed to sodium selenite (NaSe) at the concentration of 10 μM during 24 h. Values represent means ± SD of at least three experiments. *higher than control cultures at P<0.05 with one way-Anova test and Dunnett's post test for multiple comparisons.

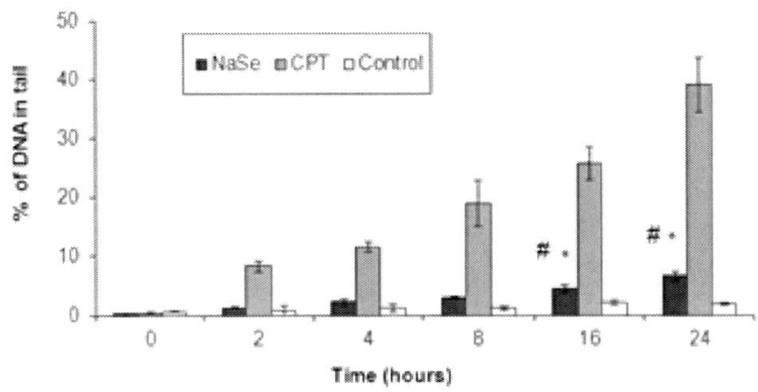

A

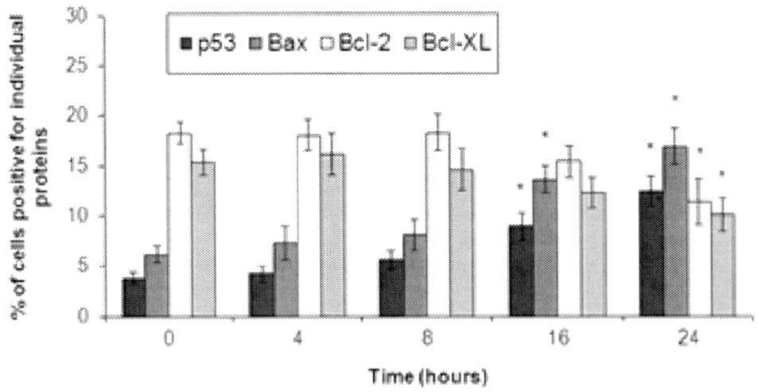

B

Figure 5. DNA damage (A) and expression of selected stress signaling proteins (B) in human colon cancer cells HCT-116 treated with sodium selenite (NaSe) at the concentration of 10 μM and camptothecin (CPT) at the concentration of 1 μM during 24 h as measured by comet assay and flow cytometry. Values represent means ± SD of at least three experiments. Values represent means ± SD of at least three experiments. #lower than cultures treated with NaSe at the concentration of 10 μM *higher than control cultures (A) higher and lower than the same cultures at the beginning of treatment (B) at P<0.05 with one way-Anova test and Dunnett's post test for multiple comparisons.

DNA damage after treatment with NaSe was next monitored using single cell electrophoresis – comet assay. As shown in Figure 5A, NaSe-induced DNA damage was first very moderate and only at later treatment periods of 16 to 24 h became significant. Conversely, in the same cells 1 μM camptothecin (topoisomerase I inhibitor) produced massive DNA damage already at 2 h of exposure.

Expression of selected proteins related to DNA damage and apoptosis in NaSe-exposed colonocytes was evaluated by flow cytometry. Data from these experiments showed an increased expression of p53 and proapoptotic Bax (later treatment interval) in the presence of decreased numbers of cells expressing antiapoptoic Bcl-2 and Bcl-XL (Figure 5B).

In addition, in the cells with altered expression of the examined proteins changes in mitochondrial density and activity were detected too although changes in mitochondrial dynamics took place generally earlier (data not shown).

3.3. MAPK Activities

The activation of specific MAPK in colon cancer HCT-116 cells exposed to NaSe was followed by ELISA-based analysis. Of three followed MAPK; i.e. ERK, p38 and JNK, NaSe consistently stimulated p38 activity which was seen raised as early as at 4 h of treatment compared to JNK activity which was increased at later periods of treatment (16 h).

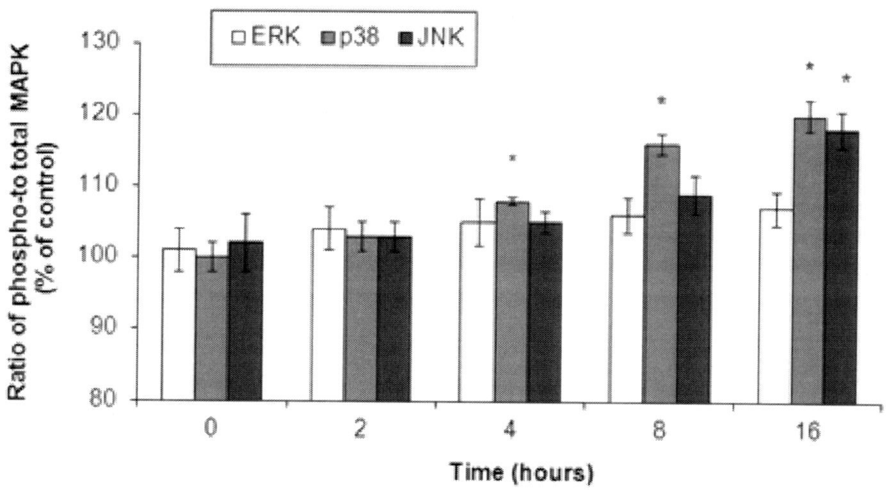

Figure 6. Phosphorylation of selected protein kinases – ERK, p38 and JNK in human colon cancer cells HCT-116 treated with sodium selenite (NaSe) at the concentration of 10 μM during 24 h. Kinase activities were measured using ELISA kits in 96-well plates. Results were normalized to micrograms of protein in the cell extract and expressed as the ratio of phospho-kinase to total kinase in the same sample. Results represent means ± SD of at least three experiments. *P<0.05 with one way-Anova test and Dunnett's post test for multiple comparisons.

On the other hand, ERK activity did not show any significant elevation during the entire course of treatment (Figure 6).

3.4. Mitochondrial Activity

In order to determine the putative contribution of mitochondria to NaSe-dependent apoptosis of HCT-116 cells, we analyzed kinetics of mitochondrial membrane potential (MMP). Results from measurements of MMP at different time points throughout the entire treatment with NaSe indicate that no decrease in MMP occurred during experiment (Figure 7).

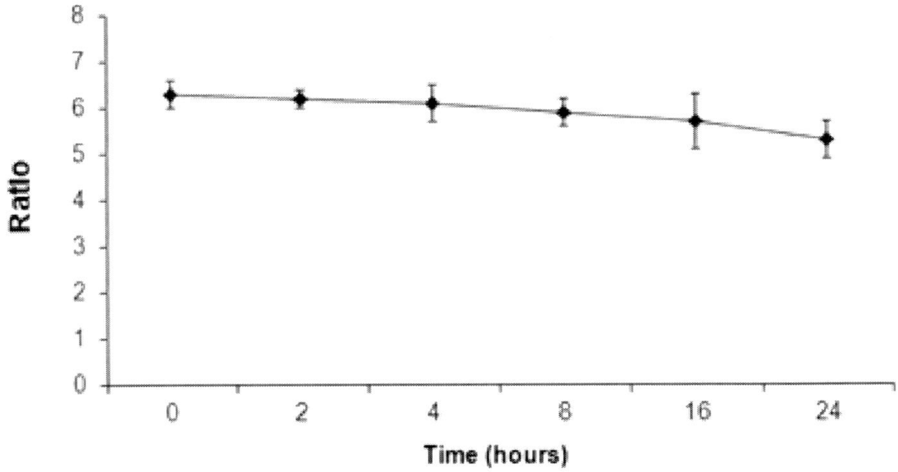

Figure 7. Kinetics of MMP loss in 10 μM sodium selenite (NaSe)-treated colon cancer cells during 24 h was measured by spectrofluorimetry. MMP loss was expressed as a ratio of (J aggregates/monomer) on y axis against individual treatment intervals (x-axis). Results represent means ± SD of at least three experiments.

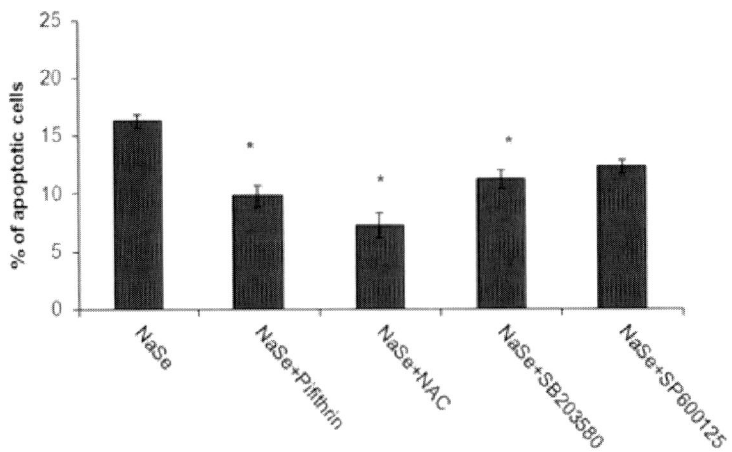

Figure 8. Effect of administered p53 inhibitor Pifithrin, antioxidant N-acetylcystein (NAC), p38 inhibitor SB203580 and JNK inhibitor SP600125 on 10 μM sodium selenite (NaSe)-induced apoptotis during 24 h of treatment.. Values represent means ± SD of at least three experiments. *lower than cultures treated with 10 μM NaSe at P<0.05 with one way-Anova test and Dunnett's post test for multiple comparisons.

3.5. Effects of Pharmacological Inhibitors on Apoptosis Signaling

Since NaSe induced proapoptotic stress signaling in colon cancer cells HCT-116, we wanted to examine the specific contribution of individual elements of this signaling to the observed apoptosis. Administration of antioxidant NAC, selective inhibitor of p53 Pifithrin and selective p38 inhibitor reduced markedly apoptosis although most potent seemed to be NAC. Convesrely, the specific inhibition of JNK decreased apoptosis of treated cells only moderately (Figure 8).

4. Discussion

Selenium exists in organic and inorganic forms and both have shown various anticancer properties ranging from a protection of normal cells and tissues against carcinogens to the active intereference with proliferation and cell death in premalignant and malignant stages [5]. These chemopreventive properties have been in particular intensively studied in colon cancer where major emphasis was paid to organic, i.e. more natural selenium forms [12, 13]. Still, recent failure of SELECT trial and advances in our knowledge concerning rather complex intracellular metabolism and speciation of organic selenium forms have spurred further scientific interest in inorganic selenium, namely in NeSe.

In the present study we employed cultured colon cancer cell line HCT-116 and exposed it to 10 μM NaSe over period of 24 h. The used cell line is often used in various cancer biology and chemoprevention studies due to its relatively stable and defined phenotype including wild type p53 and a defect in mismatch repair system. Morover, recently, several study groups used this line to characterize selected aspects of antiancer effects of NaSe. They discovered that NaSe inhibits cell proliferation by activation of JNK and suppression of β-catenin signaling [7]. Cells then die via mitochondrial apoptosis which is stimulated via NaSe-mediated Bax activation [6]. Despite these observations, however, details about timing and synchronization of NaSe-dependent apoptosis were missing, thereby substantiating our present efforts.

We noted that NaSe-dependent morphological apoptotic phenotype of HCT-116 cells starts to show only at later periods of treatment, i.e. between 16 and 24 h, respectively. Optical clarification of cytoplasm and nucleoli, detachement from substratum, cell shrinkage and membrane blebbing occurred asynchronously in individual cells with different duration in time. In addition, phosphatidylserine externalization which is often hailed as an early event in apoptosis [14] was also delayed unlike our observed earlier activation of caspase-9. Furthermore, loss of MMP which precedes cytochrome c relase and activation of caspases [15] was not detected too. Together these findings thus clearly show that at least in case of NaSe-treated HCT-116 cells, individual markers of apoptosis need to be followed in detail in order to correctly determine kinetics of the entire process and to prevent any false negative conclusions based on interpretation of raw data.

NaSe is intracellularly converted into selenide whose existence is associated with oxiative stress [6]. In our model, increased ROS levels were detectable already at early treatment intervals, however, they failed to activate a significant response until later periods of experiment. This observation could be explained by redox balance mechanisms including reduced glutathione which may have initially buffered stress-dependent changes and only after having been exhausted allowed a marked response. Thus DNA damage in NaSe-exposed cells became detectable first at 16 h of treatment unlike in case of topoisomerase inhibitor camptothecin which is known genotoxin and whose effects on DNA of HCT-116 cells were apparent already at 2 h of treatment. DNA damage response is generally associated with activation of p53-dependent pathway which may arrest exposed cells in G_0 phase of the cell cycle or signal apoptosis. The increased expression of p53, its specific phosphorylation, translocation to cytoplasm, nucleus or mitochondria and transcriptional activation of effector proapoptotic genes/proteins constitute known steps in so called intrisic apoptotic pathway [16]. Although not focusing on specific biochemical changes of p53 protein and its cellular localization (subject of our ongoing study), we verified a delayed elevation in p53 expression of NaSe-exposed cells along with almost concurrently increased abundance of Bax protein and decreased expression of antiapoptotic Bcl-2 and Bcl-XL. Temporal organization and possible interaction patterns between examined proteins suggest either direct influence of p53 or indirect effects of NaSe-mediated oxidative stress. At present we have no data to rule out either of suggested mechanisms; nevertheless it seems likely that both contribute to the certain degree which is potentially beneficial given the fact that cancer cells often harbour various mutations or defects in genetically encoded signaling pathways. Alternatively, some other mechanisms might have interefered too. One of them is activation of specific MAP stress kinases such as p38 and JNK. Both of them have been implicated in apoptosis activation and regulation in cancer cells and both of them are known to act via p53 or independently [17, 18].

We detected elevated activities of p38 and JNK in HCT-116 cells treated with NaSe although with different timing and intensity. While an early activity of p38 may not be linked to p53 expression or expression of other examined proteins, a delayed JNK activation concides with them. Still, when pretreated with pharmacological inhibitors, p38 blockage significantly prevented NaSe-induced apoptosis while JNK blockage had only a marginal effect. Thus relative importance and exact role of both kinases remains at least in the present model unclear. The concept of concurrent activation of multiple and sometimes different signals in particular study model is nowadays acknowledged [19]. It is thus only too obvious that some stimuli such as NaSe-mediated oxidative stress have very complex effects on ultimate decision over individual cells as well as their populations' fate. This is indicated by the most significant protective effect of antioxidant NAC against NaSe-induced apoptosis of HCT-116 cells.

CONCLUSION

NaSe remains a hot candidate for preclinical and clinical testing as a potential chemopreventive agent in colon cancer. Despite its toxicity, it has many benefical properties and has shown the efficacy in inhibiting growth of tumor cells and inducing their death.

Acknowledgment

This work was supported by GAUK No. 129609.

References

[1] Ip C, Hayes C, Budnick RM, Ganther HE (1991) Chemical Form of Selenium, Critical Metabolites, and Cancer Prevention. *Cancer Res.* 51:595-600.

[2] Finley JW (2006) Bioavailability of selenium from foods. *Nutr. Rev.* 64:146-151.

[3] Lippman SM, Klein EA, Goodman PJ, Lucia MS, Thompson IM, Ford LG, Parnes HL, Minasian LM, Gaziano JM, Hartline JA, Parsons JK, Bearden JD, 3rd, Crawford ED, Goodman GE, Claudio J, Winquist E, Cook ED, Karp DD, Walther P, Lieber MM, Kristal AR, Darke AK, Arnold KB, Ganz PA, Santella RM, Albanes D, Taylor PR, Probstfield JL, Jagpal TJ, Crowley JJ, Meyskens FL, Jr., Baker LH, Coltman CA, Jr. (2009) Effect of selenium and vitamin E on risk of prostate cancer and other cancers: the Selenium and Vitamin E Cancer Prevention Trial (SELECT). *JAMA* 301:39-51.

[4] Abdulah R, Miyazaki K, Nakazawa M, Koyama H (2005) Chemical forms of selenium for cancer prevention. *J. Trace. Elem. Med. Biol.* 19:141-150.

[5] Combs Jr. G, Gray W (1998) Chemopreventive Agents: Selenium. *Pharmacol. Ther.* 79:179-192.

[6] Huang F, Nie C, Yang Y, Yue W, Ren Y, Shang Y, Wang X, Jin H, Xu C, Chen Q (2009) Selenite induces redox-dependent Bax activation and apoptosis in colorectal cancer cells. *Free. Radic. Biol. Med.* 46:1186-1196.

[7] Fang W, Han A, Bi X, Xiong B, Yang W (2010) Tumor inhibition by sodium selenite is associated with activation of c-Jun NH2-terminal kinase 1 and suppression of beta-catenin signaling. *Int. J. Cancer* 127:32-42.

[8] Schroterova L, Kralova V, Voracova A, Haskova P, Rudolf E, Cervinka M (2009) Antiproliferative effects of selenium compounds in colon cancer cells: Comparison of different cytotoxicity assays. *Toxicol. In Vitro* 23: 1406-1411

[9] Kralova V, Brigulova K, Cervinka M, Rudolf E (2009) Antiproliferative and cytotoxic effects of sodium selenite in human colon cancer cells. *Toxicol. In Vitro* 23: 1497-1503

[10] Jiang C, Hu H, Malewicz B, Wang Z, Lu J (2004) Selenite-induced p53 Ser-15 phosphorylation and caspase-mediated apoptosis in LNCaP human prostate cancer cells. *Mol. Cancer Ther.* 3:877-884.

[11] Li J, Zuo L, Shen T, Xu CM, Zhang ZN (2003) Induction of apoptosis by sodium selenite in human acute promyelocytic leukemia NB4 cells: involvement of oxidative stress and mitochondria. *J. Trace. Elem. Med. Biol.* 17:19-26.

[12] Courtney E, Melville D, Leicester R (2004) Chemoprevention of colorectal cancer. *Aliment. Pharmacol. Ther.* 19:1-24.

[13] Connelly-Frost A, Poole C, Satia JA, Kupper LL, Millikan RC, Sandler RS (2006) Selenium, apoptosis, and colorectal adenomas. *Cancer Epidemiol. Biomarkers. Prev.* 15:486-493.

[14] Aubry JP, Blaecke A, Lecoanet_Henchoz S, Jeannin P, Herbault N, Caron G, Moine V, Bonnefoy JY (1999) Annexin V used for measuring apoptosis in the early events of cellular cytotoxicity. *Cytometry* 37:197-204.

[15] Kroemer G (2003) Mitochondrial control of apoptosis: an introduction. *Biochem. Biophys. Res. Commun.* 304:433-435.

[16] Roos WP, Kaina B (2006) DNA damage-induced cell death by apoptosis. *Trends. Mol. Med.* 12:440-450.

[17] Zou Y, Niu P, Yang J, Yuan J, Wu T, Chen X (2008) The JNK signaling pathway is involved in sodium-selenite-induced apoptosis mediated by reactive oxygen in HepG2 cells. *Cancer Biol. Ther.* 7.

[18] Rudolf E, Rudolf K, Cervinka M (2007) Selenium activates p53 and p38 pathways and induces caspase-independent cell death in cervical cancer cells. *Cell. Biol. Toxicol.*

[19] Zornig M, Hueber A, Baum W, Evan G (2001) Apoptosis regulators and their role in tumorigenesis. *Biochem. Biophys. Acta.* 1551:F1-37.

In: Selenium: Sources, Functions and Health Effects
Editors: Chinatsu Aomori and Megumi Hokkaido
ISBN: 978-1-61942-061-8

Chapter 10

SELENIUM AND PROSTATE HEALTH: A NEW POSSIBLE NUTRACEUTICAL CHALLENGE

Letteria Minutoli[1,*] and Herbert Marini[2]
[1]Department of Clinical and Experimental Medicine and Pharmacology, Section of Pharmacology,
[2]Department of Biochemical, Physiological and Nutritional Sciences, Section of Physiology and Human Nutrition, University of Messina, Italy

ABSTRACT

Benign prostatic hyperplasia (BPH) and prostate cancer are major sources of morbidity in older men and have an increasing impact on human health in line with the gradual aging of the population. The aetiology of those diseases is still far from being fully understood and, as a direct consequence, it is difficult to identify a rationale long-term strategy to develop an effective therapy. Nutraceuticals are components isolated or purified from food substances currently used for medical or health benefits. Several naturally derived food substances have been studied in BPH and prostate cancer in an attempt to identify alternative therapies for those diseases. Selenium (Se) plays an important role in maintaining equilibrium of a healthy organism. Recent scientific data devoted to investigating the nutraceutic effects of Se confirm a strong correlation between Se supplementation, BPH and prostate cancer. Se is an essential trace element with antioxidative, antimutagenic, antiviral and anticarcinogenic properties and exists naturally in foods, predominantly in the organic form as selenomethionine and selenocysteine. Se has several mechanisms of action, depending on its form. Recent literature suggested that Se could inhibit cell proliferation and induce cell cycle arrest of human prostate cancer cells. Moreover, Se and its derivatives can activate both the intrinsic and extrinsic pathways of apoptosis and inhibit angiogenesis. Despite the positive results obtained with Se supplementation, it is necessary to deeply verify these findings in experimental animal models and controlled clinical studies. In light of this background, our research group recently experimentally demonstrated that the Selenium,

* Corresponding author. Prof. Letteria Minutoli, MD. Department of Clinical and Experimental Medicine and Pharmacology, Section of Pharmacology, Torre Biologica 5th floor, AOU Policlinico "G. Martino ", Via C. Valeria Gazzi, 98125 Messina, Italy; Fax +39 090 2213300 Tel. +39 090 2213652; E-mail: lminutoli@unime.it

in association with Lycopene and *Serenoa repens* is helpful in reducing benign and malign prostate growth. These data confirm the positive effects of Se in prostate disorders suggesting its possible use in therapeutic management of BPH and prostate cancer.

PART I

Selenium: Introduction and General Aspects

Selenium (Se) is a micromineral needed in the diet on a daily basis, but only in very small amounts [1,2]. Plant foods are the major dietary sources of Se in most countries throughout the world; then, the concentration of Se in diet depends on the soil Se concentrations, the types of food consumed and other factors which facilitate or inhibit uptake of Se. Dietary sources of Se include brazil nuts (the most highly concentrated source of Se), fish, whole grains, wheat germ, soybean and sunflower seeds. Excellent sources of Se also include button mushrooms, shiitake mushrooms, cod, shrimp, snapper, tuna, halibut, calf's liver, and salmon [WHFoods: selenium; www.whfoods.com]. Se generally occurs in foods as selenomethionine, the organic Se analogue of the aminoacid methionine [3,4]. Selenomethionine can be incorporated into body proteins in place of methionine, and serves as a vehicle for Se storage in organs and tissues.

In human body, approximately 30 percent of tissue Se is contained in the liver, 15 percent in kidney, 30 percent in muscle, and 10 percent in blood plasma [5].

Se is usually integrated into proteins to form selenoproteins. Selenoproteins include several enzymes such as glutathione peroxidases, thioredoxin reductases, and iodothyronine deiodinases.

The role of Se in the cytosolic enzyme glutathione peroxidase was first illustrated in 1973. During stress, infection, or tissue injury, selenoenzymes may protect against the damaging effects of hydrogen peroxide or oxygen-rich free radicals [5]. The selenoenzyme thioredoxin reductase is involved in disposal of the products of oxidative metabolism [5,6]. It is a major component of a redox system with a multiplicity of functions [5,7]. Another group of selenoproteins is essential in the conversion of thyroxin, or tetraiodothyronine (T4), to its physiologically active form, triiodothyronine (T3) [5,8]. Marked changes in the T3-T4 ratio as a consequence of a reduced Se status indicate the modifying influence of Se on thyroid hormone balance in both animal models and human subjects [5].

An evaluation of Se requirements completed in 2000 by the US Institute of Medicine revised the American recommended adult intake to 55 μg/day, the level at which the enzymes with antioxidant functions are at maximum activity [9]. The minimum daily recommended dietary allowances for selenium [9] are listed below:

Birth - 6 months: 15 μg	Children 9 - 13 years: 40 μg
Infants 7 months - 1 year: 20 μg	Males/Females 14 - 18 years; 19 years and up: 55 μg
Children 1 - 3 years: 20 μg	Pregnant females: 60 μg
Children 4 - 8 years: 30 μg	Lactating females: 70 μg

Selenium: Deficiency and Supplementation

As above reported, the Se content of food is highly variable because it depends so heavily on soil conditions. In fact, some researchers have concluded that it is not possible to create a valid list of foods and their Se content for this reason. Consequently, most of the early research on Se deficiency focused on diseases in sheep, cattle, turkeys, and pigs which involved low soil concentrations of Se and insufficient amounts of Se in the forage plants eaten by these animals. Human Se deficiency is generally rare but is seen in some countries, most notably several areas of Africa, Russia, New Zealand and China, where soil concentration of Se is low [10,11]. Deficiency symptoms for Se are difficult to determine and controversial in the research literature. However, there is evidence that Se deficiency may contribute to development of a form of heart disease, hypothyroidism, and a weakened immune system [10,12,13]. It has also been reported that Se deficiency can make the body more susceptible to illnesses caused by other nutritional, biochemical or infectious stresses [10,14]. Specifically, three specific diseases have been associated with Se deficiency: Keshan Disease, which results in an enlarged heart and poor heart function, Kashin-Beck Disease, which causes osteoarthropathy, and Myxedematous Endemic Cretinism, which leads to mental retardation. Se deficiency has also been seen in people who rely on total parenteral nutrition as their sole source of nutrition [10,15,16]. Most cases of Se depletion or deficiency are also associated with severe gastrointestinal problems, such as Crohn's disease, or with gastrectomy [10,17-19]. People with acute severe illness who develop inflammation and widespread infection often have decreased levels of Se in their blood [10,20].

Indeed, people with iodine deficiency may benefit from Se supplementation. At this purpose, the SU.VI.MAX study in France, suggested that Se supplements may be protective against goiter [21].

Selenium: Toxicity

Use of Se as a nutraceutical and chemopreventive agent requires that it would be considered as a pharmacological agent and since supra-nutritional doses need to be administered, raises concerns about the safety margin and potential side effects [22]. Previous studies have shown that the choice of chemical form and dose of Se can strongly influence the observed biological effects. Inorganic Se compounds like selenite or selenate are known to produce genotoxic effects, whereas organic Se-containing compounds are better tolerated and exhibit anticancer activity [23]. Organic Se is present in the forms of selenomethionine, selenocysteine, and selenomethylselenocysteine; selenomethionine is generally considered to be the best absorbed and utilized form of Se.

High blood levels of Se (greater than 100 μg/dL) can cause selenosis, a syndrome characterized by gastrointestinal symptoms, hair loss, white blotchy nails, garlic breath odor (caused by methylated Se), fatigue, irritability, and peripheral neuropathy [24]. However, Se toxicity is generally rare. For example, the few reported cases in USA, have been associated with industrial accidents and a manufacturing error that led to an excessively high dose of Se in a supplement [10,25,26]. The Institute of Medicine of the National Academy of Sciences has set a tolerable upper intake level for Se (400 μg/day) for adults to prevent the risk of developing selenosis [9]. According to the US Department of Health and Human Services, selenium sulfide, one compound of Se used in antidandruff shampoos, but not in supplements, might cause cancer if taken internally and therefore should not be ingested. Using these shampoos is considered safe because there is no substantial absorption through intact skin [Report on Carcinogens, Twelfth Edition (2011) Table of Contents: http://ntp.niehs.nih.gov/go/roc12].

PART II

Selenium: Benign Prostatic Hyperplasia and Prostate Cancer

Benign prostatic hyperplasia (BPH) and prostate cancer are two of the most common proliferative diseases affecting the elderly male [27].

BPH is characterized by smooth muscle and epithelial proliferation primarily within the prostatic transition zone that can cause a variety of problems for patients, the most frequent being lower urinary tract symptoms (LUTS) [27]. Despite the enormous burden of BPH on public health, the pathogenesis of BPH is incompletely understood [28]. Age-related systemic/local hormonal and vascular changes appear to represent the predominant mechanism. However, an emerging body of evidence suggests that inflammation may play a key role in the development and progression of BPH [28,29]. Inflammation may contribute to tissue injury, and cytokines produced by inflammatory cells may serve to drive local growth factor production and angiogenesis [30]. As a consequence, the development of an inflammatory cascade has also suggested to have a role in prostate cancer.

On the other hand prostate cancer mostly originates in the prostate peripheral zone, but one in four arise in the transition zone where most BPH originates. In addition, the development of abnormal prostate growth is thought to involve disruption of dihydrotestosterone (DHT)-supported homeostasis between cell proliferation and cell death, and, as a result, proliferative processes predominate and apoptotic processes are inhibited [31,32]. The key role of DHT in the pathogenesis of BPH prompted the development of 5-alpha reductase inhibitors as a treatment for BPH, and potentially, for the prevention of prostate cancer [31]. Several large trials have shown the efficacy of alpha-receptor blocking medications when used alone and/or in combination with 5-alpha reductase inhibitors in BPH [33]. In addition, never data have demonstrated the benefit of anti-muscarinic medications in specific populations who suffer from bladder outlet obstruction causing storage urinary symptoms [34]. However, these therapeutic strategies are not free from side effects on sexuality and blood pressure regulation [35] and, as a direct consequence, it is difficult to identify an effective therapy without side effects.

Nutraceuticals are components isolated or purified from food substances and recent scientific data devoted to investigating the nutraceutic effects of Se confirm a strong correlation between Se supplementation, BPH and prostate cancer [36,37].

Experimental evidence suggests that Se supplementation can reduce the incidence of many types of cancer when nontoxic doses are provided to the diet of rodent species [38]. In humans, a large scale cancer prevention study demonstrated that taking a daily supplement containing 200 micrograms of Se per day could lower the risk of developing prostate, lung, and colorectal cancer [39]. Two years after these analysis, Harvard's Health Professionals

Follow-up Study tracked 33,737 men between the ages of 40 and 75 for six years [40]. Researchers found that the men with the highest Se levels at the start of the study had a 65% lower incidence of advanced prostate cancer than the men with the lowest levels. The Harvard team calculated that 159 μg/day of Se would prove protective [40,41].

A study published in 2004 (called SU.VI.MAX) reported interesting data on more than 13,000 French adults who had taken either a placebo or a combination of vitamin E, vitamin C, beta carotene, Se, and zinc [42]. After a median of more than 7 years of follow-up, among men with normal prostate-specific antigen who had taken the antioxidant supplement there was a marked statistically significant reduction in the rate of prostate cancer.

The benefit of Se was also tested in SELECT trial [43]. However, this trial was discontinued in October 2008 when a safety committee assigned to the study found that the supplements, taken alone or together for an average of 5 years, did not prevent prostate cancer.

Overall, the complex relationship between Se intake and prostate cancer risk has created a great deal of interest in understanding the mechanisms of prostate cancer potential chemoprevention by Se. As a matter of fact, Se has several mechanisms of action, depending on its form [44].

Although the precise mechanism of chemoprevention by Se remains unknown, two recent theories have evolved. First, Se metabolites have been postulated to promote apoptosis and inhibit cell proliferation selectively in cancer cells [45,46]. Second, it has been proposed that Se supplementation increases the levels of antioxidant selenoproteins and thereby prevents the DNA damage that could lead to cell transformation [47].

In light of this background, we have recently demonstrated, in a bladder outlet obstruction experimental model, that a combination of Selenium (Se), Lycopene (Ly) and *Serenoa repens* is more effective than *Serenoa repens* alone in reducing prostate inflammatory response, growth factor expression and histological features [48]. However, we still lack confirmatory data in more relevant experimental models. In this regard, testosterone administration in rats is a suitable model to investigate BPH. Prostate enlargement induced by testosterone has been used to assess the effects of potential treatments for BPH, since it reproduces adequately, although not fully, the major features of human BPH, including functional and histological changes, supporting the theory that testosterone actually produces prostate hyperplasia. Our results show a prominent growth of prostate following testosterone administration and a consequent increase in its weight, with the typical histological features of BPH [49]. The combined treatment with Se-Ly-*Serenoa repens* was more effective than *Serenoa repens* alone in preventing BPH and inhibited growth by 83%, suggesting that Selenium and Lycopene at pharmacological doses further increase *Serenoa repens* efficacy in BPH. Prostate growth inhibition by Se-Ly-*Serenoa repens* was likely stimulated via both a caspase-dependent signal, through caspase-9, and an independent mechanism involving the pro-apoptotic Bax and the anti-apoptotic Bcl-2 gene.

It was also demonstrated that Bcl-2 staining was intensified within the area of chronic inflammatory infiltrate in radical prostatectomy specimens [50]. This potential concomitant decrease in inflammation, that parallels the increase in apoptotic activity in BPH tissue, is also supported by our previous findings in the bladder-obstruction model, in which we observed a significant reduction of inflammatory infiltrate and tumor necrosis factor-α, an important BPH inflammatory marker [48] confirming the anti-inflammatory role of Se.

Growth factors and cytokines, released by inflammatory cells play a significant role in the regulation and growth of normal, hyperplastic and malignant prostatic epithelium. Moreover, prostatic cells by themselves are able to secrete inflammatory mediators and, finally, to stimulate their own growth. We found that, during testosterone-induced prostate growth, there is an over-expression of the Epidermal Growth Factor (EGF) that was prevented by treatment with Se-Ly-*Serenoa repens* combination and, specifically, Se has contributed significantly at the reduction of this growth factor [49]. As a matter of fact, EGF plays a critical role during tumorigenesis of the prostate gland [51] activating intracellular-signaling cascades leading, in turn, to activation of downstream pathways, cell proliferation, migration, adhesion, antiapoptosis, angiogenesis, and metastasis [52].

EGF and its receptor EGFR are frequently over-expressed in prostate cancer, and this is associated with a more aggressive clinical outcome [53]. Moreover, inhibition of EGFR has been shown to result in a marked decrease in Bcl-2 and a marked increase in the expression of Bax [54].

Among other growth factors involved in BPH and cancer development, a primary role is also played by VEGF that stimulates neovascularisation. VEGF is often called vascular permeability factor, since it enhances vascular leakage, an effect that contributes significantly to tumour development and metastasis [55]. Indeed, VEGF has been observed in BPH stromal cells [56], as well as in prostate cancer epithelial cells [57], where it plays a significant role in tumour growth, inducing angiogenesis. Very recently, VEGF has been indicated as possible therapeutic target to reduce prostate growth [57]. Another growth factor secreted by human prostate epithelial cells is TGF-β1. TGF-β1 is an important regulator of cell differentiation and apoptosis, and is also a potent immunosuppressive factor [58]. Accordingly, recent papers suggested that many tumors produce high levels of TGF-β1 which inhibits antigen-presenting cell activation and suppress tumor specific T cell immunity [59,60]. Consequently, the decreased expression of TGF-β1 and VEGF by sodium selenite pretreatment in human prostatic cells may contribute to the up-regulation of tumor specific immune response and, ultimately, may suppress immune escape and progression of tumors [61].

Accordingly, in our experimental models, we demonstrated a sustained anti-VEGF effect of Selenium, particularly when combined with Lycopene and *Serenoa repens*, thus pointing out to an anti-proliferative effect independent from hormone receptor status [49].

Conclusion

Benign prostatic hyperplasia and prostate cancer prevention are still under intense investigation. Many of nutraceuticals as Selenium, have therapeutic potential but it is unclear whether and how Se can contribute to eventual positive effects observed. Future studies consisting of both experimental models and well-designed clinical trials are necessary in order to better understand the effects of Selenium on prostate health.

REFERENCES

[1] Thomson CD. Assessment of requirements for selenium and adequacy of selenium status: a review. *Eur J Clin Nutr 2004*, 58:391-402.

[2] Goldhaber SB. Trace element risk assessment: essentiality vs. toxicity. *Regulatory Toxicology and Pharmacology* 2003, 38:232-242.

[3] Schrauzer GN. Commentary: Nutrition selenium supplements: Product types, quality, and safety. *J Am College of Nutr* 2001, 20:1-4.

[4] Schrauzer GN. The nutritional significance, metabolism and toxicology of selenomethionine. *Adv Food Nutr Res* 2003,47:73-112.

[5] Human Vitamin and Mineral Requirements Report of a joint FAO/WHO expert consultation Bangkok, Thailand Chapter 15:Selenium, 2002.

[6] Howie AF; Arthur JR; Nicol T; Walker SW; Beech SG; Beckett GJ. Identification of a 57-kilodalton selenoprotein in human thyrocytes as thioredoxin reductase. *J Clin Endocrinol Metab* 1998, 83:2052-2058.

[7] Mairrino M; Thomas JP; Girotti AW; Ursini F. Reactivity of phospholipid hydroperoxide glutathione peroxidase with membrane and lipoprotein lipid hydroperoxides. *Free Radic Res Commun* 1991, 12:131-135.

[8] Arthur J. Selenium biochemistry and function. In: Trace Elements in Man and Animals - 9. *Proceedings of the Ninth International Symposium on Trace Elements in Man and Animals.* Fischer PWF; L'Abbe MR; Cockel KA; Gibson RS 1997 eds. p. 1-5.Ottawa, Canada, NRC Research Press.

[9] Institute of Medicine, Food and Nutrition Board. *Dietary Reference Intakes: Vitamin C, Vitamin E, Selenium, and Carotenoids.* National Academy Press, Washington, DC, 2000.

[10] Office of Dietary Supplements National Institute of Health Dietary Supplement Fact Sheet: Selenium http://ods.od.nih.gov/factsheets/selenium

[11] Ellis DR ; Salt DE. Plants, selenium and human health. *Curr Opin Plant Biol* 2003, 6:273-279.

[12] Combs GF. Food system-based approaches to improving micronutrient nutrition: the case for selenium. *Biofactors* 2000, 12:39-43.

[13] Zimmerman MB; Kohrle J. The impact of iron and selenium deficiencies on iodine and thyroid metabolism: biochemistry and relevance to public health. *Thyroid* 2002, 12:867-878.

[14] Beck MA; Levander O; Handy J. Selenium deficiency and viral infection. *J of Nutr* 2003, 133:1463S-1467S.

[15] Levander OA. Scientific rationale for the 1989 recommended dietary allowance for selenium. *J Am Diet Assoc* 1991, 91:1572-1576.

[16] Gramm HJ; Kopf A; Bratter P. The necessity of selenium substitution in total parenteral nutrition and artificial alimentation. *J Trace Elem Med Biol* 1995, 9:1-12.

[17] Kuroki F; Matsumoto T; Lida M. Selenium is depleted in Crohn's disease on enteral nutrition. *Digestive Diseases* 2003, 21:266-270.

[18] Rannem T; Ladefoged K; Hylander E; Hegnhoj J; Jarnum S. Selenium status in patients with Crohn's disease. *Am J Clin Nutr* 1992, 56:933-937.

[19] Bjerre B; von Schenck H; Sorbo B. Hyposelaemia: Patients with gastrointestinal diseases are at risk. *J Intern Med* 1989, 225:85-88.
[20] Gartner R; Albrich W; Angstwurm MW. The effect of a selenium supplementation on the outcome of patients with severe systemic inflammation, burn, and trauma. *Bio Factors* 2001, 14:199-204.
[21] Derumeaux H; Valeix P; Castetbon K; Bensimon M; Boutron-Ruault MC; Arnaud J; Hercberg S. Association of selenium with thyroid volume and echostructure in 35- to 60-year-old French adults. *Eur J Endocrinol* 2003, 148:309-315.
[22] Reid ME; Stratton MS; Lillico AJ; Fakih M; Natarajan R; Clark LC; Marshall JR. A report of high-dose selenium supplementation: response and toxicities. J Trace Elem Med Biol. 2004, 18 :69-74.
[23] Nadiminty N; Gao AC. Mechanisms of Selenium chemoprevention and therapy in prostate cancer. Mol Nutr Food Res 2008,52:1247-1260
[24] Koller LD; Exon JH. The two faces of selenium-deficiency and toxicity are similar in animals and man. *Can J Vet Res* 1986, 50:297-306.
[25] Hathcock J. Vitamins and minerals: Efficacy and safety. *Am J Clin Nutr* 1997, 66:427-437.
[26] Raisbeck MF; Dahl ER; Sanchez DA; Belden EL; O'Toole D. Naturally occurring selenosis in Wyoming. *J Vet Diagn Invest* 1993, 5:84-87.
[27] Roehrborn CG. Male Lower Urinary Tract Symptoms (LUTS) and Benign Prostatic Hyperplasia (BPH). *Med Clin North Am* 2011, 95:87-100.
[28] Abdollah F; Briganti A; Suardi N; Castiglione F; Gallina A; Capitanio U; Montorsi F. Metabolic syndrome and benign prostatic hyperplasia: evidence of a potential relationship, hypothesized etiology, and prevention. *Korean J Urol* 2011, 52:507-516.
[29] Kramer G; Mitteregger D; Marberger M. Is benign benign prostatic hyperplasia (BPH) an immune inflammatory disease?. *Eur Urol* 2007, 51:1202-1216.
[30] Lucia MS; Lambert JR. Growth factors in benign prostatic hyperplasia: Basic science implications. *Curr Urol Rep* 2008, 9:272-278.
[31] Tindall DJ; Rittmaster RS. The rationale for inhibiting 5alpha-reductase isoenzymes in the prevention and treatment of prostate cancer. *J Urol* 2008, 179:1235-1242
[32] Hendriksen PJ; Dits NF; Kokame K; van Weerden WM; Bangma CH; Trapman J; Jenster G. Evolution of the androgen receptor pathway during progression of prostate cancer. Cancer Res 2006, 66:5012-5020.
[33] Roehrborn CG. Clinical management of lower urinary tract symptoms with combined medical therapy. *BJU Int* 2008, 102:13-17.
[34] Chapple C. Antimuscarinics in men with lower urinary tract symptoms suggestive of bladder outlet obstruction due to benign prostatic hyperplasia. *Curr Opin Urol* 2010, 20:43-48.
[35] Schulman CC. Lower urinary tract symptoms/benign prostatic hyperplasia: minimizing morbidity caused by treatment. *Urology* 2003, 62:24-33.
[36] Trottier G; Boström PJ; Lawrentschuk N; Fleshner NE. Nutraceuticals and prostate cancer prevention: a current review. *Nat Rev Urol* 2010, 7:21-30
[37] El-Bayoumy K; Das A; Narayanan B; Fiala ES; Desai D; Rao CV; Amin S; Sinha R. Molecular targets of the chemopreventive agent 1,4-phenylenebis (methylene)-selenocyanate in human non-small cell lung cancer. *Carcinogenesis* 2006, 27:1369-1376.

[38] El-Bayoumy K; Upadhyaya P; Date V; Sohn OS; Fiala ES; Reddy B. Metabolism of [14C] benzyl selenocyanate in the F344 rat. *Chem Res Toxicol* 1991, 4:560-565.

[39] Clark LC; Combs GF Jr; Turnbull BW; Slate EH; Chalker DK; Chow J; Davis LS; Glover RA; Graham GF; Gross EG; Krongrad A; Lesher JL Jr; Park HK; Sanders BB Jr; Smith CL; Taylor JR. Effects of selenium supplementation for cancer prevention in patients with carcinoma of the skin. A randomized controlled trial. Nutritional Prevention of Cancer Study Group. *JAMA* 1996, 276:1957-1963. Erratum in: *JAMA* 1997, 277:1520.

[40] Yoshizawa K, Willett WC, Morris SJ, Stampfer MJ, Spiegelman D, Rimm EB, Giovannucci E. Study of prediagnostic selenium level in toenails and the risk of advanced prostate cancer. *J Natl Cancer Inst* 1998, 90:1219-1224.

[41] Selenium and prostate cancer-Harvard Health Publications http://www.health.harvard.edu/fhg/updates/Selenium-and-prostate-cancer.shtml

[42] Hercberg S; Galan P; Preziosi P; Bertrais S; Mennen L; Malvy D; Roussel AM; Favier A; Briançon S. The SU.VI.MAX Study: a randomized, placebo-controlled trial of the health effects of antioxidant vitamins and minerals. *Arch Intern Med* 2004, 164:2335-2342. Erratum in: *Arch Intern Med* 2005, 165:286.

[43] Lippman SM; Klein EA; Goodman PJ; Lucia MS; Thompson IM; Ford LG; Parnes HL; Minasian LM; Gaziano JM; Hartline JA; Parsons JK; Bearden JD 3rd; Crawford ED; Goodman GE; Claudio J: Winquist E; Cook ED; Karp DD; Walther P; Lieber MM; Kristal AR; Darke AK; Arnold KB; Ganz PA; Santella RM; Albanes D; Taylor PR; Probstfield JL; Jagpal TJ; Crowley JJ; Meyskens FL Jr; Baker LH; Coltman CA Jr. Effect of selenium and vitamin E on risk of prostate cancer and other cancers: the Selenium and Vitamin E Cancer Prevention Trial (SELECT). *JAMA* 2009, 301:39-51.

[44] Dong Y; Lee SO; Zhang H; Marshall J; Gao AC; Ip C. Prostate specific antigen expression is down-regulated by selenium through disruption of androgen receptor signaling. *Cancer Res.* 2004, 64:19-22.

[45] Jiang C; Hu H; Malewicz B; Wang Z; Lu J. Selenite-induced p53 Ser-15 phosphorylation and caspase-mediated apoptosis in LNCaP human prostate cancer cells. *Mol Cancer Ther* 2004, 3:877-884.

[46] Ip C; Thompson HJ; Zhu Z; Ganther HE. In vitro and in vivo studies of methylseleninic acid: evidence that a monomethylated selenium metabolite is critical for cancer chemoprevention. *Cancer Res* 2000, 60:2882-2886.

[47] Diwadkar-Navsariwala V; Diamond AM. The link between selenium and chemoprevention: a case for selenoproteins. *J Nutr* 2004, 134:2899-2902.

[48] Bonvissuto G; Minutoli L; Morgia G; Bitto A; Polito F; Irrera N; Marini H; Squadrito F; Altavilla D. Effects of Serenoa Repens, Lycopene and Selenium, on pro-inflammatory phenotype activation: an in vitro and in vivo comparison study. *Urology* 2011, 77:248.e9-e16.

[49] Altavilla D; Bitto A; Polito F; Irrera N; Marini H; Arena S; Favilla V; Squadrito F; Morgia G; Minutoli L. The Combination of Serenoa Repens, Selenium and Lycopene is More Effective Than Serenoa Repens Alone to Prevent Hormone Dependent Prostatic Growth. *J Urol* 2011, 186:1524-1529.

[50] Gerstenbluth RE; Seftel AD; MacLennan GT; Rao RN; Corty EW; Ferguson K; Resnick MI. Distribution of chronic prostatitis in radical prostatectomy specimens with up-regulation of bcl-2 in areas of inflammation. *J Urol* 2002, 167:2267-2270.

[51] Itoh N; Patel U; Skinner MK. Developmental and hormonal regulation of transforming growth factor-alpha and epidermal growth factor receptor gene expression in isolated prostatic epithelial and stromal cells. *Endocrinology* 1998, 139:1369-1377.

[52] Prenzel N; Fischer OM; Streit S; Hart S; Ullrich A. The epidermal growth factor receptor family as a central element for cellular signal transduction and diversification. *Endocr Relat Cancer* 2001, 8:11-31.

[53] Di Lorenzo G; Tortora G; D'Armiento FP; De Rosa G; Staibano S; Autorino R; D'Armiento M; De Laurentiis M; De Placido S; Catalano G; Bianco AR; Ciardiello F. Expression of epidermal growth factor receptor correlates with disease relapse and progression to androgen-independence in human prostate cancer. *Clin Cancer Res* 2002, 8: 3438-3444.

[54] Xiao X; Wu J; Zhu X; Zhao P; Zhou J; Liu QQ; Zheng L; Zeng M; Liu R; Huang W. Induction of cell cycle arrest and apoptosis in human nasopharyngeal carcinoma cells by ZD6474, an inhibitor of VEGFR tyrosine kinase with additional activity against EGFR tyrosine kinase. *Int J Cancer* 2007, 121: 2095-2104.

[55] Bates DO; Curry FE. Vascular endothelial growth factor increases hydraulic conductivity of isolated perfused microvessels. *Am J Physiol* 1996, 271:H2520-2528.

[56] Jackson MW; Bentel JM; Tilley WD. Vascular endothelial growth factor (VEGF) expression in prostate cancer and benign prostatic hyperplasia. *J Urol* 1997,157: 2323-2328.

[57] Ortholan C; Durivault J; Hannoun-Levi JM; Guyot M; Bourcier C; Ambrosetti D; Safe S; Pagès G. Bevacizumab/docetaxel association is more efficient than docetaxel alone in reducing breast and prostate cancer cell growth: a new paradigm for understanding the therapeutic effect of combined treatment. *Eur J Cancer.* 2010, 46:3022-3036.

[58] Bello-DeOcampo D; Tindall DJ. TGF-ß/Smad signaling in prostate cancer. *Curr Drug Targets* 2003, 4:197-207.

[59] Thomas DA; Massagué J. TGF-beta directly targets cytotoxic T cell functions during tumor evasion of immune surveillance. *Cancer Cell* 2005, 8:369-380.

[60] Li MO; Sanjabi S; Flavell RA. Transforming growth factor-beta controls development,homeostasis, and tolerance of T cells by regulatory T cell-dependent and independent mechanisms. *Immunity* 2006, 25:455-471.

[61] Pei Z; Li H; Guo Y; Jin Y; Lin D. Sodium selenite inhibits the expression of VEGF, TGFbeta(1) and IL-6 induced by LPS in human PC3 cells via TLR4-NF-(K)B signaling blockage. *Int Immunopharmacol* 2010, 10:50-56.

In: Selenium: Sources, Functions and Health Effects
Editors: Chinatsu Aomori and Megumi Hokkaido
ISBN: 978-1-61942-061-8

Chapter 11

SELENIUM IN FOOD MATERIALS AND ITS IMPACT ON HUMAN HEALTH

Wenbiao Wu*
College of Food Science, Southwest University,
Chongqing, PRC

ABSTRACT

Since selenium was first isolated in 1817, there have been many studies on its accumulation and existing forms in food materials as well as its beneficial and toxic effects on human body. The main context of this article reviewed the positive correlation of selenium content in plant food materials with that in their growing soils, the positive correlation of selenium content in animal food materials with that in feedstuffs, the positive correlation of selenium content in food materials with that in human tissues, the association of selenium with proteins in food materials, the organic and inorganic forms of selenium in food materials, selenium deficiency and human diseases such as white muscle, or Keshan disease or other cardiovascular diseases or leading to immotile, deformed sperm and infertility, selenium supplementation including the variation of its beneficial effects caused by the variation of its forms in the supplements for curing or preventing diseases such as cancers and selenium toxicity such as losing hairs or nails to human body.

INTRODUCTION

It is able to find selenium in major human foods including animal and plant origins. The selenium content in the food originated from animals is dependent on that in feedstuffs. The selenium content in the food or feedstuffs originated from plants is dependent on that in soil or fertilization. No matter whether it is essential for plant or animal growth, or not, the selenium content in plant or animal materials increases with the increment of its concentration in soil or feedstuffs, respectively. Therefore, the selenium contents of human foods originated

* E-mail: wbwu62@yahoo.com.

from plants or animals usually varies a lot from variety to variety and from place to place, etc. For example, selenium concentration in cheese ranged from 0.022 to 0.088 μg/g wet weight (w.w.), in meat ranged from 0.064 (beef) to 0.094 (chicken) μg/g w.w., and in liver ranged from 0.307 to 0.401 μg/g w.w. (Pilarczyk et al., 2010) and in potatoes usually ranged from 0.001 to 0.270 mg/kg (Hahn et al., 1981).

Since selenium was first isolated in 1817, many studies on every aspect of selenium have been reported in the literature. For example, Reilly (2006) reported that over the past 50 years more than 100,000 scientific papers excluding popular articles and books related to selenium have been published. Among these publications, some are related to selenium in food materials and its impact on human health, which continue to appear with showing no sign of diminution today. This mass of writing scatters about different literatures and it is very necessary for an overview that takes into account the considerable amount of fresh information published over the past decade, by scientists from a wide range of specialties of which might appear to have much to do with selenium in food materials and its impact on human health.

The aim of this article is therefore to review selenium in food materials and its impact on human health. The main contents of this article cover the correlation of selenium content in plant food materials with that in their growing soils, the correlation of selenium content in animal food materials with that in feedstuffs, the correlation of selenium content in food materials with that in human tissues, the association of selenium with proteins in food materials, the other organic and inorganic forms of selenium in food materials, human diseases related to selenium deficiency and toxicity of selenium.

Correlation of Selenium Content in Plant Food Materials with that in Their Growing Soils

The samples that illustrate the relationship between the selenium content in plant food materials and that in their growing environmental soils are shown in Figure 1, Figure 2, Figure 3, Figure 4, Figure 5 and Figure 6. From these figures, it is obvious that the selenium content in food materials is highly correlated with that in their growing environmental soils. For corn seeds, tea leaves, cole leaves, potato tubers, rice and soy beans, the relative coefficients (R^2) are 0.9804, 0.9205, 0.9796, 0.9895, 0.9818 and 0.9896, respectively.

It should be noted that those figures are mainly to show the general relationship between the selenium content in plant food materials and that in their growing environmental soils. The absolute selenium content in any food material may change from place to place, or time to time, or variety to variety, etc. If the measuring method of selenium, the variety of plant (or soil), etc, vary, the slope of the regression line may vary. And also, the biological availability of the selenium in soils greatly affects the efficiency of absorbing the selenium by food plants or crops.

Selenium has not yet been proven to be essential for the growth of most food plants. It is believed that selenium is taken up by food plants from the soil solution primarily as selenate or selenite (Broadley et al., 2006). Selenate is absorbed by root cells through sulfate transporters in their plasma membranes (Terry et al., 2000; White et al., 2004). Selenium tends to replace sulphur inside plant tissues.

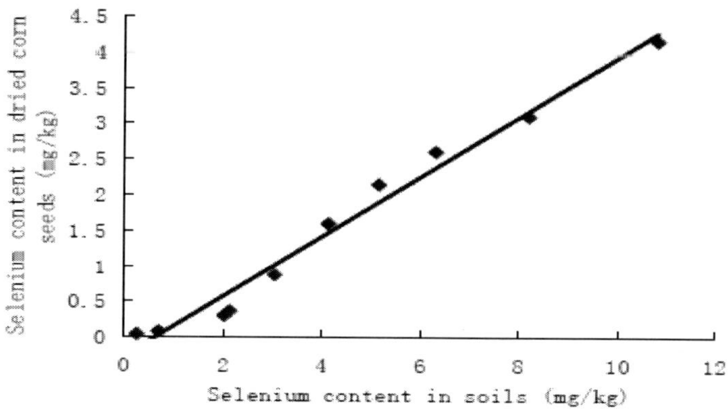

Figure 1. Correlation of selenium content in corn seeds with that in soils.

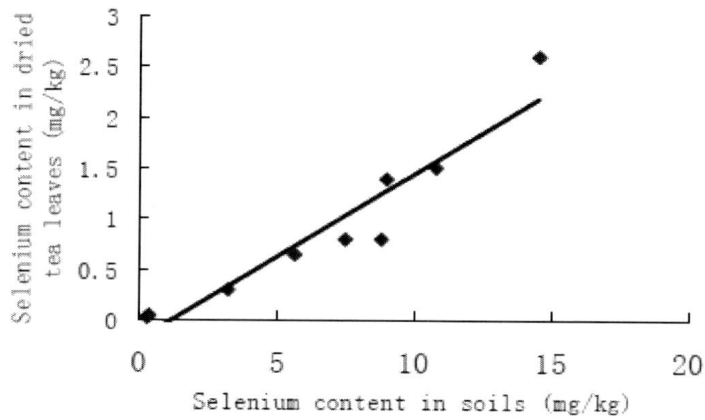

Figure 2. Correlation of selenium content in tea leaves with that in soils.

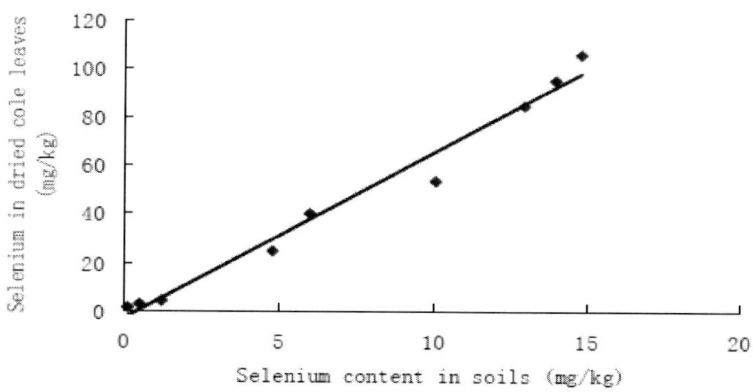

Figure 3. Correlation of selenium content in cole leaves with that in soils.

The knowledge of positive correlation of the selenium content in plant food materials with that in their growing environmental soils can be a very useful and scientific basis for producing seleniferous products that are safe to eat and avoiding excessive or inadequate selenium in the products. One aspect relating to this is that there are some areas in the world that are rich in selenium. For example, the selenium contents of soil from Enshi, Hubei, China range from 0.14 to 354 mg/kg (an area covering more than 2000 km^2). Soils in the Great Plains of the USA, Canada, South America and Russia are also seleniferous. According to the positive linear relationship between selenium content in food materials and that in their growing soils described above, the selenium content in the food materials from these areas could be too high and they could be toxic unless the bioavailability of the selenium in the soils is too low.

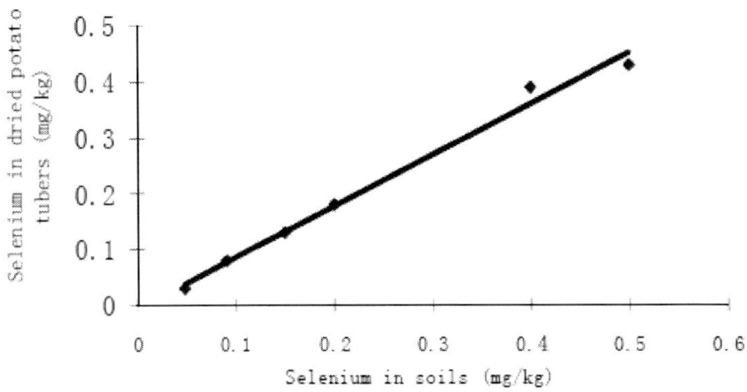

Figure 4. Correlation of selenium content in potatoes with that in soils.

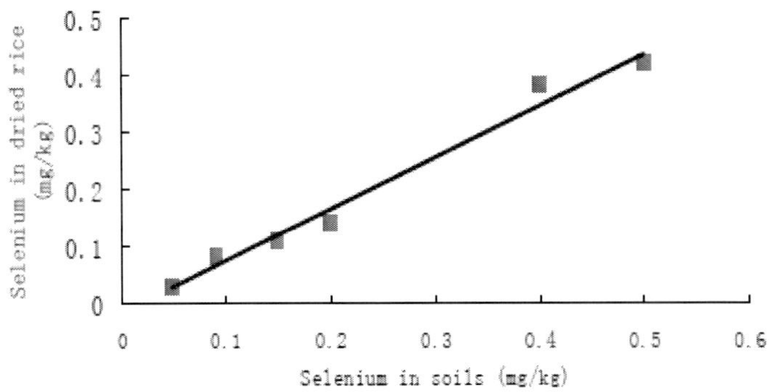

Figure 5. Correlation of selenium content in rice with that in soils.

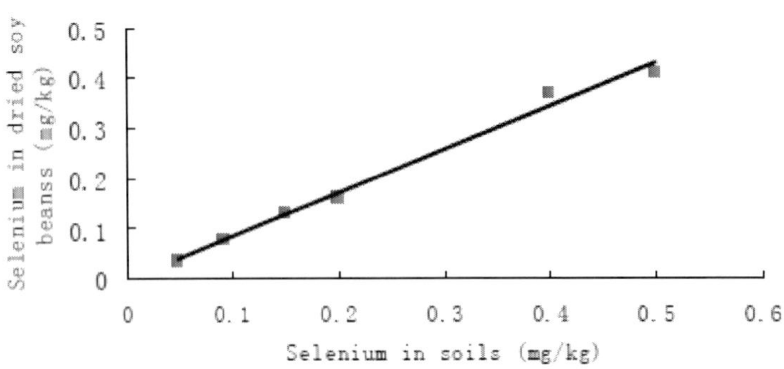

Figure 6. Correlation of selenium content in soy beans with that in soils.

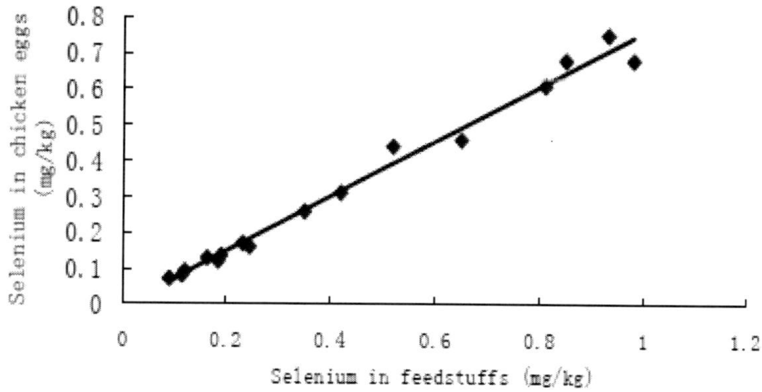

Figure 7. Correlation of selenium content in feedstuffs with that in chicken eggs.

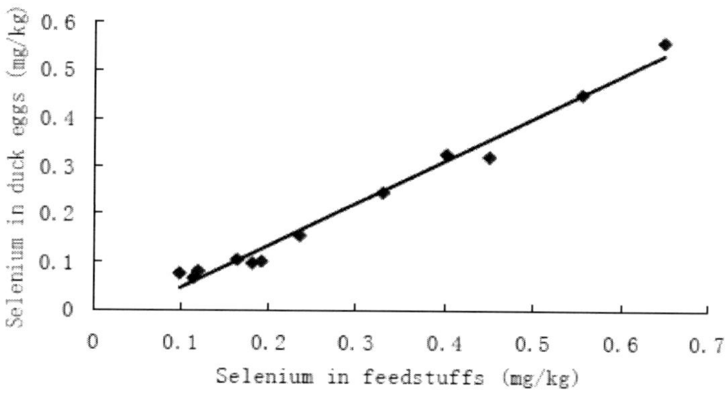

Figure 8. Correlation of selenium content in feedstuffs with that in duck eggs.

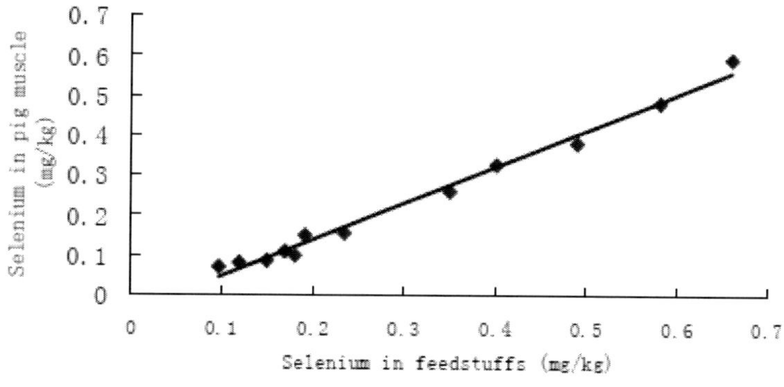

Figure 9. Correlation of selenium content in feedstuffs with that in pig muscles.

A detailed study has shown that prolonged intake of more than 3 mg of selenium daily is sufficient to invoke symptoms of poisoning (Frausto Da Silva and Williams, 1991). To render these kinds of the food materials suitable for human consumption and profitable for food industry, proper processing would be necessary. How can this problem been predicted? The well established regression curve or equation for the relationship between selenium content in a particular plant food material and that in its growing environmental soils could be very useful for predicting its dietary safety. On the other hand, selenium deficient areas have been reported throughout the world (Judson and Reuter, 1999; Spadoni et al., 2007; Hincal, 2007; Iqbal et al., 2008). In these areas, dietary supplementation of selenium is necessary and food materials enriched with selenium are accepted by many consumers. For example, those such products that have been reported widely are selenium-enriched potato (Poggi et al., 1999; Poggi et al., 2000; Turakainen et al., 2006; Ferri et al., 2007; Cuderman et al., 2008; Bordoni et al., 2008), wheat (Eurola et al., 1989; Stephen et al., 1989; Singh, 1994; Lyons et al., 2003, 2004, 2005a,b), rice (Chen et al., 2002), oat (Eurola et al., 2004) and tea (Hu et al., 2001). How can the soils been fertilized? The well established regression curve or equation for the relationship between selenium content in a particular plant food material and that in its growing environmental soils could be very useful for designing and managing the proper fertilization of soils to produce dietetically safe selenium-enriched food products.

Correlation of Selenium Content in Animal Food Materials with That in Feedstuffs

The samples that illustrate the relationship between the selenium content in feedstuffs and that in animal food products are shown in Figure 7, Figure 8 and Figure 9. From these figures, it is obvious that the selenium content in feedstuffs is highly correlated with that in animal food products. For chicken eggs, duck eggs and pig muscles, the relative coefficients (R^2) are 0.9871, 0.9840 and 0.9883, respectively.

It should be noted that those figures are mainly to show the general relationship between the selenium content in animal food materials and that in feedstuffs. The absolute selenium content in any animal food material may change from individual to individual, or variety to variety, etc. The slope of the regression line may vary with the variation of the method for measuring selenium content, feedstuffs, animals, etc. And also, the biological availability of the selenium in feedstuffs greatly affects the efficiency of absorbing the selenium by animals.

The knowledge of positive correlation of the selenium content in animal food products with that in feedstuffs can be a very useful and scientific basis for producing seleniferous animal food products that are safe to eat with avoiding excessive or inadequate selenium in the products and protecting animals from poisoning or diseases caused by selenium deficiency. For example, in selenium deficient areas, this knowledge can be applied to guide the design and practical operation of producing seleniferous animal food products that contain proper amount of selenium by the supplement of proper amount (which could be calculated by using the well established regression line or equation) of selenium in feedstuffs or in soils (or applied as foliar sprays) to produce seleniferous forage or pasture (Gupta and Gupta, 2002; Valle et al., 2002). On the other hand, in selenium rich areas, the well established regression line can be applied to predict the toxicity or dietary safety of feedstuffs to animals and the selenium content in the animal food products. The toxicity or dietary safety of the animal food products to human body can subsequently be predicted.

CORRELATION OF SELENIUM CONTENT IN FOOD MATERIALS WITH THAT IN HUMAN TISSUES

The samples that illustrate the relationship between the selenium content in food materials or daily intake and that in human tissues are shown in Figure 10, Figure 11 and Figure 12. From these figures, it is obvious that the selenium content in food materials or daily intake is highly correlated with that in human tissues. For the daily selenium intake and human blood, the relative coefficient (R^2) is 0.9580. For the selenium content in rice and that in human hairs and nails, the relative coefficients (R^2) are 0.9787 and 0.9752, respectively.

It should be noted that those figures are mainly to show the general relationship between the selenium content in food materials and that in human tissues. The absolute selenium content in any food material may change from variety to variety, etc and that in any human tissue may vary from individual to individual, etc. The slope of the regression line may vary with the variation of the method for measuring selenium content, food materials, human individualities, etc. And also, the biological availability of the selenium in food materials greatly affects the efficiency of absorbing the selenium by human body.

For experiment on rice, it was taken as a staple food for one year with the amount of vegetable intake/day unchanged. The maximum amount of selenium experimented was within the dietetically safe level. The samples for measuring selenium content in hairs and nails were selected every month for the investigation into selenium accumulation and the data presented in Figure 11 and Figure 12 were from the samples taken at the end of the experiment (1 year).

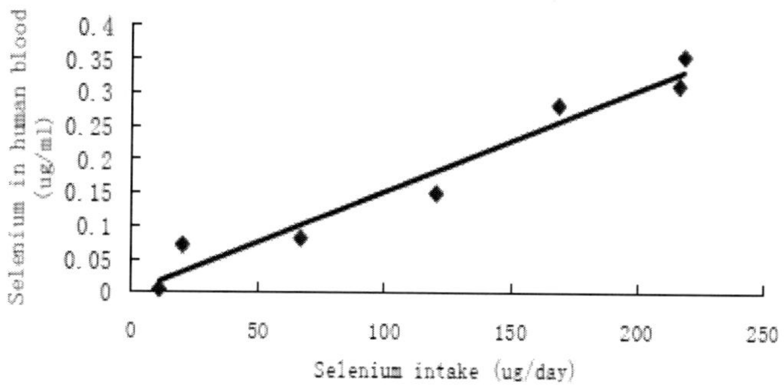

Figure 10. Correlation of selenium intake with its concentration in human blood.

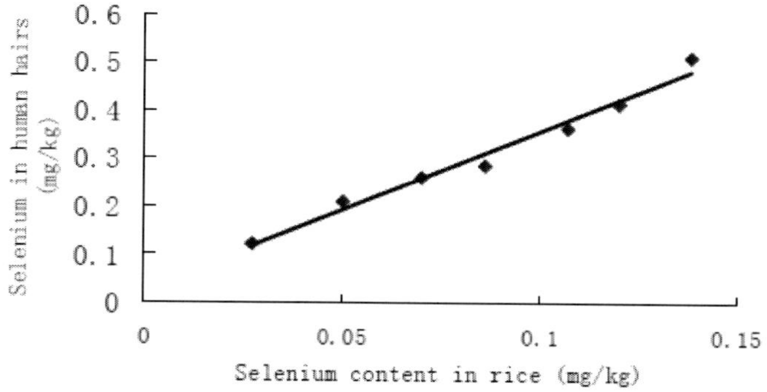

Figure 11. Correlation of selenium content in rice with that in human hairs.

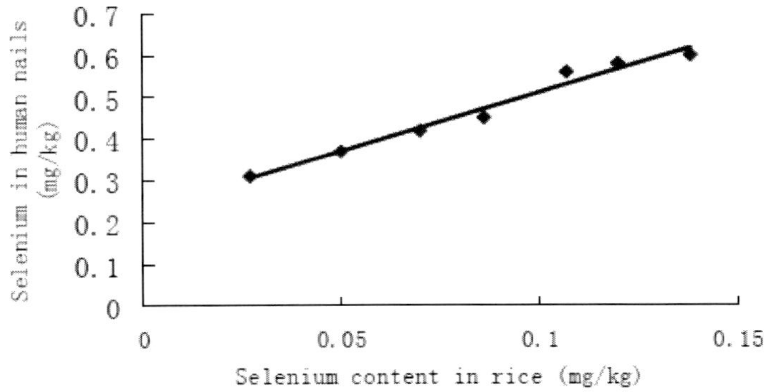

Figure 12. Correlation of selenium content in rice with that in human nails.

The knowledge of positive correlation of the selenium content in food materials with that in human tissues can be a very useful and scientific basis for predicting the biological availability of selenium and its dietary safety to human, avoiding excessive or inadequate selenium intake, and protecting human body from poisoning or diseases caused by selenium deficiency.

MOLECULAR FORMS OF SELENIUM IN FOOD MATERIALS

In human nutrition, it is not just the total selenium content in food materials that is important, but also the molecular forms of selenium are vital. The nutritional and other beneficial functions of different forms of selenium are quite different. Therefore, they should be discriminated from each other in case of applying in human nutrition. Both organic and inorganic forms of selenium exist in food materials.

Although single purified selenoprotein has not yet been isolated from land plants and identified (Broadley et al., 2006), it has been reported that proteins are closely associated with selenium in most food materials. For example, the dispersibility of selenium highly correlated with that of proteins in teas (Wu and Long, 2011) and the selenium was precipitated by precipitating proteins from potato juice (Wu, 2011). Resenfeld and Beath reported that most of selenium with its content less than 30 mg/kg in crop seeds is a constituent of protein (Spallholz et al., 1981). Approximately thirty selenoproteins have been found in mammalian tissues and about half of them have been characterised (Rayman, 2002).

Organic selenium containing compounds with small molecular weight found in both plants and animals include $H\text{-}Se\text{-}CH_2\text{-}CH(NH_2)\text{-}COOH$ (selenocysteine), $CH_3\text{-}Se\text{-}CH_2\text{-}CH_2\text{-}CH(NH_2)\text{-}COOH$ (selenomethionine), $CH_3\text{-}Se\text{-}CH_3$ (di-methyl-selenium) and $CH_3\text{-}Se\text{-}Se\text{-}CH_3$ (di-methyl-di-selenium). Those found in plants include $CH_3\text{-}Se\text{-}CH_2\text{-}CH(NH_2)\text{-}COOH$ (methyl-selenocysteine), $HOOC\text{-}CH(NH_2)\text{-}NH_2\text{-}CH_2\text{-}Se\text{-}CH_2\text{-}CH_2\text{-}CH(NH_2)\text{-}COOH$, $(CH_3)_2\text{-}Se^+\text{-}CH_2\text{-}CH_2\text{-}CH(NH_2)\text{-}COOH$ (methyl-selenomethionine), $HOOC\text{-}CH(NH_2)\text{-}CH_2\text{-}Se\text{-}Se\text{-}CH_2\text{-}CH(NH_2)\text{-}COOH$ (selenocystine) and $HOOC\text{-}CH(NH_2)\text{-}(CH_2)_2\text{-}Se\text{-}Se\text{-}(CH_2)_2\text{-}CH(NH_2)\text{-}COOH$. Those found in animal include $(CH_3)_3Se^+$ (tri-methyl-selenium) and $H_2N\text{-}CH_2\text{-}CH_2\text{-}SeO_3H$. Organic selenium containing compounds with large molecular weight include GSH-Px and selenoprotein P found in animals as well as seleno-t-RNA found in both plants and animals.

Four main oxidation states, -2 (selenide), 0 (elemental Se), +4 (selenite) and +6 (selenate) of inorganic selenium could be found in plant food materials (Gondi et al., 1992). The same author also reported that inorganic selenium is highly mobile under oxidising conditions though its mobility decreases with decreasing pH.

HUMAN DISEASES RELATED TO SELENIUM DEFICIENCY

Selenium in food materials had been treated as a toxin before it was first recognized, in the 1950s, as an essential nutrient (Schwarz and Foltz, 1957; Coultate, 1989). Selenium deficiency can cause health problems, for example, cardiovascular or muscular diseases (e.g. white muscle disease) (Campa et al., 2000; Barbaro, 2002; Natella et al., 2007; Zeng and Combs, 2008; Barretto et al., 2008), Kashin-Beck disease (an osteoarthritis disorder) (Fordyce, 2005), pancreatitis, asthma or inflammatory response syndrome (Rayman, 2000, 2002). Furthermore, immune system functioning, response to viral infection, female (e.g. reduced rates of miscarriage) and male (e.g. sperm development and function) fertility and thyroid functioning (if both selenium and iodine status are deficient) could be adversely affected by low selenium status (Rayman, 2002). It has also been reported that selenium supplementation can prevent cancer (El-Bayoumy, 2001; Duffield-Lillico et al., 2003; Klein et al., 2003; Combs, 2004). Other studies have reported the treatment of specific human disorders by dietary supplementation with selenium products (Orndahl et al., 1986; Huston et al., 1991; Peretz et al., 1992; Gartner et al., 2001; Dekok, 2005).

Selenium may exist in human body as parts of selenoproteins. The functions of these proteins may include anti-oxidative activities (e.g. glutathione peroxidase), effect on protein stability, transcription of mRNA and other biochemical activities.

TOXICITY OF SELENIUM

Selenium is also very toxic. Daily excessive intake of selenium can result in toxic damages to human body. The poisoning symptoms that have been observed include alkali disease, blind staggers, losing hairs or nails, damages to liver or kidney, etc.

Therefore, Chinese Nutrition Society recommended that the proper amount of daily selenium should be 50-250 μg and the maximum amount of intake daily should not exceed 400 μg. Nutrition Group, National Scientific Committee, USA recommended that the maximum amount of intake daily should not exceed 200 μg. The maximum amount of selenium intake daily recommended by FAO/WHO/IAEA is quite similar to those described above.

REFERENCES

Barbaro, G. (2002). Selenium deficiency and HIV-associated cardiomyopathy. *Journal of Royal Society of Medicine,* 95(1), 57.

Barretto, J.R., Silva, L.R., Leite, M.E., Boa-Sorte, N., Pimentel, H., Purificação, A.C., Gildasio, C., Fontes, M.I.M.M. and Amorim, T. (2008). Poor zinc and selenium status in phenylketonuric children and adolescents in Brazil. *Nutrition Research,* 28(3), 208-211.

Bordoni, A., Danesi, F., Malaguti, M., Nunzio, M.D., Pasqui, F., Maranesi, M. and Biagi, P.L. (2008). Dietary selenium for the counteraction of oxidative damage: fortified foods or supplements? *British Journal of Nutrition,* 99, 191-197.

Broadley, M.R., White, P.J., Bryson, R.J., Meacham, M.C., Bowen, H.C., Johnson, S.E., Hawkesford, M.J., McGrath, S.P., Zhao, F.J., Breward, N., Harriman, M. and Tucker, M. (2006). Biofortification of UK food crops with selenium. *Proceedings of the Nutrition Society*, 65, 169–181.

Campa, A., Shor-Posner, G. and Baum, M. (2000). Selenium status and the human immunodeficiency virus. *Journal of American Diet Association,* 100(4), 418.

Chen, L., Yang, F., Xu, J., Hu, Y., Hu, Q., Zhang, Y. and Pan, G. (2002). Determination of selenium concentration of rice in China and effect of fertilization of selenite and selenate on selenium content of rice. *Journal of Agricultural and Food Chemistry*, 50, 5128–5130.

Combs, G.F. (2004). Status of selenium in prostate cancer prevention. *British Journal of Cancer,* 91, 195–199.

Coultate, T.D. (1989). *FOOD The Chemistry of Its Components*, 2nd Ed.. Pp. 287-288. Royal Society of Chemistry, Cambridge.

Cuderman, P., Kreft, I., Germ, M., Kovacevic, M. and Sibilj, V. (2008). Selenium Species in Selenium-enriched and Drought-exposed Potatoes. *Journal of Agricultural and Food Chemistry,* 56(19), 9114-9120.

Dekok, H.J. (2005). Case Report: The medical treatment of obstructive sleep apnoea syndrome (OSAS) with Selenium. *Medical Hypotheses,* 65(4), 817-818.

Duffield-Lillico, A.J., Slate, E.H., Reid, M.E., Turnbull, B.W., Wilkins, P.A., Combs, G.F. Jr, Park, H.K., Gross, E.G., Graham, G.F., Stratton, M.S., Marshall, J.R. and Clark, L.C. (2003). Nutritional Prevention of Cancer Study Group. Selenium supplementation and secondary prevention of nonmelanoma skin cancer in a randomized trial. *Journal of the National Cancer Institute,* 95(19), 1477-1481.

El-Bayoumy, K. (2001). The protective role of selenium on genetic damage and on cancer. *Mutation Research,* 475, 123-139.

Eurola, M., Ekholm, P., Ylinen, M., Koivistoinen, P. and Varo, P. (1989). Effects of selenium fertilization on the selenium content of selected Finnish fruits and vegetables. *Acta Agriculturae Scandinavica*, 39, 345–350.

Eurola, M., Hietaniemi, V., Kontturi, M., Tuuri, H., Kangas, A., Niskanen, M. and Saastamoinen, M. (2004). Selenium content of Finnish oats in 1997–1999: effect of cultivars and cultivation techniques. *Agricultural and Food Science*, 13, 46–53.

Ferri, T., Favero, G. and Frasconi, M. (2007). Selenium speciation in foods: Preliminary results on potatoes. *Microchemical Journal,* 85(2), 222-227.

Fordyce, F. (2005). Selenium deficiency and toxicity in the environment. In Essentials of Medical Geology, pp. 373–415 [O Selinus, B Alloway, J Centeno, R Finkelman, R Fuge, U Lindh and P Smedley, editors]. London: Elsevier.

Frausto Da Silva, J.J.R. and Williams, R.J.P. (1991). *The Biological and Chemistry of the Elements- the Inorganic Chemistry of Life*. Pp.461. Oxford: Clarendon Press.

Gartner, R., Albrich, W. and Angstwurm, M.W. (2001). The effect of a selenium supplementation on the outcome of patients with severe systemic inflammation, burn and trauma. *Biofactors,* 14(1-4), 199-204.

Gondi, F., Panto, G., Feher, J., Bogye, G. and Alfthan, G. (1992). Selenium in Hungary – the rock-soil-human system. *Biological Trace Element Research*, 35, 299–306.

Gupta, U.C. and Gupta, S.C. (2002). Quality of animal and human life as affected by selenium management of soils and crops. *Communications in Soil Science and Plant Analysis*, 33, 15–18.

Hahn, M.H., Kuennen, R.W., Caruso, J.A. and Fricke, F.L. (1981). Determination of Trace Amounts of Selenium in Corn, Lettuce, Potatoes, Soybeans, and Wheat by Hydride Generation/Condensation and Flame Atomic Absorption Spectrometry. *Journal of Agricultural and Food Chemistry*, 29, 792-796.

Hu, Q., Pan, G., An, X. and Ding, R. (2001). Physiological function of Se-enriched tea fertilised with sodium selenite and naturally high-Se tea in rats. *Journal of Science of Food and Agriculture*, 81(2), 202-204.

Huston, R.K., Jelen, B.J. and Vidgoff, J. (1991). Selenium supplementation in low-birthweight premature infants: relationship to trace metals and antioxidant enzymes. *Journal of Parenteral and Enteral Nutrition,* 15(5), 556-559.

Klein, E.A., Lippman, S.M., Thompson, I.M., Goodman, P.J., Albanes, D., Taylor, P.R. and Coltman, C. (2003). The selenium and vitamin E cancer prevention trial. *World Journal of Urology,* 21(1), 21-27.

Lyons, G., Ortiz-Monasterio, I., Stangoulis, J. and Graham, R. (2005a). Selenium concentration in wheat grain: Is there sufficient genotypic variation to use in breeding? *Plant and Soil*, 269, 269–380.

Lyons, G., Stangoulis, J. and Graham, R. (2003). High-selenium wheat: biofortification for better health. *Nutrition Research Reviews*, 16, 45–60.

Lyons, G.H., Stangoulis, J.C.R. and Graham, R.D. (2004). Exploiting micronutrient interaction to optimize biofortification programs: The case for inclusion of selenium and iodine in the Harvest-Plus program. *Nutrition Reviews*, 62, 247–252.

Lyons, G.H., Stangoulis, J.C.R. and Graham, R.D. (2005b). Tolerance of wheat (Triticum aestivum L.) to high soil and solution selenium levels. *Plant and Soil*, 270, 179–188.

Natella, F., Fidale, M., Tubaro, F., Ursini, F. and Scaccini, C. (2007). Selenium supplementation prevents the increase in atherogenic electronegative LDL (LDL minus) in the postprandial phase. *Nutrition, Metabolism and Cardiovascular Diseases,* 17(9), 649-656.

Orndahl, G., Sellden, U., Hallin, S., Wetterquist, H., Rindby, A. and Selin, E. (1986). Myotonic dystrophy treated with selenium and vitamin E. *Acta Medica Scandinavica,* 219(4), 407-414.

Peretz, A., Neve, J., Duchateau, J. and Famaey, J.P. (1992). Adjuvant treatment of recent onset rheumatoid arthritis by selenium supplementation: preliminary observations. *British Journal of Rheumatology,* 31(4), 281-282.

Pilarczyk, B., Tomza-Marciniak, A., Mituniewicz-Małek, A., Wieczorek-Dąbrowska, M., Pilarczyk, R., Wójcik, J., Balicka-Ramisz, A., Bąkowska, M. and Dmytrów, I. (2010). Selenium content in selected products of animal origin and estimation of the degree of cover daily Se requirement in Poland. *International Journal of Food Science and Technology*, 45(1), 186–191.

Poggi, V., Arcioni, A., Filippini, P. and Pifferi, P.G. (2000). Foliar Application of Selenite and Selenate to Potato (Solanum tuberosum): Effect of A Ligand Agent on Selenium Contents of Tubers. Journal of Agricultural and Food Chemistry, 48(10), 4749-4751.

Poggi, V., Pifferi, P.G., Bordoni, A. and Biagi, P.L. (1999). Plant-derived foods supplied with selenium: the potato [in Italian]. Industrial Alimemt, 38, 1107-1113.

Rayman, M.P. (2000). The importance of selenium to human health. Lancet, 356, 233–241.

Rayman, M.P. (2002). The argument for increasing selenium intake. Proceedings of the Nutrition Society, 61, 203–215.

Reilly, C. (2006). Selenium in Food and Health, 2nd ed., Springer-Science+Business Media, New York, USA.

Schwarz, K. and Foltz, C.M. (1957). Selenium as an integral part of factor 3 against dietary necrotic liver degeneration. *Journal of American Chemical Society*, 79, 3292-3293.

Singh, B.R. (1994). Effect of selenium-enriched calcium nitrate, top-dressed at different growth-stages, on the selenium concentration in wheat. *Fertilizer Research*, 38, 199–203.

Spallholz, J.E., Martin, J.L. and Ganther, H.E. (1981). *Selenium in Biology and Medicine*. Pp.179. Avi., Westport, C.T..

Stephen, R.C., Saville, D.J. and Watkinson, J.H. (1989). The effects of sodium selenate applications on growth and selenium concentration in wheat. *New Zealand Journal of Crop and Horticultural Science*, 17, 229–237.

Terry, N., Zayed, A.M., de Souza, M.P. and Tarun, A.S. (2000). Selenium in higher plants. *Annual Review of Plant Physiology and Plant Molecular Biology*, 51, 401–432.

Turakainen, M., Hartikainen, H., Ekholm, P. and Seppanen, M.M. (2006). Distribution of selenium in different biochemical fractions and raw darkening degree of potato (Solanum tuberosum L.) tubers supplemented with selenate. *Journal of Agricultural and Food Chemistry*, 54, 8617-8622.

Valle, G., McDowell, L.R., Prichard, D.L., Chenoweth, P.J., Wright, D.L., Martin, F.G., Kunkle, W.E. and Wilkinson, N.S. (2002). Selenium concentration of fescue and bahia grasses after applying a selenium fertilizer. *Communications in Soil Science and Plant Analysis*, 33, 1461–1472.

White, P.J., Bowen, H.C., Parmaguru, P., Fritz, M., Spracklen, W.P., Spiby, R.E., Meacham, M.C., Mead, A., Harriman, M., Trueman, L.J., Smith, B.M., Thomas, B. and Broadley, M.R. (2004). Interactions between selenium and sulphur nutrition in Arabidopsis thaliana. *Journal of Experimental Botany*, 55(404), 1927–1937.

Wu, W.B. (2011). Concentration of selenium by precipitating proteins from potato juice. *International Journal of Food Science and Technology,* 46(2), 402–405. DOI:10.1111/j.1365-2621.2010. 02501.x.

Wu, W.B. and Long, X.J. (2011). Correlation of Dispersibility of Proteins with that of Selenium in Teas. *Biological Trace Element Research,* 142(2), 137-142. DOI: 10.1007/s12011-010-8753-7.

Zeng, H. and Combs, G.F. (2008). Selenium as an anticancer nutrient: roles in cell proliferation and tumor cell invasion. *The Journal of Nutritional Biochemistry,* 19(1), 1-7.

In: Book Title
Editor: Chinatsu Aomori and Megumi Hokkaido
ISBN: 978-1-61942-061-8

Chapter 12

STRUCTURAL, ELECTRONIC, VIBRATIONAL AND OPTICAL PROPERTIES OF AMORPHOUS SELENIUM: A FIRST PRINCIPLES MOLECULAR DYNAMICS SIMULATION

J. A. Reyes-Retana*
Instituto de Investigaciones en Materiales,
Universidad Nacional Autónoma de México,
Apartado Postal, México, D.F., México

Abstract

A new approach is used to generate amorphous selenium structures by an *ab initio*-molecular dynamics method. Crystalline cubic supercells start with 64, 100, 150, 216 and 512 atoms and with the experimental densities of 4.25 g cm^{-3} and 4.45 g cm^{-3}. The samples are amorphized using $DMol^3$ from the suite in Material Studio 3.2® by heating the periodic structures to just below the melting point (the *undermelt-quench* approach) and then cooling them down to 0 K. The structures are relaxed by annealing and quenching, and finally a geometry optimization is carried out. The structural properties: radial distribution functions g(r), bond angle distributions and dihedral angle distributions; electronics properties: electronics density of states; vibrational properties; vibrational density of states and the optical properties: tauc approximation are reported. It is found that the amorphous structure, for both densities, is mainly formed by chains but not at all linear, there are some ring-like structure although not closed, in this chapter gives the first quantitative ratio of chains and rings.

PACS 31.15.A-, 81.05.Gc, 63.50.-x

Keywords: Amorphous selenium, *ab initio* molecular dynamics, electronic density of states, vibrational density of states

*E-mail address: amorpho@gmail.com

1. Introduction

The selenium (**Se**), the sulphur (**S**) and the tellurium (**Te**) are the chalcogen elements. Se was the first photoconductor, it was discovered in 1873 by W. Smith [1]. He used it as resistor rods for submarine telegraphic cables when he realized that the resistance depended on the illumination. The first selenium solar cell was reported by Charles Fritts and it was available from 1920 until 1960 when the silicon devices came. But the most important application was the xerography [2, 3], this continue until the late 80's when the organic semiconductors arose. The Se photoconductivity, in the amorphous phase (**a-Se**), is still interesting. Nowadays, it is used for imaging applications, such as as an avalanche photoconductor in ultrahigh sensitivity vidicon tubes (HARPICONs) [4, 5]); and as an X-ray photoconductor in direct conversion X-ray detectors [6, 7].

The chalcogen elements are those which belong to the subgroup VIB of the periodic table. These elements are: Sulfur, Selenium and Tellurium. The subgroup VIB have also the Oxygen and Polonium. The chalcogen are the basic elements of the chalcogenide alloys, in which at least have one chalcogen and an electropositive element or radical. The chalcogenide name come from the Greek: $\chi\alpha\lambda\kappa o\zeta = copper$, $\gamma\varepsilon\nu\varepsilon\sigma = born$ y $\varepsilon\iota\delta o\zeta = type$, ascribed first to the minerals which contained copper mixed with Sulfur, Selenium or Tellurium

1.1. Selenium (Se)

The Selenium is the element with Z=34 in the periodic table, its atomic mass is 78.96. The selenium abundance in the Earth shell is 8×10^{-5} $at.\%$. Pure selenium is very scarce. Selenium is obtained from different minerals (more than 40), for example, $FeSe_2$, $NiSe_2$, $CoSe_2$, CuSe, ZnSe, Fe_3CuSe_4, $As_2(S,Se_2)_2$, Bi_2Se_3, HgSe, AgSe, PbSe, CdSe, Ag_4SeS, $PbSeO_4$, etc.

The mixed selenium consist of six stables isotopes: ^{74}Se (1%); ^{76}Se (9%); ^{77}Se (7.5%); ^{78}Se (23.5%); ^{80}Se (50%);^{82}Se (9%).

The configuration of the valence electrons of selenium is $4s^2 4p^4$. The oxidations states are: -2, 0, +2, +4, +6, and the hybridization sp^3 is less stable than the sulphur [8].

1.2. Crystalline Selenium

The principal features of the crystalline phases of selenium are based that the atoms built the *trans* configuration (see fig. 1(b))[1]. All allotropic forms of selenium have configurations of short order (first neighbors) in the range distance of 2.32 and 2.40 Å.

1.2.1. Hexagonal (gray or metallic) Selenium (Se_γ)

This is the stablest form of selenium. The cell is made up by parallel helicoidal chains. Each atom has two neighbors on the chain at the distance of 2.30 and 2.37 Å and four neighbors on the range of 3.42 and 3.48 Å. The helix's radius is 0.984 Å. The bonds within the chains

[1] According IUPAC: "Stereoisomeric olefins or cycloalkanes (or hetero-analogues) which differ in the positions of atoms (or groups) relative to a reference plane: in the cis-isomer the atoms are on the same side, in the trans-isomer they are on opposite sides." [9]

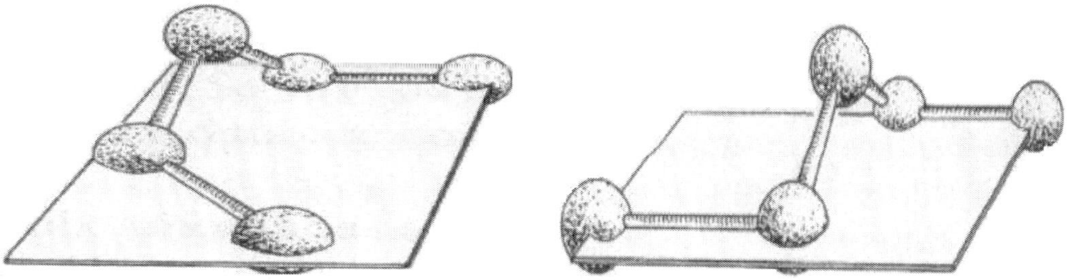

(a) cis (b) trans

Figure 1. *Cis* and *trans* configurations.

are covalent and between them the molecular forces are Van der Waals type with a weak metallic component. The hexagonal unit cell contains three atoms. The lattice parameters measured at 18 °C are: a=4.35 Å, c=4.96 Å, c/a=1.14, space group $P3_121$ [10]. The bond angle is 103.1 ± 0.2 ° and the torsion (dihedral) angle is 100.7 ± 0.1 °.

1.2.2. Monoclinic Selenium Se_α

This allotropic form is obtained by the slow evaporation of a saturated dissolution of selenium in carbon disulphide. Usually the monoclinic Se_α is found together with monoclinic Se_β. Th space group of Se_α is: $P2_1/n - C_2n^5$ and the lattice parameters are: a=9.05 Å, c=11.60 Å y $\beta = 90.82°$. Se_α is built by 8-members rings.

In this configuration the selenium atoms are in the corners of two superposed squares, rotated 45° and are separated such as the selenium ring are not in the same plane. Each atom in the ring have four neighbors at the mean distance of 3.7 Å. The difference between Se_α and Se_β is the ring packing. The mean bond angle of monoclinic Se is 105.7° and the mean torsion angle is 101.3°.

1.2.3. Monoclinic Selenium Se_β

This is a prismatic, transparent and dark-red crystal. The unit cell has the following parameters: a=12.58 Å, b=8.07 Å, c=9.31 Å, $\beta = 93.13°$ (space group $P2_1/aC_{2h}^5$). Both monoclinic crystals are obtained by heating slowly the amorphous Selenium films with the thickness of 600-800 Å, Se_α is appeared at 35-40°C and around 65°C Se_β is detected.

1.2.4. Rhombohedral Selenium

It was discovered in 1980 [11, 12]. The crystalline Selenium is obtained from carbon sulphide solution saturated in Selenium. The corresponding space group is $C_{3i}^2 - R_3$. The unit cell has a six-members ring. The lattice parameters are: a=11.36 Å, c=4.43 Å. The rhombohedral Selenium changes to Se_γ at temperatures above 105 °C (irreversible process).

1.2.5. Cubic Selenium

Andrievski *et al.* [8] showed that through heating thin films of amorphous Selenium at 150-160° C, it can be produced a creeping crystallization and then held them at 160° C become a new crystal form, face centered cubic with the lattice parameters of a=5.76 Å. This crystal was named α-cubic Selenium. They also showed that by a local heating occurs another crystalline form with lattice parameter: a=2.79 Å, it was called β-cubic Selenium. All allotropic forms of Selenium transform in hexagonal Selenium (Se_γ) through a heating of 180-220 °C.

2. Amorphous Selenium

The non-crystalline Selenium (vitreous or amorphous) is a dark gray solid. Both are built with disordered chains and rings of divalent Selenium. Some of the values for the amorphous selenium, such as covalent distance, dihedral and plane angles, and the distance to the second neighbors, seems to be very congruent with hexagonal Selenium form. Between the chain, the forces are Van der Waals type. The density of amorphous selenium (**a-Se**) has 10% less than the crystalline one, therefore the packed structural units are not close packing.

Typically it can be found that a-Se consists of a mixed of two structural types: large helicoidal chains and disordered eight-member rings, with covalent bonds and Van der Waals forces. Although some of the evidence of the rings existence had been questioned and the structural details had not yet presented [13]. In this chapter, a quantitative study of the structure of chains and rings is presented.

2.1. Experimental radial distribution functions

First of all, a bibliographic revision for the amorphous selenium topology is described. In 1967 Henninger *et al.* [14] reported a neutrons and X-ray diffraction study for amorphous selenium. The distances from the first to the fifth neighbor were given: de 2.33, 3.7, 4.6, 5.8, y 7.3 Å respectively. They concluded that the amorphous selenium structure is conformed by random chains. In table 1 the distances of the neighbors are reported and in the figure 2(a) the radial distribution functions (**RDF**) are reproduced. X-ray diffraction studies carried out by Chang *et al* [15] showed the distances to the next neighbors at 2.3, 3.6, 4.1, 4.7, 5.3 y 5.8 Å. The RDFs are shown in figure 2(b). In this paper, the author did not mention how the structure is built.

In 1968 Kaplow *et al.* [16] prepared three different amorphous selenium samples: cast amorphous selenium, a-Se-Cast, vapor deposited amorphous selenium at 298 K, a-Se-298 and vapor deposited amorphous selenium at 77 K, a-Se-77. They also described the crystalline structure for the hexagonal selenium. The correlations peaks are shown in table 2 and the experimental RDFs in figure 3(a). They concluded that the amorphous structure are mainly conformed with slightly disordered eight-member rings and some of this rings could open enough to mimic the trigonal symmetry or very deformed chains.

According with the second peak position in the RDFs, Richter [17] established three different forms of amorphous selenium (Se(I), Se(II), Se(III)), these samples were prepared

Table 1. Summary for the first five peaks distribution of the RDF obtained experimentally by Henninger *et al.* [14].

RDF	Peaks (Å)				
	First	Second	Third	Fourth	Fifth
X-ray	2.33	3.78	4.63	5.80	7.30
X-ray	2.32	3.70	4.65	5.78	7.42
Neutrons	2.31	3.72	4.58	5.80	7.33
Neutrons	2.33	3.71	...	5.74	7.25

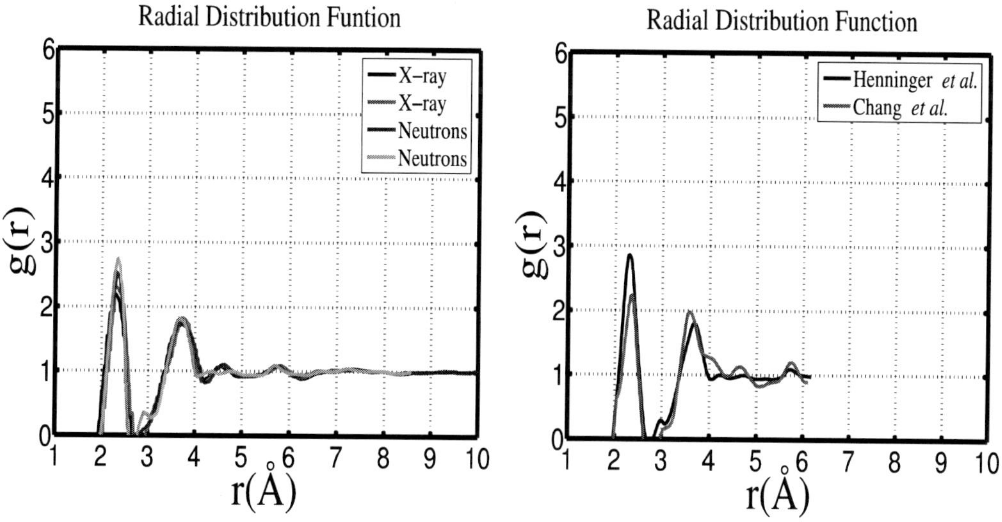

(a) Henninger *et al.* (b) Chang *et al.*

Figure 2. Experimental radial distribution functions (a) obtained by Henninger *et al.* [14] with X-ray and neutrons at different energies, (b) obtained by Chang *et al.* [15] and compared with Henninger *et al.*

by vapor deposition. He also reported two a-Se samples that were melted at 543 and 723 K. The RDFs are described in figures 3(b) and 3(c) and their features are resume in table 3. The author focus on the fact that a-Se had chain structure.

In 1973 Renninger and Averbach [18] measured the RDFs for the amorphous As-Se alloys and pure a-Se. They varied the As-Se concentration from 0 at % to 50 at %. One of the results was that while the concentration of As grew from 0 to 50 at %, then the first peak of the RDF shifted from 2.34 to 2.40 Å. The a-Se distance values for the first and second neighbors were reported as 2.34 and 3.71 Å and the coordination numbers as 2.07 and 3.48 atoms respectively. The RDF is shown in figure 4(a).

There is a second study for the amorphous selenium by neutron diffraction, in this work, Hansen *et al.* [19] showed two RDFs measured at liquid nitrogen and room temperature. In figure 3 of the reference [19] the RDF can be seen before and after a correction, then a shoulder between the first and second peaks despaired. This characteristic peak was also

Table 2. Experimental correlations for amorphous selenium first peaks reported by Kaplow *et al.* [16]. C_1 y C_2 are the coordination numbers for the fist and second neighbors respectively

RDF	Peaks(Å)		Coordination number	
	First	Second	C_1	C_2
a-Se-Cast	2.34	3.75	2.0	6.4
a-Se-298	2.34	3.75	2.0	6.3
a-Se-77	2.31	3.70	2.0	6.6
Hexagonal	2.32	3.55	2.0	5.8

Table 3. First neighbors distribution for the experimental samples reported by Richter [17].

RDF	Peaks (Å)			
	First		Second	Third
Se(I)	2.35	...	3.68	5.75
Se(II)	2.35	...	3.78	5.75
Se(III)	2.35	3.0	3.55	5.60
Fundido 543 K	2.36	3.09	3.58	...
Fundido 723 K	2.30	2.85	3.67	...

observed in the RDFs reported by Henninger *et al.*, Chang *et al.* and Richter [14, 15, 17]. This peak situated around 3 Å is an important part of the discussion for the topologic properties. The values for the first and second peak are: 2.32 y 3.69 Å. They did not mentioned anything about the structure type (chains or rings). In figure 4(b) the RDFs are shown.

Bellissent [20] measured, by neutrons diffraction, the short order in amorphous alloys of Se-Te and pure a-Se. The first and second peaks positions for a-Se are situated at 2.32 and 3.72, the coordination number for the first neighbor is 2. The experimental RDF for a-Se is in figure 4(c). The author did not mention which kind of structure were embedded in the amorphous material.

Another procedure to obtain amorphous materials is by mechanical milling. Trigonal selenium powder was placed under an As gas atmosphere, using a planetary ball mill apparatus (Fritsch Pulversette 5) for 0, 5, 10 and 50 h. Fukunaga *et al.* [21] carried out the experimental method described above, they also annealed the 50 h of milling sample to compared structure differences. The first and second peaks distance for 50 h of milling were situated at 2.36 and 3.96 Å, while after the annealing these values moved to 2.37 and 3.70 Å respectively. The figure 4(d) shows the RDFs. The authors concluded that for the mechanical milling the amorphous structure was conformed with chains, while for the annealing sample, this is conformed with chains and rings, but the proportion was not mentioned.

In 1994 Inui *et al.* [22] reported neutron and X-ray scattering diffraction studies for the a-Se. In this paper the authors highlighted the detection of the lone pair and also the first distances at 2.30 and 3.70 Å, see figure 5(a). Then on 1999, in a second X-ray diffraction,

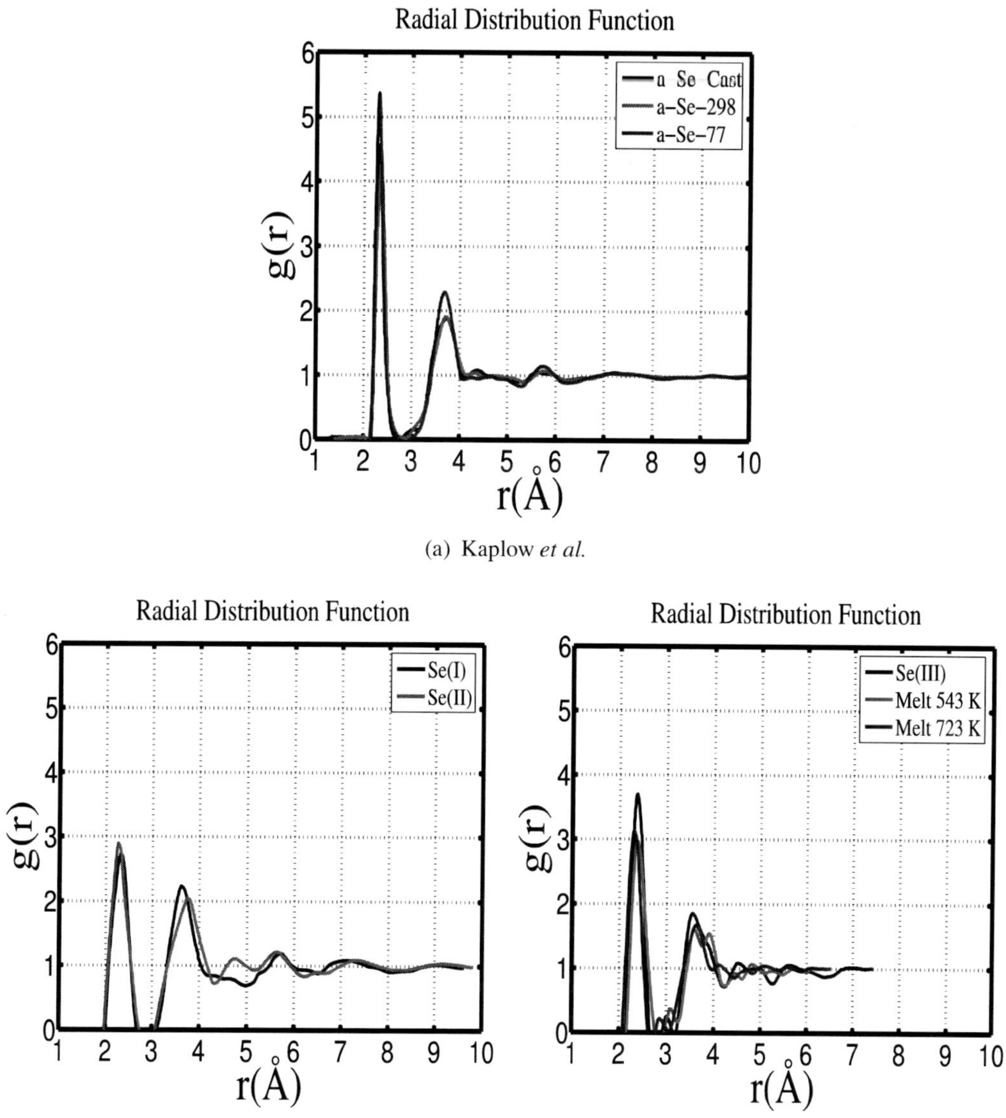

(a) Kaplow *et al.*

(b) Richter

(c) Richter

Figure 3. (a) Experimental radial distribution functions obtained by Kaplow *et al.* [16]. Experimental RDFs for (b) Se(I), Se(II) y (c) Se(III), melted selenium at 543 and 723 K, obtained by Richter [17].

Inui *et al.* [23] presented the amorphous selenium dilatation for differents temperatures and pressures; in their work they reported the RDF for a-Se at 300° C and 10 bars. The values for the first and second neighbors were: 2.35 and 3.70 Å. the authors suggested a chain-like structure for the amorphous sample. In figure 5(a) the experimental RDFs are shown.

The second mechanical milling study was made by Lima *et al.* [24], the authors reported the influence of aging on the thermal and structural properties of amorphous selenium and also the molecular chains into rings-like (Se_8) structure. The experimetal RDF can be seen in figure 5(b) where the first peak is situated at 2.37 Å and the second at 3.85 Å. At the

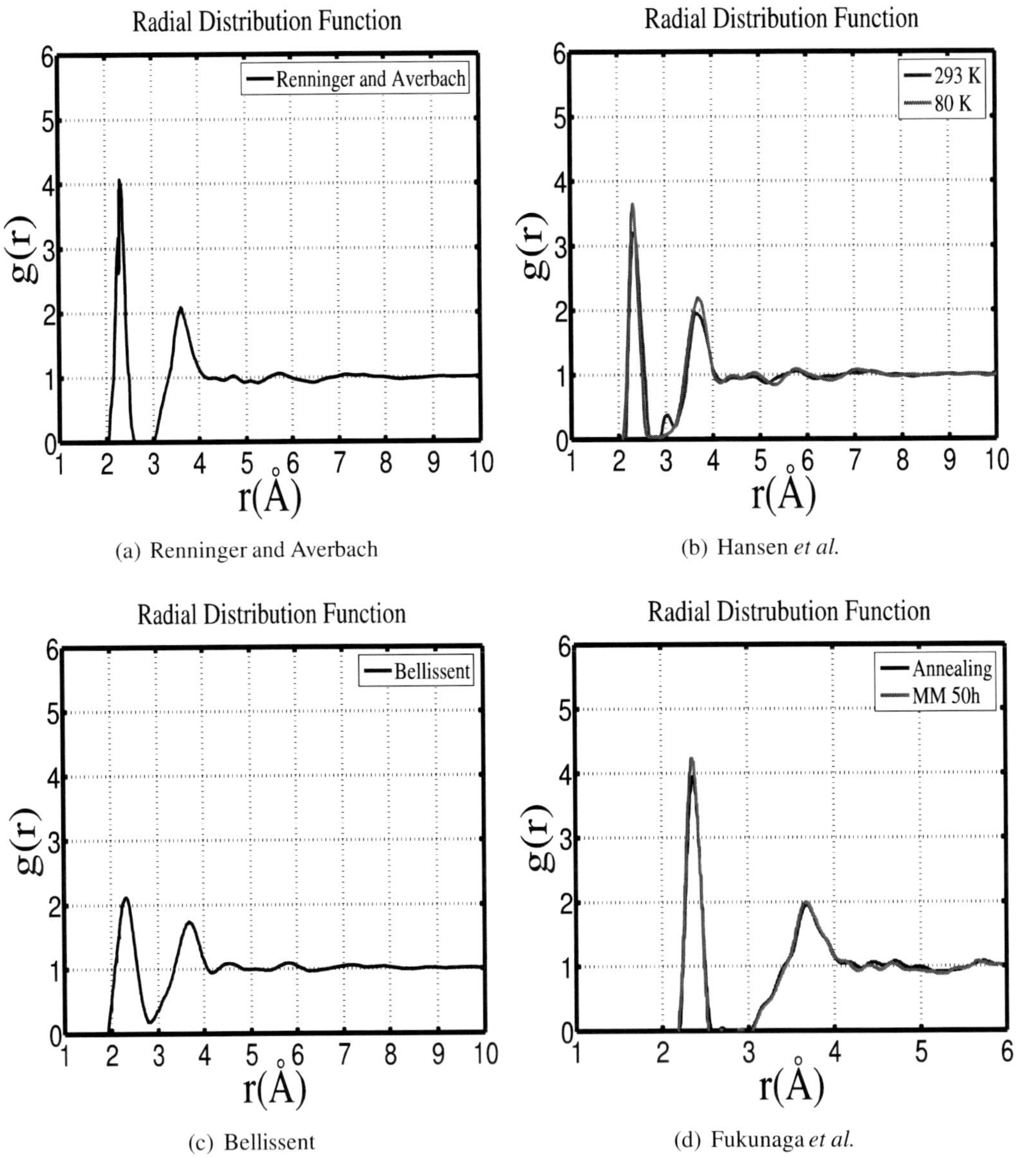

Figure 4. Experimental radial distribution functions for a-Se (a) obtained by Renninger and Averbach [18], (b) obtained by Hansen *et al.* [19], (c) obtained by Bellissent [20] and (d) obtained by Fukunaga *et al.* [21].

same time, Brüning *et al.* [25] published the X-ray diffraction results for a-Se, they first analyzed the diffraction when the melt was cooling down, the same sample was annealing and analyzed it one more time. The figure 5(c) shows the RDFs and the corresponding values for the first two neighbors were: 2.35 and 3.96 Å. They reported a random chain structure for a-Se.

The last experimental RDF that is presented was made by Jóvári *et al.* [26], using neutron diffraction. The first and second neighbors were situated at 2.32 and 3.70 Å (see figure

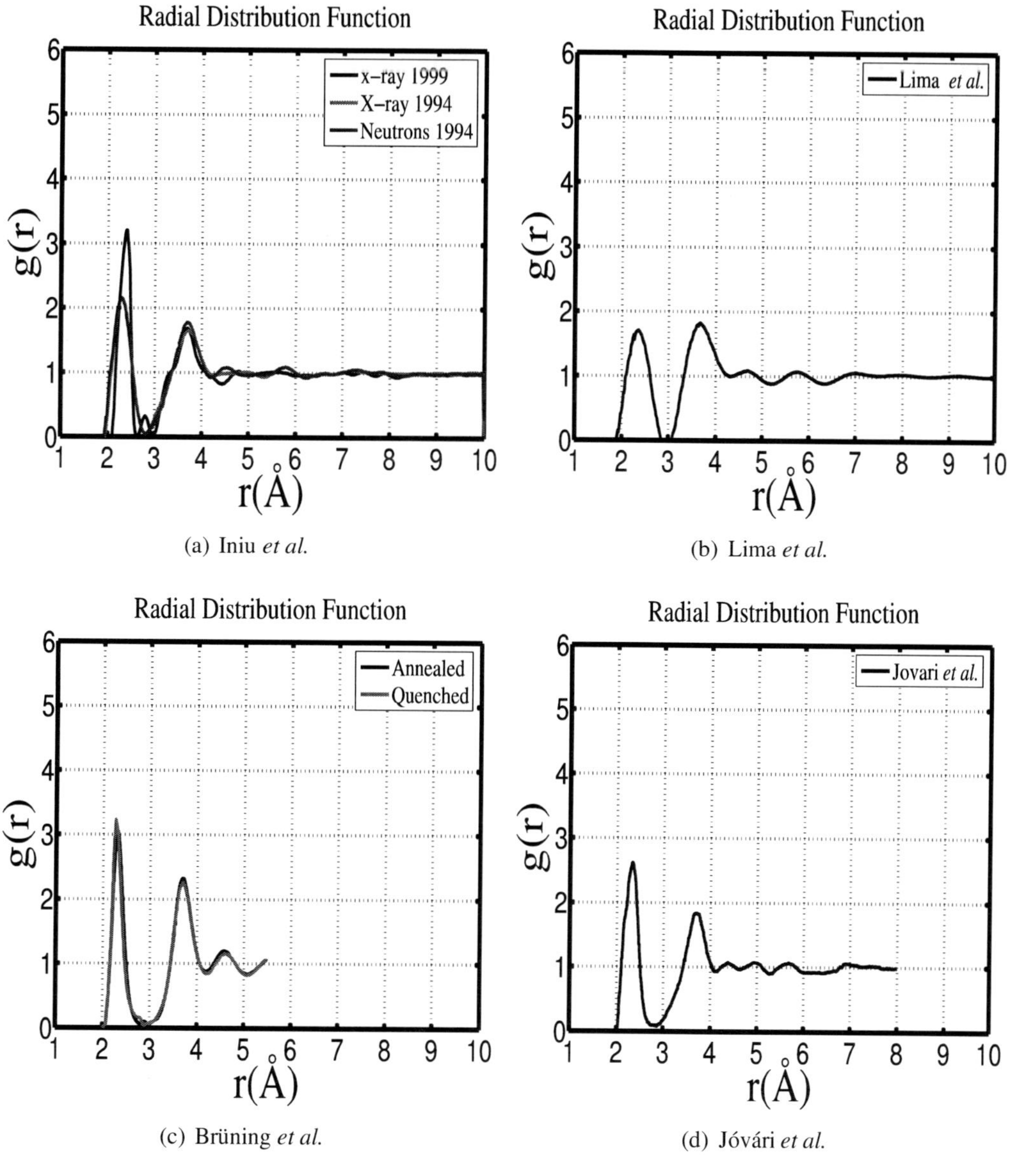

(a) Iniu *et al.* (b) Lima *et al.* (c) Brüning *et al.* (d) Jóvári *et al.*

Figure 5. Experimental radial distribution functions for a-Se (a) obtained by Inui *et al.* [22,23], (b) obtained by Lima *et al.* [24], (c) obtained by Brüning *et al.* [25] and (d) obtained by Jóvári *et al.* [26].

5(d)). They concluded that even how the amorphous selenium is prepared, its structure is mainly conformed with molecular chains and the existences of eight members rings is improbable.

2.2. Computational radial distribution functions

In this section the radial distribution functions obtained by computational methods are presented. In 1968 Kaplow *et al.* [16] built a 100 atoms model in which they used Monte Carlo process to perturb a monoclinic selenium structure (Se_{α}) until reproduce the experimental radial distribution functions that they measured, see figure 3(a).

In 1982 Corb *et al.* [27, 28] made two models to reproduce the RDF obtained by X-ray diffraction. In their two models they began with different structures but both had the following restrictions:

- The molecules are deformed helicoidal chains where the bond and dihedral angle distributions are centred in 105° and 102° respectively.
- The molecules are packed such as they minimized the Van der Waals attractions.
- The molecular compact packed is achieved allowing that the dihedral angle magnitud varied slightly around 102° and also to allow that the dihedral angle phase varied randomly.
- The electronic orbital Coulomb repulsions were simulated such as the distances less than 3 Å are forbidden.

The authors also inserted a little percent of close eight member-rings (Se_8). They mentioned that the ratio between *cis* (figure 1(a)) and *trans* (figure 1(b)) is not unique for the a-Se and then this can not been used for the structure description.

The first *ab initio* study for amorphous selenium was published in 1990 by Hohl and Jones [29], they also included the liquid selenium description. They began with a 64 atoms embedded in a cubic cell, the sample was heated until 2500 K and after several thermic process it was held at 350 K. Figure 6(a) shows the RDF. The values for the first and second peak that they reported are: 2.43 and 3.70 Å, a little different to the experimental values (2.32 Å). In the RDF of this first principle study, appeared a shoulder around 3 Å, the authors associated it a selenium atoms with coordination three, but in their next paper [30] argued that this peak had an ambiguos description.

In 1996 Olischleger *et al.* [31] presented a intermolecular potential for the selenium description. This potential is based in small molecular structures of Se (Se_3, Se_4, Se_5, Se_6, Se_7, Se_8) calculated by density functional theory (DFT) [32]

The second *ab initio* study for amorphous selenium was made by Zhang and Drabold in 1998 [33]. They used 64 atoms in a cubic cell where the atoms were put randomly. The values for the first and second peaks were: 2.30 and 3.65 Å (see figure 6(b)). They also described that the shoulder at 3 Å in the RDF was due to the distances between a ring and a chain in the final configuration. This configuration consisted in a long chain and a six member-ring (Se_6). In their following paper [34], they reported a 216 atoms cell simulation to research the light influence on the amorphous selenium. Figure 6(b) shows the RDF and the values for the first and second peaks are: 2.39 and 3.75 Å.

Bichara *et al.* [35] used simulations of tight binding combined with Monte Carlo (TB-MC) process to search the chain structure of the a-Se, they conclude that the percent of the coordination two did not exceed 70%. In 1999 Simuzu *et al.* [36] performed TB-MC for

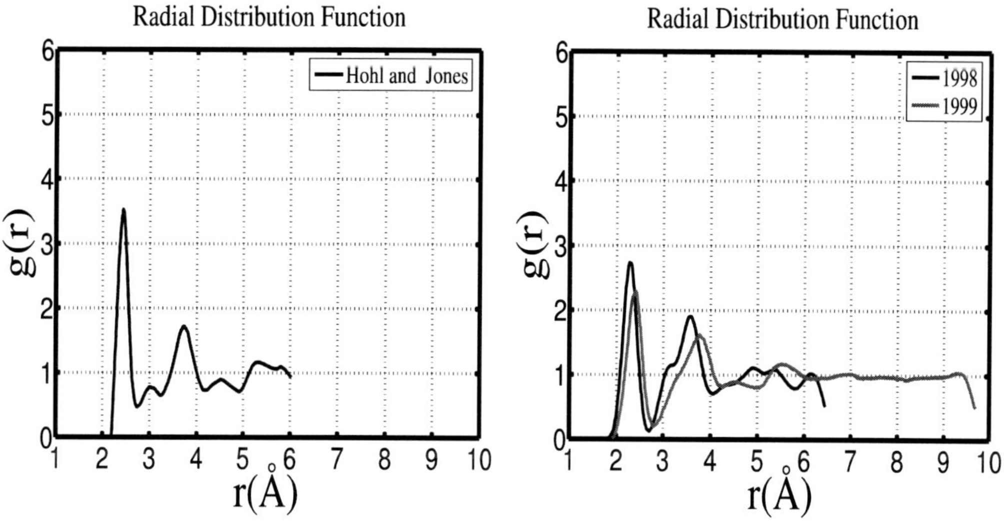

(a) Hohl and Jones

(b) Zhang and Drabold

Figure 6. First principles radial distribution functions obtained by (a) Hohl and Jones and by (b) Zhang and Drabold.

64, 216 and 512 atoms, they reported that the first minimum of the RDF was situated at 2.75 Å, see figure 7(a). The values of the first and second peaks are obtained from the RDF, these are: 2.24 and 3.50 Å.

Cabrion and Schober [37] used 2000 atoms and the potential described by Olischleger *et al.* [31] to simulate the a-Se structure. In this study, they calculated the coordination number for the first neighbors, this values corresponded to 2.1 atoms. The principal purpose was to give a values for the vitreous transitions temperature, which was at $T_g \approx 300$ K. Figure 7(b) shows the RDF and the first and second peaks are situated at 2.35 and 3.64 Å. They concluded that as liquid as well as the amorphous were conformed with chains and rings with the coordination number mention above, also found that the chains and rings were connected. In their next work, Cabrion and Schober varied the pressure on the sample and therefore the vitreous transition temperature grew.

Hegedüs *et al.* [40,41] also used Olischleger potential to perform a classical molecular dynamics and to describe the cooling process of the melt and the evaporation. They built four a-Se models, one for the rapid quench and three for the evaporation simulation with different deposition energies (bombarding): 0.1, 1 and 10 eV. They only reported the first distance values and it is the same for the four models: 2.37 Å. Hegedüs *et al.* analyzed the volume changes due to the light incident, in this tight binding molecular dynamics (TB-MD) study they showed one more time the shoulder around 3 Å, which they correlated with the distance of the selenium in two different chains.

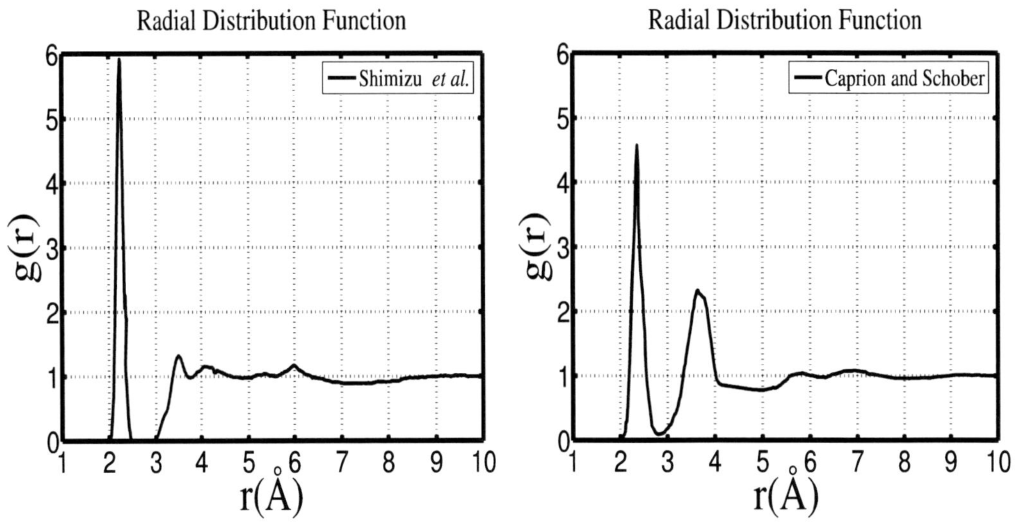

(a) Shimizu *et al.* (b) Caprion and Schober

Figure 7. Radial distribution functions obtained (a) through TB-MC by Shimizu *et al.* and (b) through classical methods by Caprion and Schober [37,38].

3. Description for the Amorphous Selenium Cells

The number of the amorphous selenium simulations within a cubic cell are increasing from 64, 150, 216 until 512 atoms. The starting cell for the amorphous process is a cubic cell diamond-like with periodic conditions. The density theory functional theory (DFT) is used [43], with the local density approximation (LDA) [44]. The energy calculation is computed using Harris functional [45,46] with differents orbital bases set: *dn* and *dnp* (*double numerical and double numerical with polarization*). Two options for the core treatment are used: *dspp* (*DFT semi-core pseudopotencials*) and *none* (all electron are treatment as valence electrons). There are two densities for the amorphous selenium, the first density is the experimental 4.25 g/cm^3 [19] and the other is the calculated 4.45 g/cm^3 [42]. For all simulations the cutoff is 5 Å and the *time step* is taken as 16 fs. Table 4 shows all the parameter and the label for each sample.

The label description for the amorphization process is the following: at the beginning the element's name, Se; then the number of atoms (64, 150, 216 or 512), immediately a capital letter: A if the density es 4.45 g/cm^3 or B if the density is 4.25 g/cm^3, obital bases set (dn or dnp) and finally the core treatment (dspp or none). At the end of the every label is the capital letter A which means the amorphization process.

The *undermelt* process for the amorphization is used [47]. The starting temperature is 300 K and through lineal heating of 100 steps the samples is reached to 480 K (0.113$\times 10^{15}$K/s) just below the melting point (494 K), then they are cooling down to 0 K in 266 steps (-0.113$\times 10^{15}$K/s). At this moment the samples are subjected to a annealing process: 60 steps at 300 K and then 66 steps of cooling down (-0.113$\times 10^{15}$K/s), see figure 8. After the amorphization process a geometry optimization is carried out.

The table 5 shows the optimization serie; the label description for these simulations

Table 4. Simulation parameters for the amorphizations with 64, 150, 216 and 512 atoms.

Number of atoms	Density (g/cm^3)	Basis	Pseudo-potentials	Sample
64	4.25	Dn	Dspp	Se64B_DnDsppA
		Dnp	Dspp	Se64B_DnpDsppA
			None	Se64B_DnpNoneA
	4.45	Dnp	Dspp	Se64A_DnpDsppA
			None	Se64A_DnpNoneA
150	4.25	Dnp	Dspp	Se150B_DnpDsppA
	4.45		Dspp	Se150A_DnpDsppA
216	4.25	Dn	None	Se216B_DnNoneA
		Dnp	Dspp	Se216B_DnpDsppA
			None	Se216B_DnpNoneA
	4.45	Dnp	Dspp	Se216A_DnpDsppA
512	4.45	Dn	Dspp	Se512A_DnDsppA

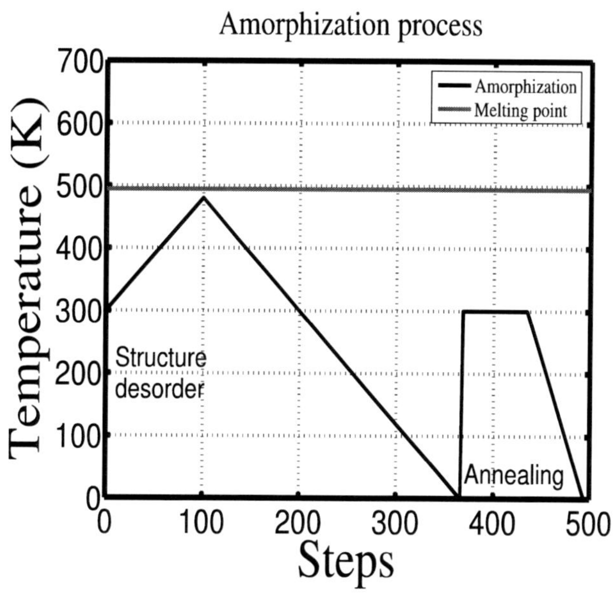

Figure 8. Amorphization process.

is the same that for the amorphization process except that there not capital letter at the end. Now all this sample are analyzed for topological, vibrational, electronic and optical properties.

Table 5. Sample labels for the disorder (amorphization process, see table 4) and the corresponding optimization.

Disorder	Basis	Pseudopotential	Optimization
Se64B_DnDsppA	Dn	Dspp	Se64B_DnDspp
Se64B_DnpDsppA	Dnp	Dspp	Se64B_DnpDspp
Se64B_DnpNoneA	Dnp	None	Se64B_DnpNone
Se64A_DnpDsppA	Dnp	Dspp	Se64A_DnpDspp
Se64A_DnpNoneA	Dnp	None	Se64A_DnpNone
Se150B_DnpDsppA	Dnp	Dspp	Se150B_DnpDspp
Se150B_DnpDsppA	Dn	Dspp	Se150B_DnDspp
Se150B_DnpDsppA	Dnp	None	Se150B_DnpNone
Se150A_DnpDsppA	Dnp	Dspp	Se150A_DnpDspp
Se150A_DnpDsppA	Dnp	None	Se150A_DnpNone
Se216B_DnNoneA	Dn	Dspp	Se216B_DnDspp
Se216B_DnpDsppA	Dnp	Dspp	Se216B_DnpDspp
Se216B_DnpNoneA	Dnp	None	Se216B_DnpNone
Se216A_DnpDsppA	Dnp	Dspp	Se216A_DnpDspp
Se512A_DnDsppA	Dn	Dspp	Se512A_DnDspp

4. Topological Properties

In this section the topological properties of the amorphous selenium are presented. First to all the radial distribution functions (RDF) are presented, the first and second neighbors distances are given and also the comparison with the experiment data. Then the bond and dihedral bond distribution are shown. In this part of the chapter, the study of existence embedded in the structure is described as well as the percent within these final structures are given.

4.1. Radial Distribution Functions

The radial distribution functions for all amorphous samples are calculated, therefore the first and second peak are given. A very important contribution of this chapter, is the shoulder explanation around 3 Å, which was called r_α. The characteristic mentioned above is associated to a crystal phase: α-cubic [48]. It is congruent with section 1.2.5., where the author reported that by heating the amorphous selenium samples, these became a α-cubic form.

Figure 9 shows the RDFs for 64 atoms simulations, within the box, the coordination number is presented. The first and second neighbors details are reported in table 6. It is worth to mention, how many electrons are taken into account when *dspp* is used. The approximation above joins $1s^2 2s^2 2p^6 3s^2 3p^6$ in a effective potential within the Hamiltonian, which the $3d^{10} 4s^2 4p^4$ (16 electrons) act directly in the electron-electron and ion-electron interactions. In case of did not use any pseudopotential (*none*), all electrons act in the Hamiltonian, *i. e.* 34 electrons per selenium atom; $1s^2 2s^2 2p^6 3s^2 3p^6 3d^{10} 4s^2 4p^4$

In the first simulation serie or a-Se, the RDF, obtained by *dnp* basis set and the used of

Table 6. r_1 is the first neighbor distance, r_α α-cubic distance and r_2 second neighbor distance for 64 atoms samples

Sample	r_1 (Å)	r_α (Å)	r_2 (Å)
Se64B_DnDspp	2.43	3.03	3.73
Se64B_DnpDspp	2.35	3.05	3.75
Se64B_DnpNone	2.35	2.85	3.65
Se64A_DnpDspp	2.43	2.93	3.83
Se64A_DnpNone	2.34	3.04	3.54

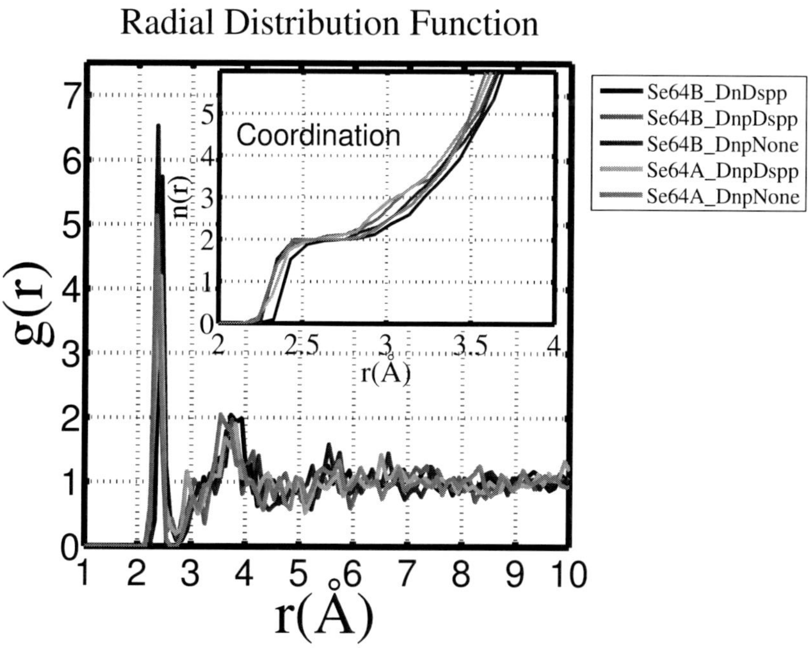

Figure 9. RDFs for the 64 atoms samples, with different basis set and pseudopotentials.

all electrons, shows similar behavior to the corresponding experimental part. The sample simulation with density 4.25 g/cm^3, *dspp* pseudopotential also gives a good RDF which fits satisfactory to the experiments. However, the samples which were treated with *dn* basis set, density 4.25 g/cm^3 reached the disorder but did not reproduced the first and second distances in comparison with samples mention above (*dnp* basis set and *none* psudopotential).

In computational simulation field, it is common to used the smoothing tool to eliminate the noise which is produced by the low number of atoms. In this *ab initio* study did not use the tool described above, instead the number of atoms were increased from 64 to 512 to observe how the different properties curves were smoothed.

Figure 10 shows the RDFs for 150 atoms simulations, within the box, the coordination number is presented. The first and second neighbors details are reported in table 7. These results give us the following information: when a sample is amorphized with *dn* basis set, *dspp* pseudopotential and after it is optimized with a completer basis set (*dnp*) then the topo-

Table 7. r_1 first neighbor distance, r_α α-cubic distance and r_2 second neighbor distance for 150 atoms samples.

Sample	r_1 (Å)	r_α (Å)	r_2 (Å)
Se150B_DnDspp	2.43	2.93	3.83
Se150B_DnpDspp	2.35	3.15	3.75
Se150B_DnpNone	2.36	3.06	3.76
Se150A_DnpDspp	2.32	2.92	3.72
Se150A_DnpNone	2.34	3.04	3.74

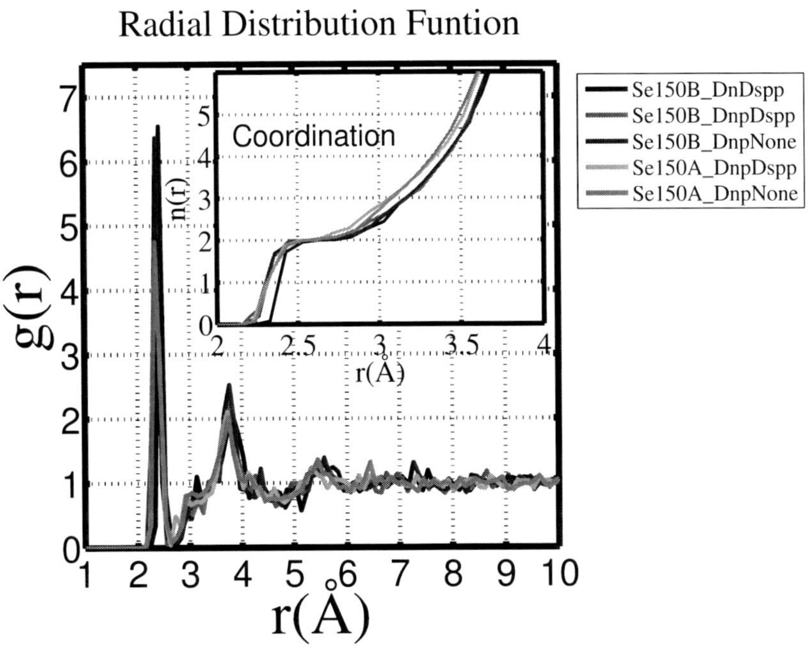

Figure 10. RDFs for the 150 atoms samples, with different basis set and pseudopotentials.

logical properties are in a good agrement with the experimental and also the computational time has been reduced.

With only two amorphization process (150B_DnpDsppA and 150A_DnpDsppA), five different final disordered structures are obtained, three geometry optimization with the density 4.25 g/cm^3 and two with 4.45 g/cm^3.

The 150B_DnDspp is very similar to the sample 64 atoms with the same parameters, this suggests that the polarizations involved an important factor for the amorphous selenium description. The geometry optimization with *dnp* basis set, *dspp* pseudopotential generates a RDF which is a good physic representation for the atomic model.

The mention above is valid for both densities. The RDF curves with 150 atoms are smoother that the 64 atoms curves, which is due to the number of atoms increment.

Figure 11 shows the RDFs for 216 atoms simulations, the box within describes the coordination number. The next neighbor distances are given in table 8. In the case of

Table 8. r_1 first neighbor distance, r_α α-cubic distance and r_2 second neighbor distance for 216 atoms samples.

Sample	r_1 (Å)	r_α (Å)	r_2 (Å)
Se216B_DnDspp	2.44	3.04	3.84
Se216B_DntoDnp	2.34	3.04	3.74
Se216B_DnpDspp	2.35	3.05	3.65
Se216B_DnpNone	2.35	2.85	3.65
Se216A_DnpDspp	2.35	3.05	3.75

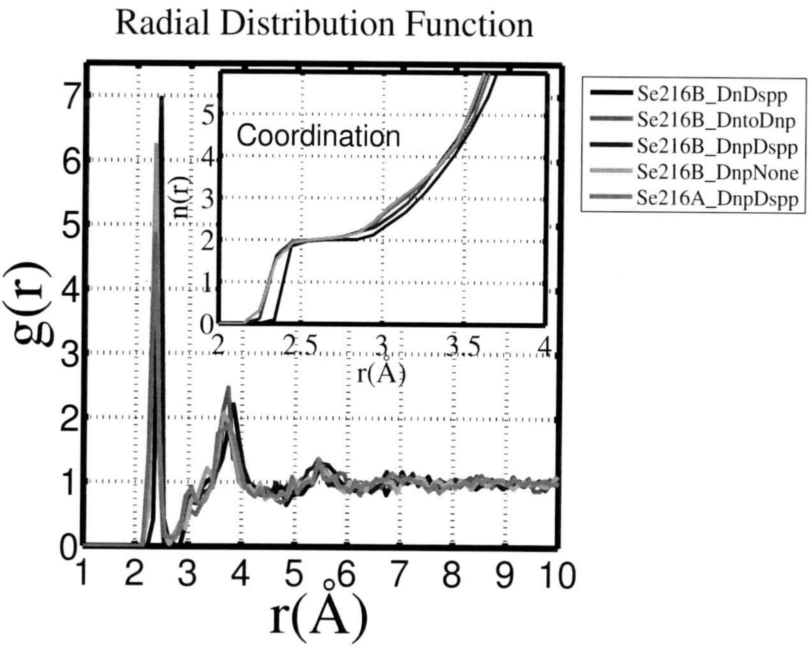

Figure 11. RDFs for the 150 atoms samples, with different basis set and pseudopotentials.

the amorphization process Se216B_DnNoneA, the final structure is taken and optimized with *dnp* basis set and *dspp* pseudopotential (Se216B_DntoDnp), which produced a good agrement with the experiment data. This amorphization-optimization path leads to a less computational time that if *dnp* basis set were taken from the beginning.

Figure 12 shows RDFs for 512 atoms simulations, the box within describes the coordination number. The first and second neighbors details are reported in table 9. There are no more simulations at this moment for this number of atoms, because the expensive computational time. But it can been seen that while the number of atoms are incremented then the topological properties are smoothed. A completest work with more that 512 atoms is carrying out and will be published soon.

In summary, the amorphous selenium structures with 64, 150, 216 and 512 atoms within a supercell are generated, the parameters for the amorphization and optimization process are the *dnp* basis set and *dspp* pseudopotential. An alternative is first an amorphization process with *dn* basis set and *dspp* pseudopotential then the geometry optimization with completer basis set (*dnp*) and dspp, which the results are very similar when all electron are used.

Table 9. r_1 first neighbor distance, r_α α-cubic distance and r_2 second neighbor distance for 216 atoms samples.

Sample	r_1 (Å)	r_α (Å)	r_2 (Å)
Se512A_DnDspp	2.43	3.13	3.83

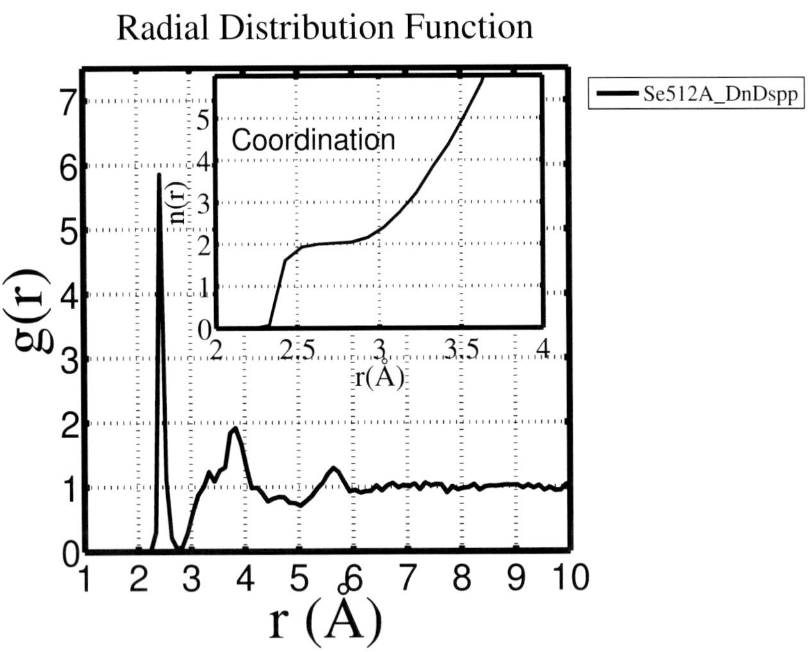

Figure 12. RDF for the 512 atoms sample, basis set *dn* y pseudopotentiales *dspp*.

4.2. Bond angle distribution

From the first and second distance, the average bond angle can be computed with the following equation [49]:

$$\theta = 2arcsin\frac{r_2}{2r_1} \tag{1}$$

where r_1 is the first distance and r_2 is the second. Another way to report the bond angle is by counting each bond angle and then making the distribution. It is needed to know the criterio to decide when two atoms are joined. In this chapter, the first minimum of the RDF is taken. Table 10 shows the maximum bond length value for each sample, the average bond angle calculated by the equation 1 and the maximum for each distribution.

Figures 13 and 14 show the bond angle distributions (BAD) for each amorphous sample. It can be seen that the average bond angle for te hexagonal crystalline (103.2°) and

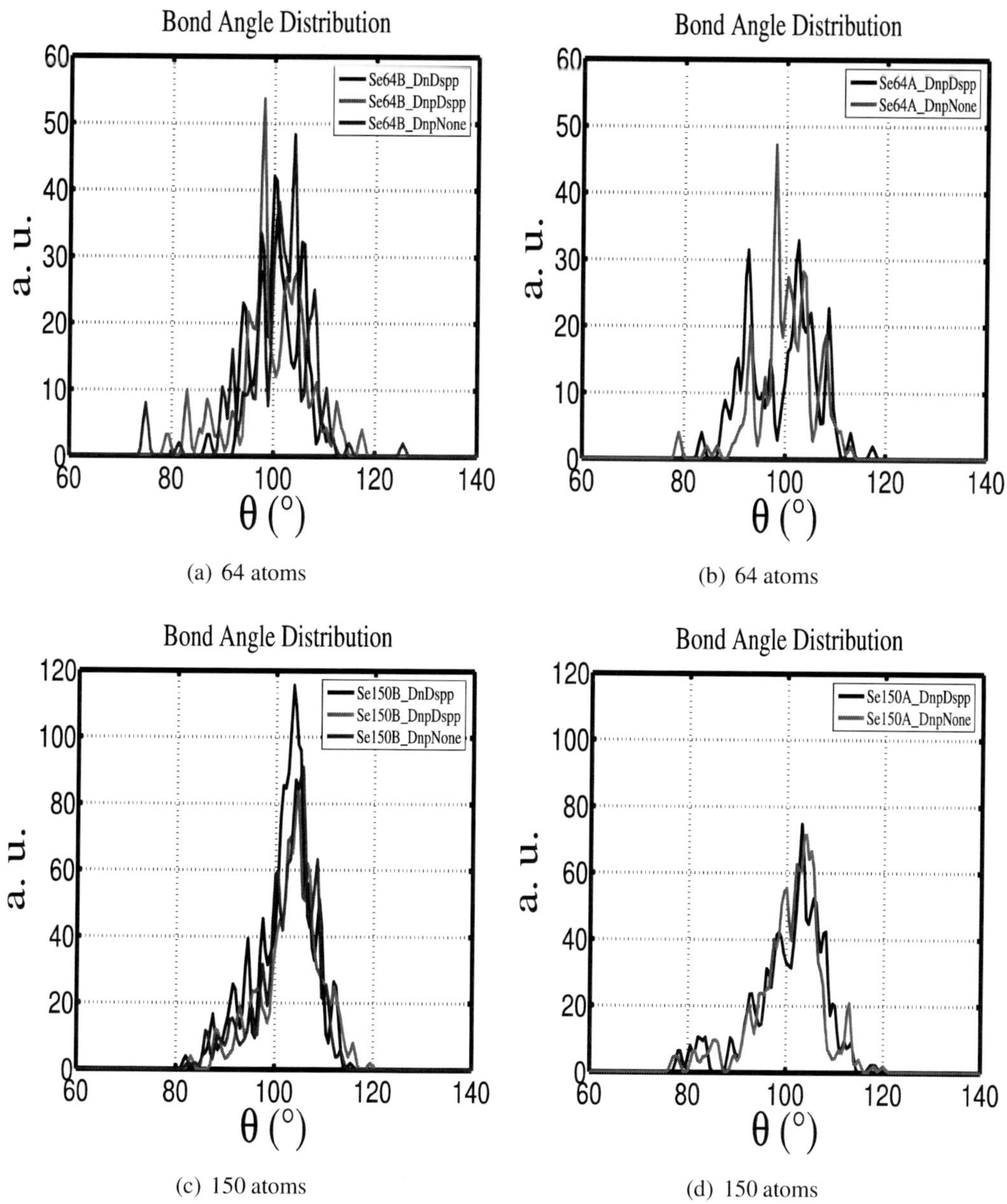

Figure 13. Bond angle distributions for the amorphous samples.

monoclinic (105.7°) structure selenium are embedded in all distributions. Therefore, at this point it is not clear what structure is dominated in the amorphous phase, neither rings or chains.

4.3. Dihedral angle distribution

The dihedral angle is defined as the angle conformed between two planes, the first plane is built by three atoms chain and the second plane is built by another three atoms chain,

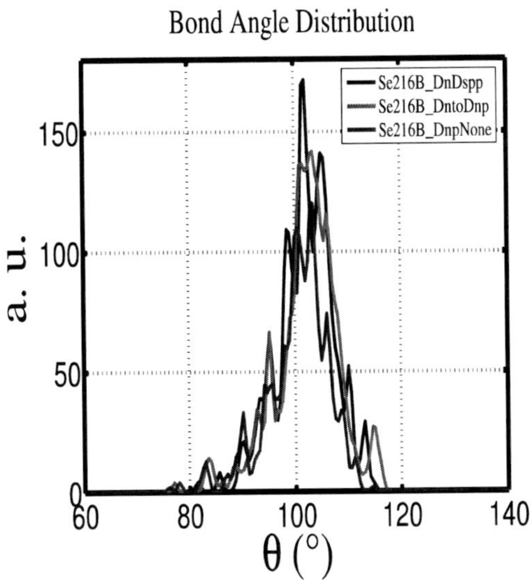

(a) 216 atoms

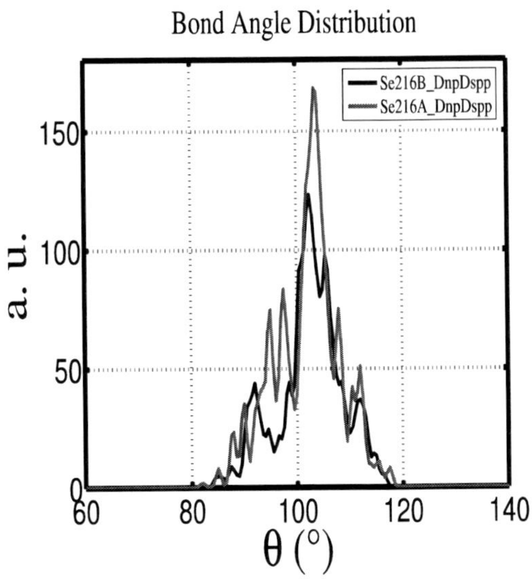

(b) 216 atoms

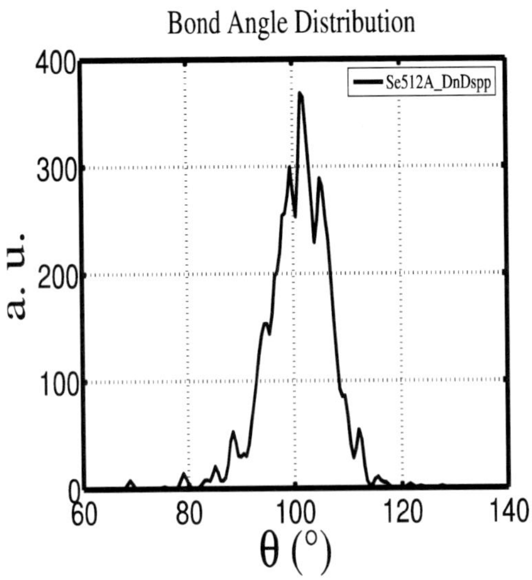

(c) 512 atoms

Figure 14. Bond angle distributions for the amorphous samples.

Table 10. Bond angle distribution for all amorphous samples, where r_e is the maximum bond length values, θ is the bond given for the equation 1 and the θ_{max} is the BAD maximum value.

Sample	r_e (Å)	θ	θ_{max}
Se64B_DnDspp	2.83	100.26°	100.0°
Se64B_DnpDspp	2.75	105.85°	98.0°
Se64B_DnpNone	2.55	101.90°	104.0°
Se64A_DnpDspp	2.73	104.23°	105.5°
Se64A_DnpNone	2.64	98.30°	98.0°
Se150B_DnDspp	2.73	104.10°	105.0°
Se150B_DnpDspp	2.75	105.85°	104.0°
Se150B_DnpNone	2.76	105.61°	105.5°
Se150A_DnpDspp	2.62	106.59°	103.0°
Se150A_DnpNone	2.64	106.10°	103.5°
Se216B_DnDspp	2.84	103.80°	101.5°
Se216B_DntoDnp	2.64	106.10°	103.5°
Se216B_DnpDspp	2.55	101.90°	100.5°
Se216B_DnpNone	2.65	101.90°	103.0°
Se216A_DnpDspp	2.75	105.85°	103.5°
Se512B_DnDspp	2.73	104.01°	101.5 °

which shares the first two atoms with the former chain. The criterio to have two atoms connected is the described in the last subsection (r_e), see table 10. Figures 15 and 16 show the dihedral angles distribution. The values for the hexagonal and monoclinic crystalline structures (101.3° and 100.7°) are embedded in the distributions. With this information is hard to decide what kind of form predominate for the selenium amorphous structures. But it is evident te possibility to find chains as well as rings but also α-cubic structures as Andrievsky *et al.* had reported.

4.4. Ring distribution

In the bibliographic revision, there are only two types of forms which can be found into the amorphous selenium: disordered chains or random oriented rings. But there are not a quantitative study. In this subsection the percent of ring structure embedded in the amorphous phase is presented. The criterio for the maximum bond was defined above.

Table 11 shows the rings distribution for the different simulations, also the percent of disordered chains and rings are given. In general, it is found that the amorphous selenium is mainly conformed with disordered chains and low percent (10 %) of random oriented rings with different members.

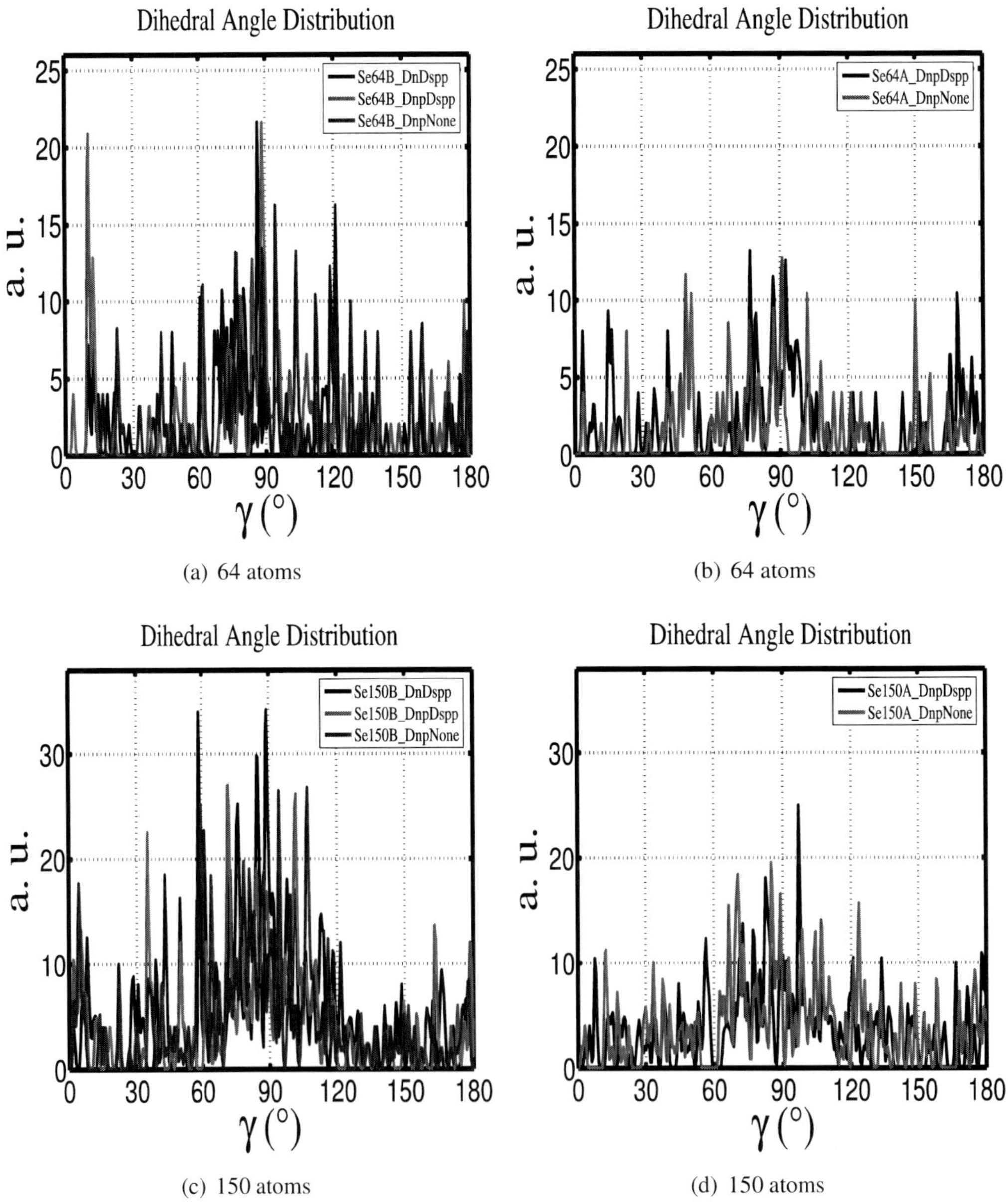

Figure 15. Dihedral angle distributions for the amorphous samples.

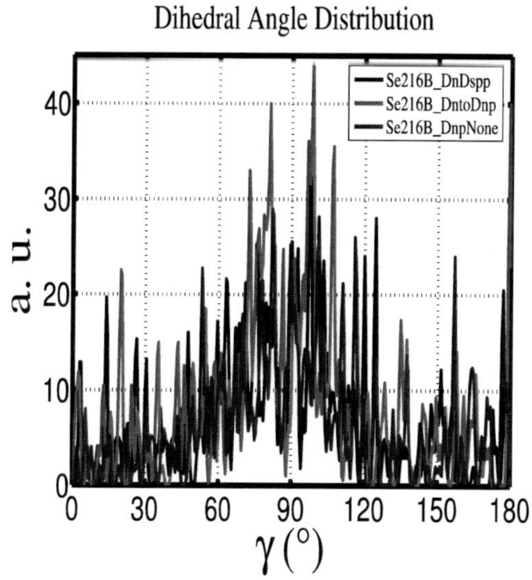

(a) 216 atoms

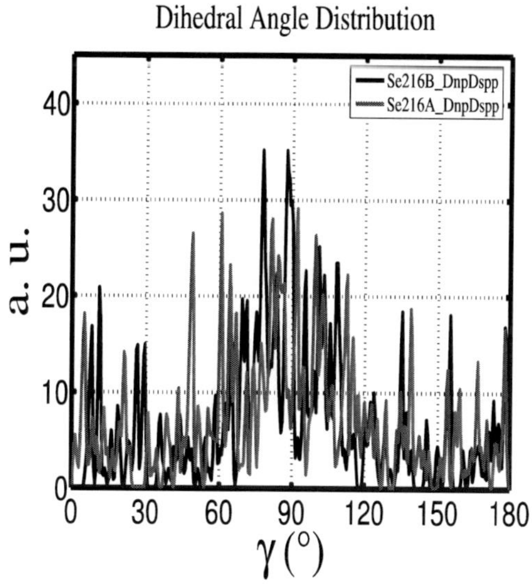

(b) 216 atoms

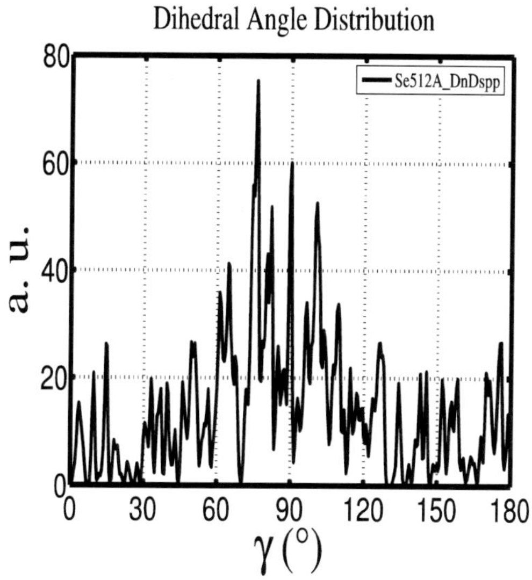

(c) 512 atoms

Figure 16. Dihedral angle distributions for the amorphous samples.

Table 11. Ring distribution for amorphous selenium, also the percent of the chains and rings are presented.

Sample	Rings distribution	Type percent ring (%)	chain(%)
Se64B_DnDspp	There is not rings	0	100
Se64B_DnpDspp	There is not rings	0	100
Se64B_DnpNone	There is not rings	0	100
Se64A_DnpDspp	There is not rings	0	100
Se64A_DnpNone	There is not rings	0	100
Se150B_DnDspp	1 of 7 members	4.7	95.3
Se150B_DnpDspp	1 of 4 and 1 of 7 members	7.3	92.7
Se150B_DnpNone	1 of 7 members	4.7	95.3
Se150A_DnpDspp	1 of 7 members	4.7	95.3
Se150A_DnpNone	1 of 7 members	4.7	95.3
Se216B_DnDspp	2 of 5, 1 of 6 and 2 of 7 members	13.9	86.1
Se216B_DntoDnp	3 of 5 and 1 of 7 members	10.2	89.8
Se216B_DnpDspp	1 of 6 members	2.8	97.2
Se216B_DnpNone	2 of 6 and 1 of 7 members	8.8	91.2
Se216A_DnpDspp	1 of 5 and 1 of 8 members	6	94
Se512B_DnDspp	2 of 3, 2 of 5, 3 of 6 and 1 of 7 members	8	92

5. Vibrational Properties

A very important dynamical property is the solid vibrational spectrum. If there is not free electrons, the heat is transported by te vibrational excitations and the heat capacity is obtained by the allowed vibrational modes [49]. When the harmonic approximation is used, each atom in the solid is like a harmonic oscillator, the vibrational frequencies are computed from the eigenvalues of the Hessian matrix weighted with the atomic mass [51].

5.1. Vibrational density of states

The first experiment study of the vibrational density of states (vDOS) was made by Gompf in 1979 [52]. In this neutron inelastic diffraction study, the author compared the trigonal, vitreous, and red amorphous selenium vibrational density of states, see figure 17. Even that there was a dispersion relation for the crystalline selenium (see ref 1 in [52]), there was not a model to explain it physically. Through the comparison between crystalline and amorphous form, the author tried to give a model for the disordered structure vibrations.

He described the amorphous selenium as disordered helicoidal chains with the following distances: the first distances are 2.31 Å for crystal, 2.32 Å for vitreous and amorphous, the second distance is 3.69 Å for all of them. He also reported the distance between chains as 3.46, 3.69 and 3,86 Å for the crystal, vitreous and amorphous, respectively.

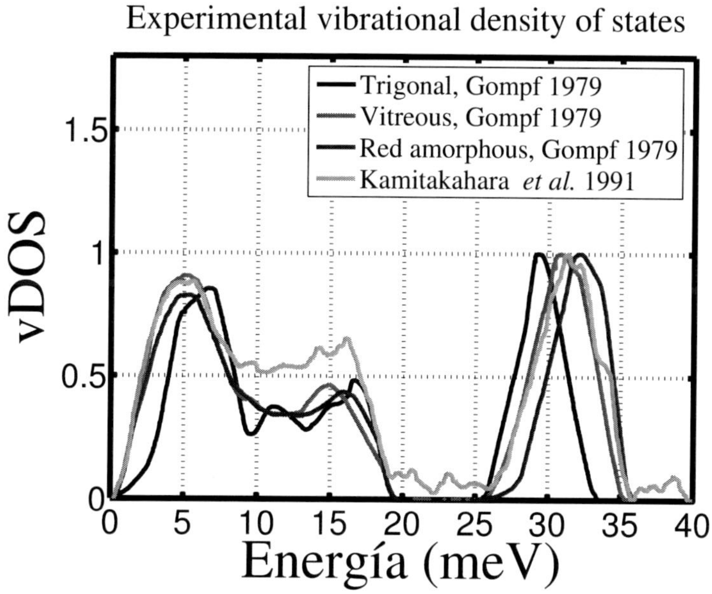

Figure 17. Experimenta vDOS reported by Gompf [52] (trigonal, vitreous and red amorphous) and Kamitakahara [53].

In the Gompf model, there are four peaks due to torsion (~ 5 meV), libration (~ 11 meV), bending (~ 16 meV) and dilatation (~ 30 meV). The first peak which corresponds to the torsion is smoother in the disordered structures. The next local maximum only can be seen in the crystalline form. He said, that these oscillation modes are rotational modes around the helix axes where the restitute forces were caused by the neighbors chains. The bending modes were situated around 16 meV, which is observed again smoother for the disordered systems (vitreous and amorphous). The peak decrement in the disordered material can be explain with the theoretical model development by Meek [54], who calculated the vibrational spectrum for different dihedral angles in the a-Se. At last, the peak in the range of 30 meV, it is caused to the stretching modes. In Genesky and Newell [55] model is discussed that if weak forces are introduced between polymeric chains then the hight frequencies at the end of the spectrum are shifted to low frequencies.

There is other neutron diffraction study for the Se-Ge and Se-As-Ge alloys made by Kamitakahara *et al.* [53], in which they included the vDOS for a-Se, see figure 17. The vibrational spectrum description is the following: from 0 to 19 meV is the floppy peak, follow by a gap, but it seems that there were some states within this gap. There is a sharped peak around 31 meV. They consider the range of 0-19 meV as vibrations occurred by the bending bonds, the range 25-35 meV as vibrations occurred by the stretching bonds, and the range of 35-40 meV they did not mention anything.

When the final structure is obtained, it means at the end of the geometry optimization,

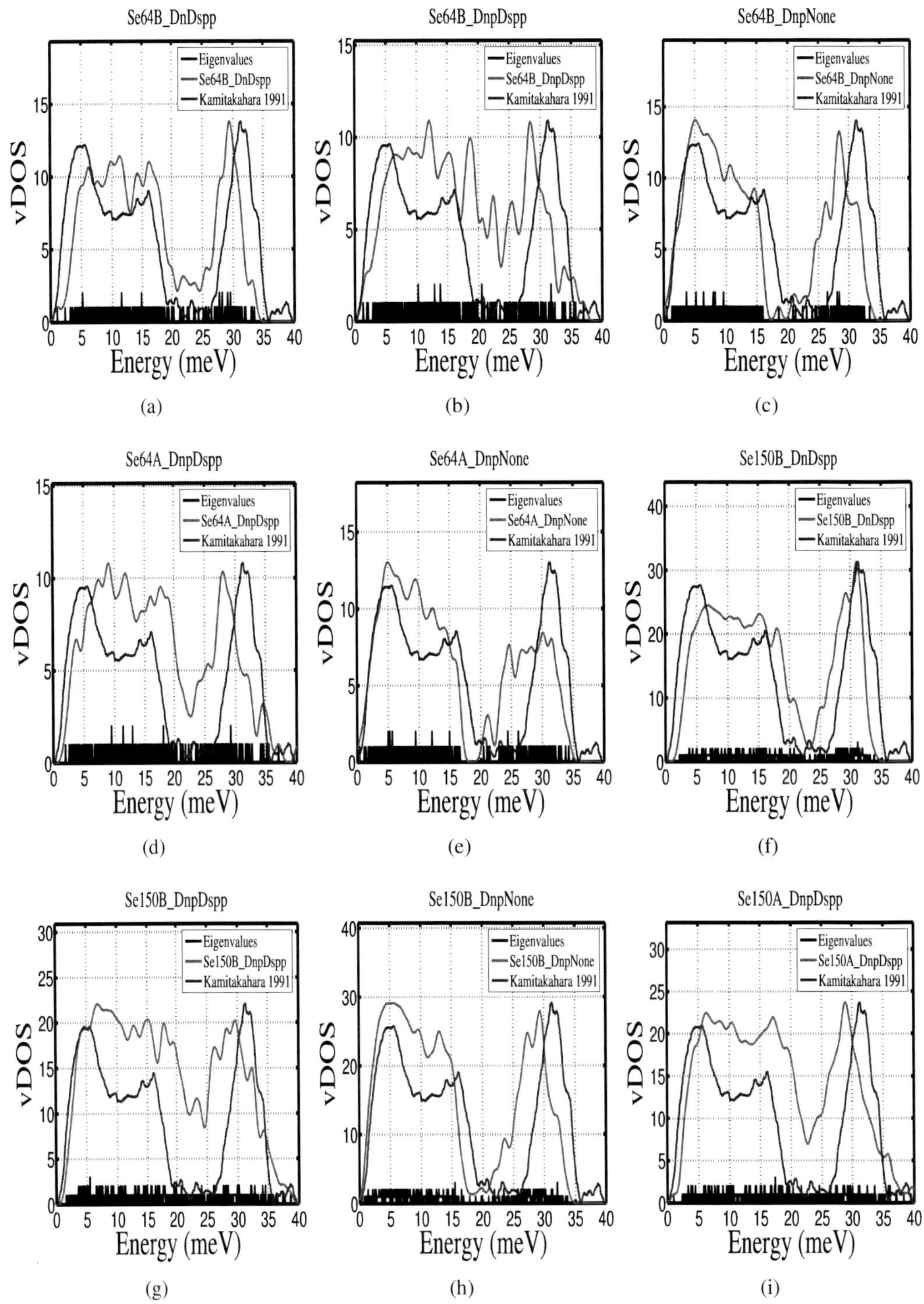

Figure 18. Vibrational density of states for the simulated samples with 64 and 150 atoms and compared with the experiment [53].

Table 12. Coordination defect distribution for the amorphous samples using the criterio of maximum bond value r_e.

Sample	Coordination(%)			
	1	2	3	4
Se64B_DnDspp	10.94	78.12	10.94	0
Se64B_DnpDspp	1.57	84.37	14.06	0
Se64B_DnpNone	6.25	87.50	6.25	0
Se64A_DnpDspp	9.37	78.13	12.50	0
Se64A_DnpNone	7.81	84.38	7.81	0
Se150B_DnDspp	8.00	84.00	8.00	0
Se150B_DnpDspp	6.00	84.67	8.67	0.66
Se150B_DnpNone	6.67	83.33	9.34	0.66
Se150A_DnpDspp	10.00	80.00	10.00	0
Se150A_DnpNone	10.00	80.00	10.00	0
Se216B_DnDspp	8.80	81.48	9.72	0
Se216B_DntoDnp	8.33	83.34	8.33	0
Se216B_DnpDspp	10.18	81.49	8.33	0
Se216B_DnpNone	7.87	84.26	7.87	0
Se216A_DnpDspp	6.94	81.49	11.57	0
Se512B_DnDspp	13.67	72.66	13.67	0

each atom is moved along the three axis to get the Hessian matrix. Once the Hessian matrix is diagonalized, each element in the diagonal corresponds to one vibrational mode. In the vibrational density of states are the allowed modes.

Figures 18 and 19 show all vDOS for the different cells and parameters. It is worth to mention that the time to compute the vibrational spectrum for the 512 is to large, and the calculations are still running.

The simulated vDOS can reproduce the floppy and stretching peaks. There are not a clear gap, but it can be seen a decrement states in the range of 19 to 25 meV. To explain it a unidimensional crystal with two different masses. In a unidimensional crystal with two different masses exists a gap in the vibrational spectrum [56]. It can be changed the effect of two different masses with the effect of two different springs, *i. e.*, two different bond forces types can be found in the selenium; weak forces (Van der Waals) which interact between chains for the hexagonal crystal or between rings for the monoclinic selenium, the strong forces or covalent bonds are the responsable to keep join the atoms in the chains such as the rings, depending on the structure. In figure 17 can be seen a clear gap for the crystalline selenium, which may be associated to the separation between forces that act in the solid. The first vDOS peak mention above can be attributed to weak forces follow by a gap and then a sharped peak due to the covalent bonds. In the disordered systems, there are a defects distribution and therefore a new types of bond forces are risen (floating or dangling bonds), see table 12. This produced a new states for the vDOS spectrum, which are situated within the corresponding crystalline gap for amorphous systems.

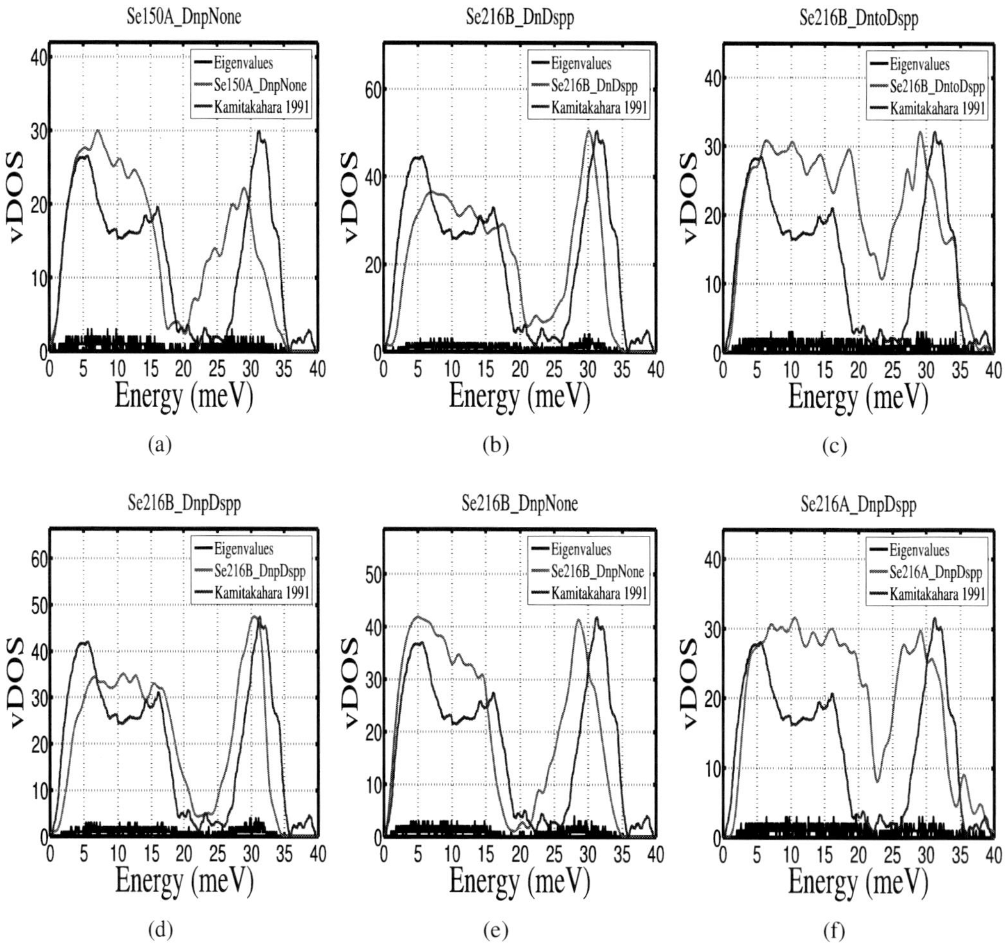

Figure 19. Vibrational density of states for the simulated samples with 150 and 216 atoms and compared with the experiment [53].

6. Electronic Properties

The experimental determination of the electronic density of states (eDOS) is obtained by different methods: soft X-ray emission and absorption, optical uv and X-ray photoemission, optical spectroscopy, etc [57]. Besides, there are other techniques to calculate the band gap for a semiconductor, in which the density of states is very small. All these methods have been applied on amorphous Si and Ge, which are an obligatory test for theory and experiment. The mentioned above demonstrated, as a whole, the eDOS is remarkably similar to the corresponding crystal. In particular, the optical spectroscopy shows the evidence of band gap, which are very similar to those of crystalline Si and Ge.

The band theory is not valid anymore for the relation E *vs* $\mathbf{k}$ in the amorphous materials but the eDOS it is. In contrast with the crystal gap, the amorphous band gap have states within as result of the disorder and the defects, which makes the characterization a little bit difficult. Other manner to get the band region is through the optical gap value. However, the *ab initio* calculation allows to have all the electronic states and the occupation, therefore the *lowest unoccupied molecular orbital* (LUMO) and the *highest occupied molecular orbital* (HOMO) are given, see table 13.

Figures 20 and 21 show the electronic density of states for all amorphous selenium simulations. The orbital *s*, *p* bonding, the lone pair and *p* antibonding are represented on the eDOS. The region between -15 to -10 eV belongs mainly to the orbital *s*, then -5 to 2.5 eV corresponds to *p* bonding, from -2.5 to 0 eV associates to the lone pair and from 0 to 5 eV the *p* antibonding orbitals are situated. As mentioned before, the coordination defects is less tan 20% of the coordination distribution. The dangling a floating bonds have almost the same percent in the coordination distribution. Only on the 150 atoms simulation a tetra-fold is founded, which can be attributed to the α-cubic structure and also it is representes on the radial function distribution around 3 Å [58].

Table 13. HOMO and LUMO values and the difference between them for all amorphous selenium simulations.

Sample	HOMO (eV)	LUMO (eV)	LUMO-HOMO (eV)
Se64B_DnDspp	-5.597	-4.514	1.083
Se64B_DnpDspp	-5.546	-4.547	0.999
Se64B_DnpNone	-5.748	-4.554	1.194
Se64A_DnpDspp	-5.657	-4.579	1.078
Se64A_DnpNone	-5.646	-4.516	1.130
Se150B_DnDspp	-5.508	-4.513	0.995
Se150B_DnpDspp	-5.556	-4.645	0.911
Se150B_DnpNone	-5.647	-4.634	1.013
Se150A_DnpDspp	-5.587	-4.652	0.935
Se150A_DnpNone	-5.617	-4.652	0.965
Se216B_DnDspp	-5.519	-4.662	0.857
Se216B_DntoDnp	-5.575	-4.647	0.928
Se216B_DnpDspp	-5.642	-4.656	0.986
Se216B_DnpNone	-5.649	-4.627	1.022
Se216A_DnpDspp	-5.552	-4.627	0.925
Se512B_DnDspp	-5.436	-4.788	0.648

Venderbilt and Joannopoulus [59–61] computed the pure and defected crystalline structures eDOS by tight-binding and self-consistent pseudopotentials calculations. They reported how the defects (uni- and three- fold bonds) influence on the eDOS when is compared with the trigonal selenium. In summary, they found states within the gap, see figure 4 and 8 of the reference [61].

Zhang and Drabold [33] reported the eDOS for 64 atoms amorphous selenium cell. They mentioned a long chain with uni and three-fold bonds defects, also a six-member ring. The authors located the defects states on the valence and conduction edges, with a electronic difference of 2.2 eV. Figure 22 shows the comparison with the corresponding 64 atoms simulations.

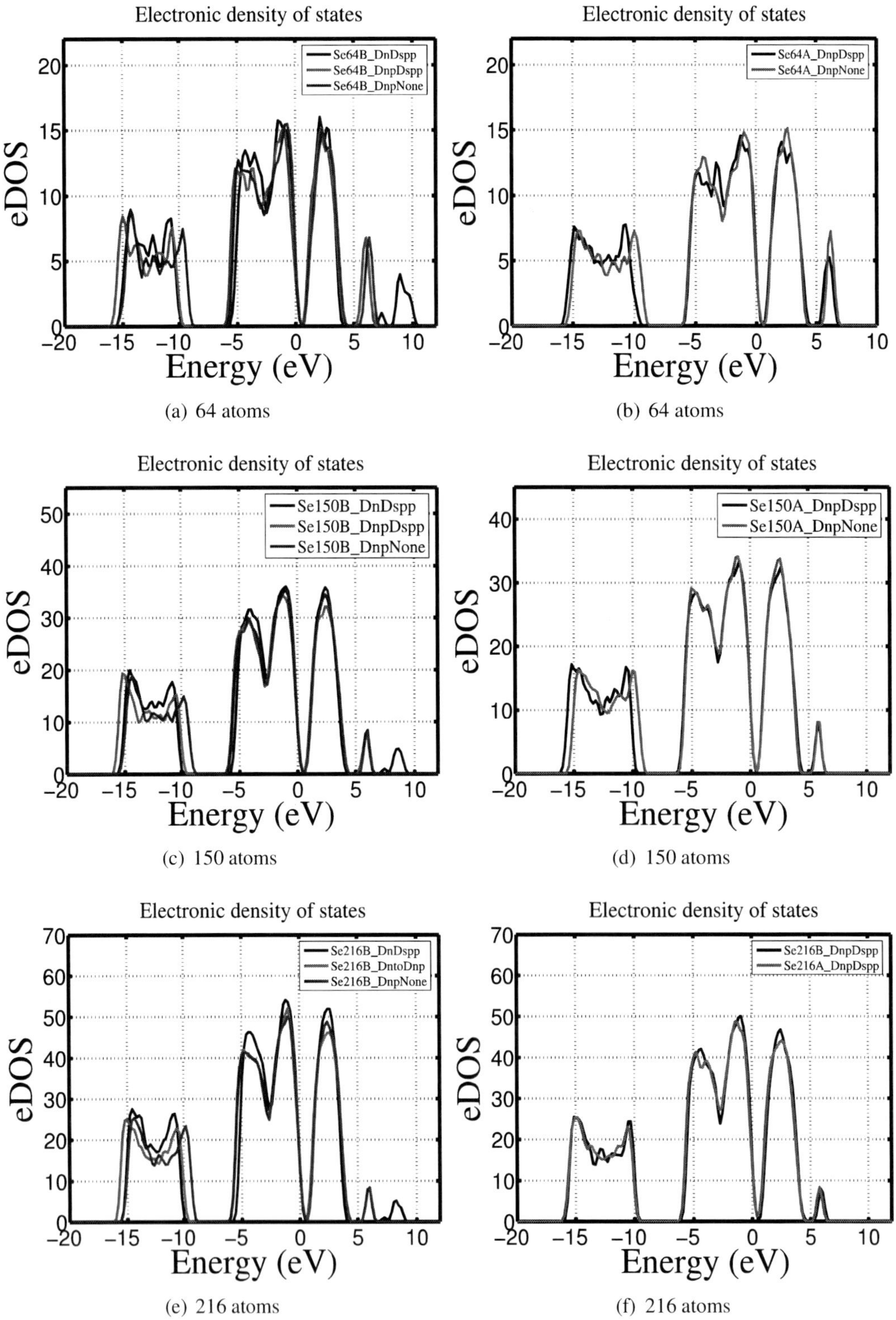

Figure 20. Electronic density of states for all amorphous selenium simulations.

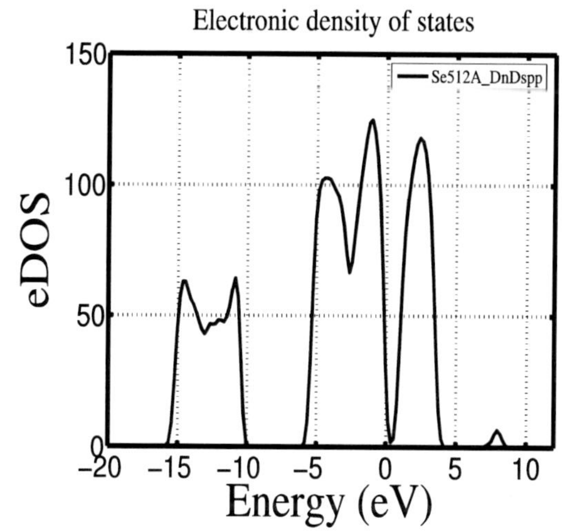

Figure 21. Electronic density of state for the 512 atoms sample.

It is worth to mention, that the scissors operation was not used, it is helpful to fit the electronic difference between valance and conduction with the experimental value. Table 2.14 reports the difference between HOMO-LUMO. An alternative method to obtain the gap is through the optical properties.

7. Optical Properties

The semiconductor band gap can be determined by electronic or optical methods and the results are in general different. In a crystal exists two principal process in which a photon can be absorbed. The first is *direct* process, it consists when the photon energy is absorbed by a electron and promoted to the conduction band, producing a hole. The second is an *indirect* process in which the photon absorbed is assisted by a phonon [62]. In the amorphous materials, the experimental optical methods allows to find a gap E_g^0 insensible to the localized states and the tails in the bands.

7.1. Tauc approximation

The materials interaction with the light gives a powerful tool to obtain the electronic and vibrational structure in a solid. In a first order optical absorption process just only one elemental excitation is carried out, where the absorbed photon energy is the same that the produced excitation. An optical gap empirical definition E_0 can be given by observation [63], in which for the amorphous semiconductor, the interband region can be described with the following relation [64], is the following:

$$\alpha \approx const \times (h\nu - E_0)^2 \tag{2}$$

A justification for this relation, according Zallen [64]; in a crystal photon energy $h\nu$ can promoted a transition from a state with energy E to a empty state with energy $E + h\nu$ only if the initial and final state have the same wave number k and satisfaces the selection

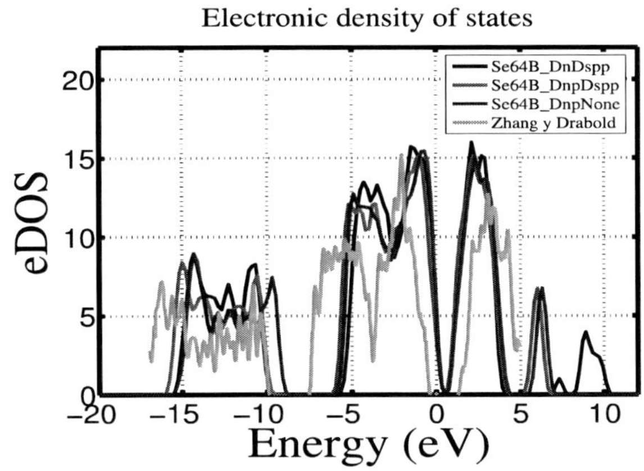

(a) 64 atoms

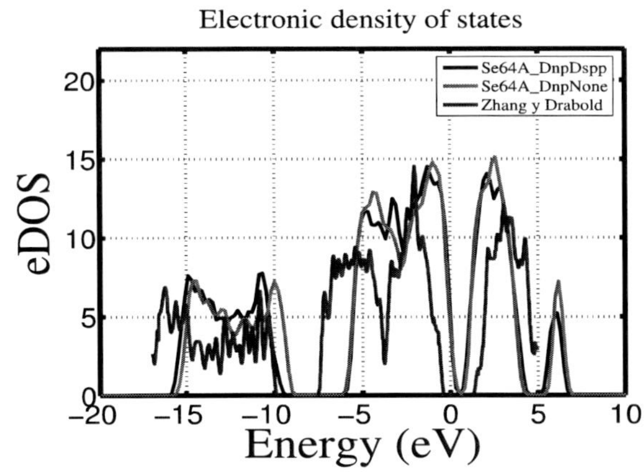

(b) 64 atoms

Figure 22. EDOS comparison with those reported by Zhang and Drabold [33].

rules. So, there are only few contributions in the optical absorbtion between all electronic states pairs separated with energy $h\nu$. But in the amorphous systems these rules are not applied. If it is assumed that there are extended states, all the full pair states (filled in the energy E and empty in the energy $E + h\nu$) could participate in the optical process. With the restriction above and the approximation of the matrix elements over energy range close to the electronic threshold are constant, the absorptive coefficient may be written:

$$\alpha = const. \times \int n_v(E) n_c(E + h\nu) dE, \tag{3}$$

where $n_v(E)$ and $n_c(E)$ are the occupied and unoccupied eigenstates respectively. If it is assumed that, as in crystals, the densities of extended states around the band edges are given by $n_v(E) \sim (E_v - E)^{1/2}$ and $n_c(E) \sim (E - E_c)^{1/2}$. Then $\alpha \sim (h\nu - E_0)^2$ with $E_0 = E_c - E_v$. Now the $\alpha^{1/2}$ vs $h\nu$ graphic yields a reasonably good straight-line fit to the absorption edge of an amorphous, the extrapolation $h\nu$ at which $\alpha^{1/2} = 0$ provides a convenient experimental benchmark for the optical bandgap E_0. This is known as *Tauc approximation*.

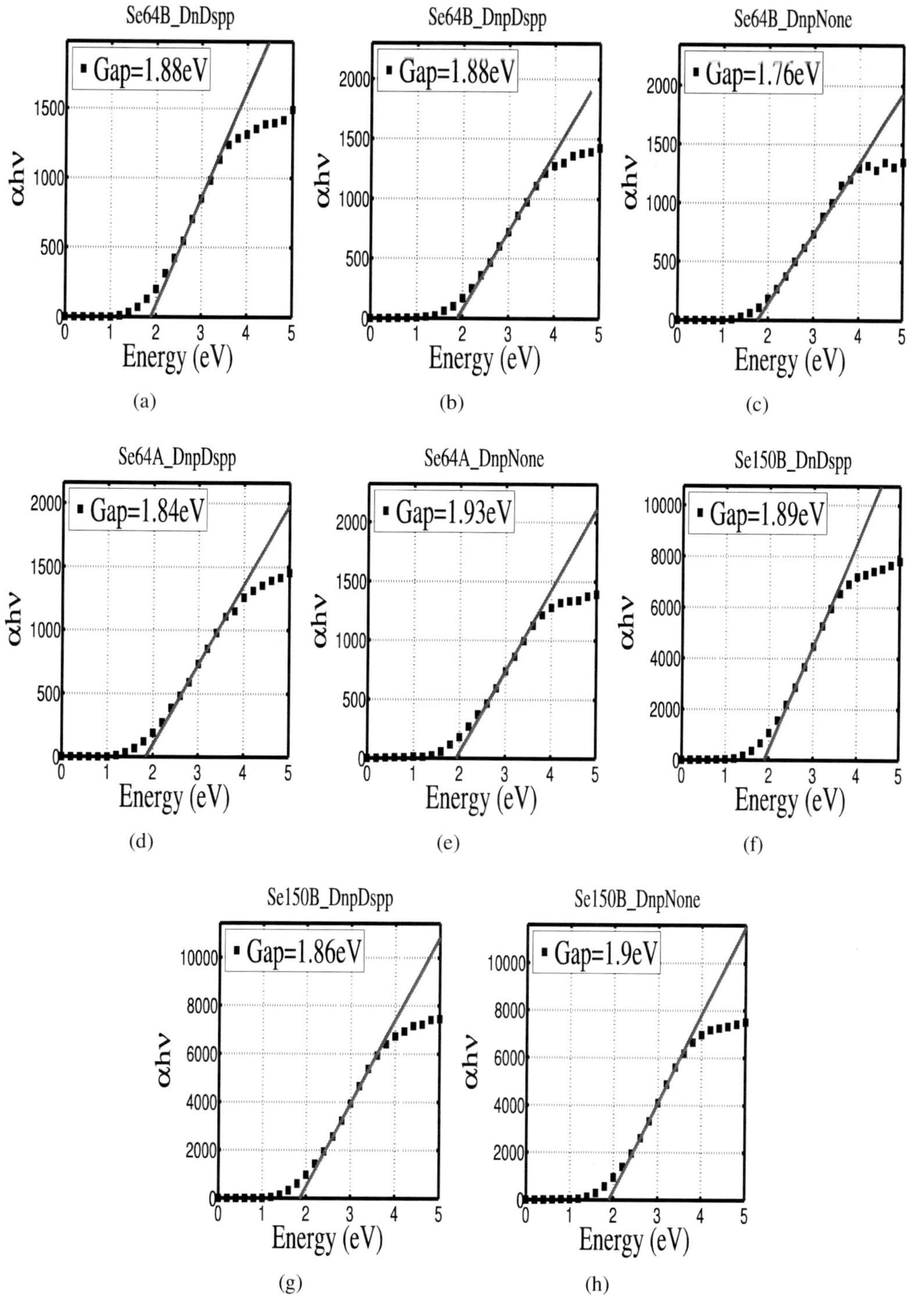

Figure 23. Optical band gap for the amorphous selenium simulations with 64 and 150 atoms using Tauc approximation.

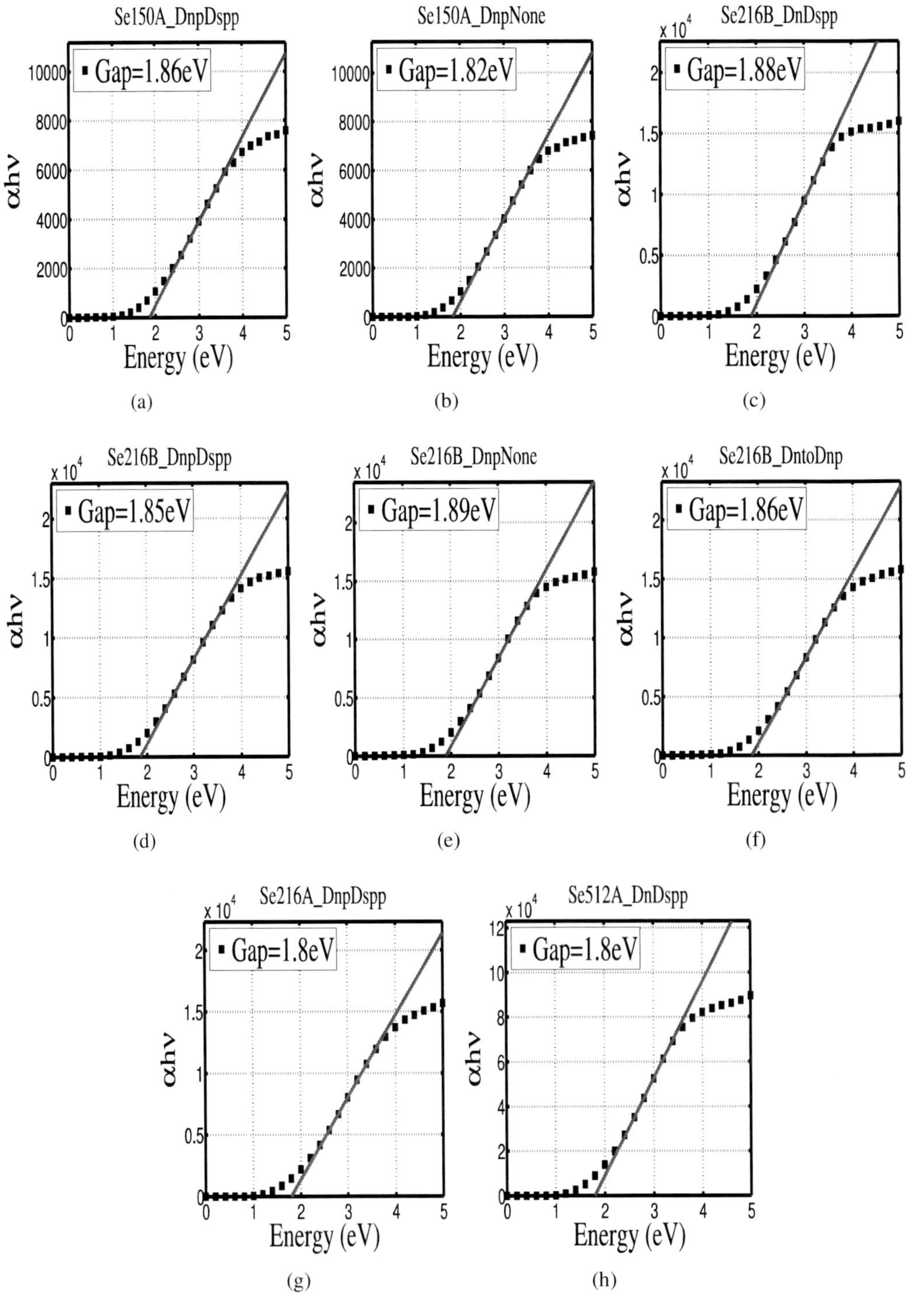

Figure 24. Optical band gap for the amorphous selenium simulations with 150, 216 and 512 atoms using Tauc approximation.

The behavior at the interbans regions is computed and linear fit is carried out (Tauc fit) on the $\alpha h\nu$ vs $h\nu$ graphics, according with the suggested by Elliott ([49] p.329). This linear fit corresponds at the low energies region, such as the maximum number of points were taken account [65]. Some experiments values for the optical band gap are: 1.76, 1.82, 1.84 eV reported by Roy *et al.* [66], 1.65 eV for black amorphous, 1.8 eV for monoclinic and 2.05 for a-Se reported by Zienab [67], 2.09 reported by Bindu *et al.* [68]. For a different amorphous selenium evaporation, Audiere *et al.* [69] presented the following values: 1.97, 1.80, 2.03, 2.07 eV and at last Chaudhuri *et al.* [70,71] gave 1.82 and 1.76 eV for the optical band gap, see table 14.

Table 14. Experimental optical ban gaps summary for amorphous selenium prepared by different methods, see text.

Experimental optical gaps	
Roy *et al.* [66]	1.76, 1.82 y 1.84 eV
Zienab [67]	1.65, 1.80, 2.05 eV
Bindu *et al.* [68]	2.09 eV
Audiere *et al.* [69]	1.97, 1.80, 2.03, 2.07 eV
Chaudhuri *et al.* [70,71]	1.82, 1.76 eV

Figures 23 and 24 show the results for the optical band gap. The basis set have an special influences on the optical properties. In summary, it can be seen that the best values are given when *dnp* basis set and all electron parameters are used.

8. Conclusion

In this chapter a serie of supercells with different number of atoms were simulated by *ab initio* studies. The topological properties show a characteristic peak around 3 Å and it is concluded that no more than 10 % of rings are embedded in the amorphous structure. The vibrational properties clarify the states in the gap for the amorphous samples and the defects distribution is given. The two-fold coordination is at least 80 % of the coordination distribution, see table 12. The electronic properties provide the electronic density of states which reproduce the features for the *s*, *p* bonding and the lone pair, see figures 20, 21 and 22. From the electronic eigenvalues *Tauc approximation* is calculated and the resulting values are in a good agrement with the experimental data, see table 14.

Acknowledgements

JARR acknowledges the financial support of CONACyT during his PhD studies. I offer my sincerest gratitude to my supervisor, Dr Ariel Valladares, who has supported me throughout my thesis with his patience and wisdom. The calculations were carried out in the computer center of DGTIC-UNAM. M.T. Vázquez and O. Jimenéz provided the information requested.

References

[1] W. Smith, *Effect of Light on Selenium during the passage of an Electric Current*, Nature **7** 303 (1873)

[2] J. Mort, *Anatomy of Xerography* (McFairland and Co., London, UK, 1989)

[3] J. Mort, *Xerography: A Study in Innovation and Economic Competitiveness*, Physics Today **47** April 32 (1994)

[4] S. O. Kasap and J. A. Rowlands, in *Insulating and semiconducting glasses* edited by P. Boolchand (World Scientific, Singapore, 2002), p.781.

[5] J. A. Rowlands and S. O. Kasap, *Amorphous semiconductors usher in digital X-ray imaging*, Physics Todays **50** November 24 (1997)

[6] G. Belev and S. O. Kasap *Amorphous selenium as an X-ray photoconductor*, J. Non-Cryst. Solids **345** 484 (2004)

[7] M. Hoheisel, L. Bätz, T. Mertelmeier, J. Giersch, A. Korn *Modulation Transfer Function of a Selenium-Based Digital Mammography System*, Ieee Transactions on Nuclear Science **53** 1118 (2006)

[8] Mihai A. Popescu *Non-Crystalline Chalcogenides* (Kluwer Academic Publishers, New York, 2000) pp. 9-13.

[9] A. D. McNaught and A. Wilkinson, *Compendium of Chemical Terminology The Gold Book* Second Edition (Blackwell Science, 1997) p 1093.

[10] Otfried Madelung, *Semicondutors: Data handbook* 3er edition, (Springer, Berlin, 2004)

[11] Y. Miyamoto, *Growth of Trigonal Selenium from* CS_2 *Solution*, Jap. J. Appl. Phys. **16** 2257 (1977)

[12] Y. Miyamoto, *Structure and Phase Transformation of Rhombohedral Selenium Composed*, Jap. J. Appl. Phys. **19** 1813 (1980)

[13] N.F. Mott and E.A. Davis, *Electronic processes in non-crystalline materials* (Oxford University Press, Cambridge, 1971)

[14] E. H. Henninger, R. C. Buschert, L. Healton, *Atomic Radial Distribution in Amorphous Selenium by X-Ray and Neutron Diffraction*, J. Chem. Phys. **46** 586 (1967)

[15] R. Chang and P. Romo, *The Atomic Radial Distribution Function in Amorphous Selenium, Acta Cryst.* **23** 700 (1967)

[16] R. Kaplow, T. A. Rowe, B. L. Averbach, *Atomic Arrangement In Vitreous Selenium, Phys. Rev.* **168** 1068 (1968)

[17] H. Richter, *Different Forms of amorphous selenium*, J. Non-Cryst. Solids **8** 388 (1972)

[18] A. L. Renninger and B. L. Averbach, *Atomic Radial Distribution Functions of As-Se Glasses*, Phys. Rev. B **8**, 1507 (1973)

[19] F. Yssing Hansen, T. Steen Knudsen, K. Carneiro, *Structure of amorphous selenium studied by neutron diffraction*, J. Chem. Phys. **62** 1556 (1975)

[20] R. Bellissent, *Short range order in the disordered states of selenium-tellurium mixtures*, Nucl. Instr. and Meth. **199** 289 (1982)

[21] T. Fukunaga, M. Utsumi, H. Akatsuka, M. Misawa, U. Mizutani, *Structure of amorphous Se prepared by milling*, J. Non-Cryst. Solids **205** 531 (1996)

[22] M. Inui, K. Maruyama, S. Takeda, S. Tamaki, Y. Waseda, *Electron Charge Distribution in Amorphous Se*, J. Phys. Soc. Jpn. **63** 1378 (1994)

[23] M. Inui, K. Tamura, Y. Oh'ishi, I. Nakaso, K. Funakoshi, W. Utsumi *X-ray diffraction measurements for expanded fluid-Se using synchrotron radiation*, J. Non-Cryst. Solids **250** 519 (1999)

[24] J. C. de Lima, T. A. Grandi, R. S. Biasi, *Influence of aging on the thermal and structural properties of amorphous selenium prepared by ball milling*, J. Non-Cryst. Solids **286** 93 (2001)

[25] R. Brüning, E. Irving, G. LeBlanch, *Reverse Monte Carlo study of structural relaxation in vitreous selenium*, J. Appl. Phys. **89** 3215 (2001)

[26] P. Jóvári, R. G. Delaplane, L. Pusztai, *Structural models of amorphous selenium*, Phys. Rev B **67** 172201 (2003)

[27] W. D. Wei, B. W. Corb, B. L. Averbach, *A correlation model of amorphous selenium*, J. Non-Cryst. Solids **53** 19 (1982)

[28] B. W. Corb, W. D. Wei, B. L. Averbach, *Atomics models of amorphous selenium, J. Non-Cryst. Solids* **53** 29 (1982)

[29] D. Hohl and R. O. Jones, *First principles MD simulation of liquid and amorphous selenium*, J. Non-Cryst. Solids **117/118** 922 (1990)

[30] D. Hohl and R. O. Jones, *First-Principles Molecular-Dynamics Simulation of Liquid and Amorphous Selenium, Phys. Rev B* **43** 3856 (1991)

[31] C. Oligschleger, R. O. Jones, S. M. Reimann, H. R. Schober, *Model interatomic potential for simulations in selenium*, Phys. Rev. B. **53** 6165 (1996)

[32] D. Hohl, R. O. Jones, R. Car, M. Parrinello, *The structures of selenium Clusters - Se_3 to Se_8*, Chem. Phys. Lett. **139** 540 (1987)

[33] X. Zhang and D. A. Drabold, *Evidence for valence alternation, and a new structural model of amorphous selenium*, J. Non-Cryst. Solids **241** 195 (1998)

[34] X. Zhang and D. A. Drabold, *Direct Molecular Dynamic Simulation of Light-Induced Structural Change in Amorphous Selenium*, Phys. Rev. Lett. **83** 5042 (1999)

[35] C. Bichara, A. Pellegatti, J. P. Gaspard, *Chain structure of liquid selenium investigated by tight-binding Monte Carlo simulation*, Phys. Rev. B **49** 6581 (1994)

[36] F. Shimizu, H. Kaburali, T. Oda, Y. Hiwatari, *Chain structure of liquid and amorphous selenium: tight-binding molecular-dynamics simulation*, J. Non-Cryst. Solids **250** 433 (1999)

[37] D. Caprion and H. R. Schober, *Structure and relaxation in liquid and amorphous selenium*, Phys. Rev B **62** 3709 (2000)

[38] D. Caprion and H. R. Schober, *Computer simulation of liquid and amorphous selenium*, J. Non-Cryst. Solids **326** 369 (2003)

[39] J. Hegedüs, K. Kohary, S. Kugler, *Comparative analysis of different preparation methods of chalcogenide glasses: molecular dynamics structure simulations*, J. Non-Cryst. Solids **338-340** (2004) 283.

[40] J. Hegedüs and S. Kugler, *Molecular dynamics structure simulations of amorphous selenium prepared by different methods*, J. Optoelec. Adv. Mat. **7** 1923 (2005)

[41] J. Hegedüs, Kohary, D. G. Pettifor, K. Shimakawa, S. Kugler, *Photoinduced Volume Changes in Amorphous Selenium*, Phys. Rev. Lett. **95** 206803 (2005)

[42] Pál Jóvári and László Pusztai *Structure of disordered forms of selenium close to the melting point*, Phys. Rev B **64** 014205 (2001)

[43] B. Delley,*An all-electron numerical method for solving the local density functional for polyatomic molecules*, J. Chem. Phys. **92** (1990) 508, B. Delley, *From molecules to solid with DMol^3 approach*, J. Chem. Phys. **113** 7756 (2000). DMol^3 is implemented in the Materials Studio suite ©.

[44] S.S.H. Vosko, L. Wilk, M. Nusair, *Accurate spin-electron liquid correlation energies for local spin density calculations: a critical analysis*, Can. J. Phys. **58** 1200 (1980)

[45] J. Harris, *Simplified method for calculating the energy of waekly interaction fragments*, Phys. Rev. B **31** 1770 (1985)

[46] Z. Lin and J. Harris, *A localized-basis scheme for molecular dynamics*, J. Phys. Condens. Matter **4** 1055 (1992)

[47] Ariel A. Valladares, *A new approach to the ab initio generation of amorphous semiconductors structures. Electronic and vibrational studies*, in Glass Materials Research Progress, edited by Jonas C. Wolf and Luka Lange, Nova Science Publisher 2008 pp. 1-64.

[48] A. Popov, *Atomic structure and structural modificacion of glass*, in: *Semiconducting Chalcogenide Glass I*, edited by R. Fairman and B. Ushkov, (Elsevier Academic Press, New York, 2004), p. 68.

[49] S. R. Elliott, *Physics of amorphous materials* Second Edition (Longman Scientific & Technical, John Wiley & Sons, Inc. New York, 1990) pp. 186-190.

[50] F. Alvarez and A. A. Valladares, *Optical gaps of ab initio generated random networks for a-SiN$_x$ alloys*, Appl. Phys Lett. **80**, 58 (2002)

[51] I. Levine, *Quantum chemistry*, 5th ed. (Prentice Hall, New Jersey, 2000)

[52] F. Gompf, *Comparison of the Phonon Density of States of Trigonal, Vitreous and Red Amorphous Selenium*, Proc. Int. Conf. on The Physics of Se and Te (Springer-Verlang, Berlin, 1979) pp. 64-67

[53] W. A. Kamitakahara, R. L. Cappelletti, P. Boolchand, B. Halfpap, F. Gompf, D. A. Neumann, H. Mutka, *Vibrational Densities of States and Network rigidity in Chalcogenide Glasses*, Phys. Rev. B **44** 94 (1991)

[54] P. E. Meek, *Vibrational spectra of disordered chains and the structure of amorphous Se and Te*, Phil. Mag. **34** 767 (1976)

[55] S. M. Genesky and G. F. Newell, *Vibration Spectrum and Heat Capacity of a Chain Polymer Crystal*, J. Chem. Phys **26** 486 (1957) y referencias contenidas.

[56] N. W. Ashcroft and N. David Mermin *Solid State Physics*, Brooks Cole, New York (1976)

[57] B. Kramer and D. Weaire in, *Amorphous Semiconductors* edited by M. H. Brodsky, second edition, Springer-Verlag, Berlin (1985) pp. 9-39.

[58] J.A. Reyes-Retana and Ariel A. Valladares, *Structural properties of amorphous selenium: An ab initio molecular-dynamics simulation*, Computational Materials Science **47** 934 (2010)

[59] D. Vanderbilt and J. D. Joannopoulos, *Calculations of defects states in amorphous selenium*, Phys. Rev. Lett. **42** 1012 (1979)

[60] D. Vanderbilt and J. D. Joannopoulos, *Bonding coordination defects in selenium* in *The physics of selenium and tellurium*, edited by E. Gerlach and P. Gross (Springer, New York, 1979) p. 203.

[61] D. Vanderbilt and J. D. Joannopoulos, *Theory of defects states in glassy selenium, Phys. Rev. B* **22** 2927 (1980)

[62] A. P. Sutton, *Electronic Structure in Materials* (Clarendon Press, Oxford 1993)

[63] D. L. Wood and Tauc, J. *Weak Absortion Tails in Amorphous Semiconductors*, Phys. Rev. B **5** 3144 (1972)

[64] R. Zallen *The Physics of Amorphous Solids* Wiley, New York (1983)

[65] F. Alvarez and A. A. Valladares, *First-principles simulations of atomic networks and optical properties of amorphous* SiN_x *alloys*, Phys. Rev. B **68** 205203 (2003)

[66] R. Roy, V. S. Choudhary, M. K. Patra, A. Pandya, *Effect of annealing temperature on the electrical and optical properties of nanocrystalline selenium thin films*, J. Optoelec. Adv. Mat. **8** 1352 (2006)

[67] S. E. Zienab Mandouh, *Effect of againg on some physical properties of selenium thin films*, Fizika A2 **1** 35 (1993)

[68] K Bindu, M Lakshmi, S Bini, C Sudha Kartha, K P Vijayakumar, T. Abe, Y. Kashiwaba, *Amorphous selenium thin films prepared using chemical bath deposition: optimization of the deposition process and characterization*, Semicond. Sci. Technol. **12** 270 (2002)

[69] J. P. Audiere, C. Mazieres, J. C. Carballes, *Non-crystalline Se thin films from controlled vapor- preparation, crystallization and optical properties*, J. Non-Cryst. Solids **27** 411 (1978)

[70] S. Chaudhuri, S. K. Biswas, A. Choudhury, K. Goswami, *Amorphous to cristalline transition of selenium thin films of different thicknesses*, J. Non-Cryst. Solids **46** 171 (1981)

[71] S. Chaudhuri, S. K. Biswas, A. Choudhury, K. Goswami, *Variation of optical gap of thick amorphous selenium film heat treatment*, J. Non-Cryst. Solids **54** 179 (1983)

INDEX

B

C

D

H

I

J

K

L

M

N

O

P

Q

R

S

T

U

V

W

Y

Z